# Cognitive Neuroscience Robotics B

Masashi Kasaki • Hiroshi Ishiguro • Minoru Asada
Mariko Osaka • Takashi Fujikado
Editors

# Cognitive Neuroscience Robotics B

## Analytic Approaches to Human Understanding

Springer

*Editors*
Masashi Kasaki
Graduate School of Letters
Kyoto University
Kyoto, Japan

Minoru Asada
Graduate School of Engineering
Osaka University
Osaka, Japan

Takashi Fujikado
Graduate School of Medicine
Osaka University
Osaka, Japan

Hiroshi Ishiguro
Graduate School of Engineering Science
Osaka University
Osaka, Japan

Mariko Osaka
Division of Cognitive Neuroscience Robotics
Institute for Academic Initiatives
Osaka University
Osaka, Japan

ISBN 978-4-431-54597-2 ISBN 978-4-431-54598-9 (eBook)
DOI 10.1007/978-4-431-54598-9

Library of Congress Control Number: 2016934599



Cover Illustration by Takashi Ikeda

Printed on acid-free paper

This Springer imprint is published by Springer Nature
The registered company is Springer Japan KK

# Preface

A variety of technologies are developing and making our society mechanized and computerized at an unprecedented pace, while the overall effects of this development on the human brain are not considered. Some examples of this development include exposure to a massive amount of information via computer networks and the widespread uses of cell phones and automobiles. At least some consequences of the development are not necessarily beneficial. It is highly plausible that our society, as it is today, places an excessive cognitive burden on the brains of children, the elderly, and even adults in the prime of life (see Fig. 1). In order to lead the development of technologies for every member of society in a healthy direction, it is necessary to develop new Information and Robot Technology (IRT) systems that can provide information and services on the basis of the understanding of higher human brain functions (or functions of the "cognitive brain").

In order to deeply understand the human brain functions and develop new IRT systems, Osaka University established the Center of Human-Friendly Robotics Based on Cognitive Neuroscience in 2009, with funding from the Global Center of Excellence (GCOE) Program of the Ministry of Education, Culture, Sports, Science and Technology, Japan. The Center integrates the following world-class research programs and studies at Osaka University, Advanced Telecommunications Research Institute International (ATR), and National Institute of Information and Communications Technology (NICT):

- World-famous human–robot interaction studies: Graduate School of Engineering and Engineering Science, Osaka University, ATR Intelligent Robotics and Communication Laboratories
- Japan's largest-scale program in cognitive psychology: Graduate School of Human Sciences, Osaka University
- World-recognized pioneering studies in brain science and brain machine interface: Graduate School of Medicine, Osaka University, and ATR Computational Neuroscience Laboratories, and NICT

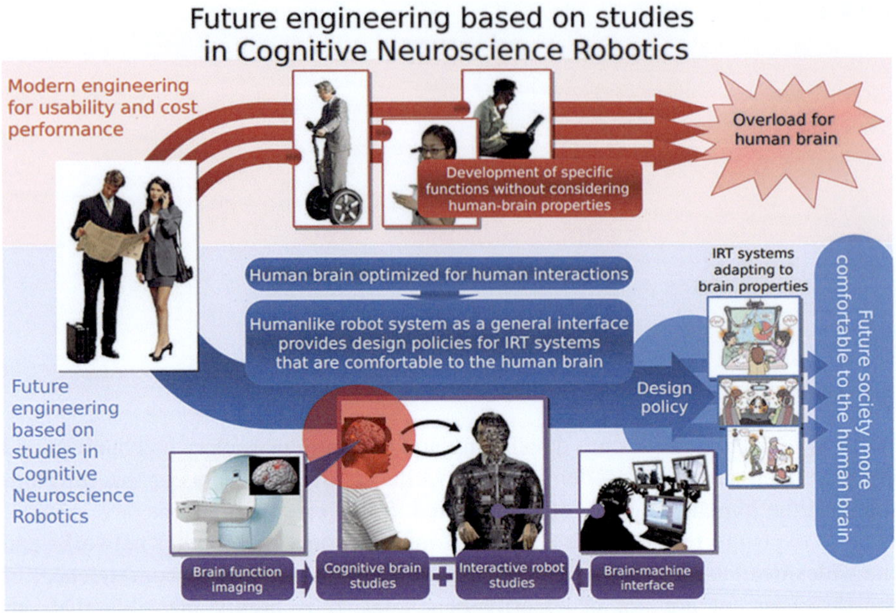


**Fig. 1** Traditional engineering vs. future engineering

The Center pursues a new research and education area in which humanities and sciences closely collaborate with each other. Thus, the Center has reorganized education and research at the graduate schools of Osaka University and provided students and researchers a place to engage in the new research and education area. This new area is named cognitive neuroscience robotics.[1]

In more detail, cognitive neuroscience robotics addresses three interrelated research tasks, among others. The first task is to explore how higher brain functions [e.g., consciousness, memory, thinking, emotion, *kansei* (feeling), and so on] are involved in the use of IRT systems, by measuring brain activities with brain-imaging technology. This requires interdisciplinary studies between cognitive and brain sciences. The second related task is to investigate higher brain functions on the basis of brain functional imaging studies on brain function disabilities and studies on brain–machine interfaces (BMI). This requires interdisciplinary studies between brain science and engineering. The third task is to develop prototypes of human-friendly IRT systems and new hypotheses about the cognitive brain by combining studies relevant to the other tasks. In short, cognitive neuroscience robotics, with new technologies at hand, will establish a new understanding of the cognitive brain and develop prototype systems, to solve the problems with modern

[1] The Center finished its proposed research under the funding from the GCOE program in 2014. The Center was then integrated into the Division of Cognitive Neuroscience Robotics, Institute for Academic Initiatives, Osaka University.

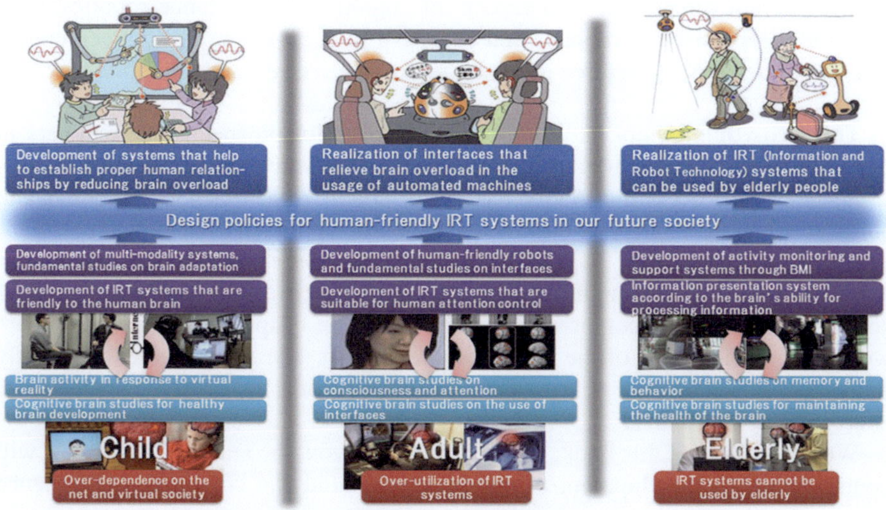


**Fig. 2** Solutions by systems based on cognitive neuroscience robotics

mechanized society. Figure 2 shows a typical example of each of the three tasks of cognitive neuroscience robotics.

The Center consists of four interdisciplinary education and research groups. They are organized into a unified five-year education and research program dedicated to the tasks stated above.

- The Group for Establishment of Cognitive Neuroscience Robotics encompasses all research activities in the Center. It establishes the direction of the Center's education and research and aims to systematize the new area of cognitive neuroscience robotics, through scientific and philosophical considerations.
- The Group for Interdisciplinary Studies in Cognitive and Brain Sciences aims to reveal higher brain functions (the cognitive brain) with brain-imaging technology.
- The Group for Interdisciplinary Studies in Brain Science and Engineering develops brain–machine interfaces that directly connect the human brain with IRT systems.
- The Group for Development of Cognitive Brain Systems develops prototypes of future IRT systems that do not cause the overload of the human brain, as opposed to the existing IRT systems.

These interdisciplinary research groups include prospective researchers, engineers, and entrepreneurs. The Center offers them a graduate minor program of cognitive neuroscience. This program provides them with basics of cognitive neuroscience robotics and prepares them to address and accommodate the needs of the future society. Students enrolled in the minor program of cognitive neuroscience are required to take two courses: "synthetic approach to human understanding" and "cognitive brain science." Synthetic approach to human understanding and cognitive brain science are two aspects of cognitive neuroscience robotics, seen from the perspectives of robotics and cognitive science, respectively. Each course consists of a series of lectures given by representative researchers in the research groups.

This two-volume book is written as a textbook for prospective researchers in cognitive neuroscience robotics. Volume A, *Synthetic Approaches to Human Understanding*, covers the robotics aspect of cognitive neuroscience robotics and corresponds to the content of the course "synthetic approach to human understanding"; Volume B, *Analytic Approaches to Human Understanding*, covers the cognitive science aspect of cognitive neuroscience robotics and corresponds to the content of the course "cognitive brain science." The chapters of each volume are written by the lecturers of the corresponding course. The two volumes are jointly designed for young researchers and graduate students to learn what cognitive neuroscience robotics is.

We, the editors of this book, strongly hope that you, the reader of this book, will contribute to the development of our society by studying cognitive neuroscience robotics.

Lastly, we would like to convey our appreciation and gratitude to all authors of the individual chapters of this two-volume book. The main editor, Masashi Kasaki, read every chapter and provided detailed feedback to each author. His contribution to the book deserves special mention here.

Japanese Society for the Promotion of Science Postdoctoral Fellow, Graduate School of Letters, Kyoto University
Guest Associate Professor, Division of Cognitive Neuroscience Robotics, Institute for Academic Initiatives, Osaka University
Masashi Kasaki

Leader of the GCOE Center of Human-Friendly Robotics Based on Cognitive Neuroscience
Professor, Graduate School of Engineering Science, Osaka University
Hiroshi Ishiguro

Director of the Division of Cognitive Neuroscience Robotics, Institute for Academic Initiatives
Professor, Graduate School of Engineering, Osaka University
Minoru Asada

Guest Professor, Division of Cognitive Neuroscience Robotics, Institute for Academic Initiatives, Osaka University
Mariko Osaka

Professor, Graduate School of Medicine, Osaka University
Takashi Fujikado

# Contents

# Main Contributors

**Takashi Fujikado** Graduate School of Medicine, Osaka University, Osaka, Japan

**Masayuki Hirata** Graduate School of Medicine, Osaka University, Osaka, Japan

**Osamu Imura** Graduate School of Human Sciences, Osaka University, Osaka, Japan

**Aya Nakae** Immunology Frontier Research Center, Osaka University, Osaka, Japan

**Masayuki Nakamichi** Graduate School of Human Sciences, Osaka University, Osaka, Japan

**Yasuo Nakayama** Graduate School of Human Sciences, Osaka University, Osaka, Japan

**Mariko Osaka** Division of Cognitive Neuroscience Robotics, Institute for Academic Initiatives, Osaka University, Osaka, Japan

**Youichi Saitoh** Graduate School of Medicine, Osaka University, Osaka, Japan

**Eku Shimosegawa** Graduate School of Medicine, Osaka University, Osaka, Japan

**Kazumitsu Shinohara** Graduate School of Human Sciences, Osaka University, Osaka, Japan

**Naoki Tani** Graduate School of Medicine, Osaka University, Osaka, Japan

**Yasunobu Yasoshima** Graduate School of Human Sciences, Osaka University, Osaka, Japan

**Yoshichika Yoshioka** Immunology Frontier Research Center, Osaka University, Osaka, Japan

Center for Information and Neural Networks, National Institute of Information and Communications Technology, Osaka, Japan

# Chapter 1
# Perceptual and Cognitive Processes in Human Behavior

**Kazumitsu Shinohara**

**Abstract** In this chapter, major models of perception and cognition are reviewed with a focus on human information processing and some related research paradigms. The main focus is on the psychological models of perception and cognition that experimental psychology has developed to explicate the psychological functions of human information processing. By adopting these models, many studies have clarified the roles of the brain activities that are closely related to the components of the models. The models are then important for understanding how we perceive and recognize surrounding environments and how we decide to behave in response to them.

**Keywords** Skill-Rule-Knowledge based model (SRK model) • Selective attention • Cocktail party phenomenon • Divided attention • Spotlight • Orientation • Useful field of view • Visual search • Feature integration theory • Coherence theory • Attentional resources • Multitasking • Working memory

## 1.1 Introduction

It is necessary for humans to adaptively behave in a variety of environments. Adaptive behaviors include acquiring information from surrounding environments, activating information stored in memory, processing acquired or activated information, and executing a behavior. Perception and attention are the first stage of adaptive behavior. In this chapter, major models of perception and cognition are reviewed with a focus on human information processing and some related research paradigms.

Obviously, perception and cognition fully depend on a wide variety of brain activities. Many researches have contributed to clarify the relationships between psychological phenomena and brain activities, resulting in a huge amount of neuropsychological findings in this domain. Compared to higher cognitive processes, such as thinking and intending, perceptual and cognitive processes are

K. Shinohara (✉)
Graduate School of Human Sciences, Osaka University, Suita, Osaka, Japan
e-mail: sinohara@hus.osaka-u.ac.jp



M. Kasaki et al. (eds.), *Cognitive Neuroscience Robotics B*,
DOI 10.1007/978-4-431-54598-9_1

simple; it is relatively easy to study the relationships between these psychological processes and brain activities under strictly-controlled experimental conditions.

In this chapter, however, only a few findings about brain activity are discussed. The main focus of the chapter is on the psychological models of perception and cognition that have been developed in experimental psychology. These models conceptually pertain to the psychological function of human information processing, although they do not specify how brain activities are related to psychological processes. Most of these models have been developed on the basis of behavioral and neuropsychological findings. By adopting these models, many studies have clarified the roles of the brain activities that are closely related to the components of the models. It is recommended that the reader refer to related neuropsychological studies if detailed information on neuropsychological evidence for the models is needed.

## 1.2 Human Information Processing

### *1.2.1 Automatic and Intentional Control of Behavior*

In the study of cognitive psychology, human behavior has been regarded as a sequence of information processing: information processing involves obtaining information from the surroundings, understanding the situation on the basis of information, making a decision about how to act, and executing a behavior.

Many psychological models of human behavior distinguish between two modes of behavior: the automatic mode and the intentional and effortful mode. One of the most famous models is the dual process theory, and it assumes System 1 and System 2 (Stanovich and West 2000). According to Kahneman (2002), System 1 is fast, automatic, effortless, associative and difficult to control or modify, while System 2 is slower, serial, effortful, and deliberately controlled. When we are awake, System 1 and System 2 are corporately working. System 1 continuously generates suggestions for System 2, and System 2 often accepts these suggestions with little or no modification, leading to smooth performance with minimum effort. When System 1 encounters some difficulties, System 2 is called upon for detailed and specific processing (Kahneman 2011).

Whether the mode of behavior is automatic or intentional, there are many psychological processes underneath any given behavior. To understand human information processing more analytically, it is necessary to use the model of human behavior that divides psychological processes into specific components, and to design a study that can pinpoint the function of each component in an adequate research method. In this section, a model frequently used in the human factors research, Rasmussen's Skill-Rule-Knowledge based model (SRK model; Rasmussen 1983, 1986, 1987), is described (Fig. 1.1). The SRK model is mainly used for analyzing and classifying human errors. In the human factors research,

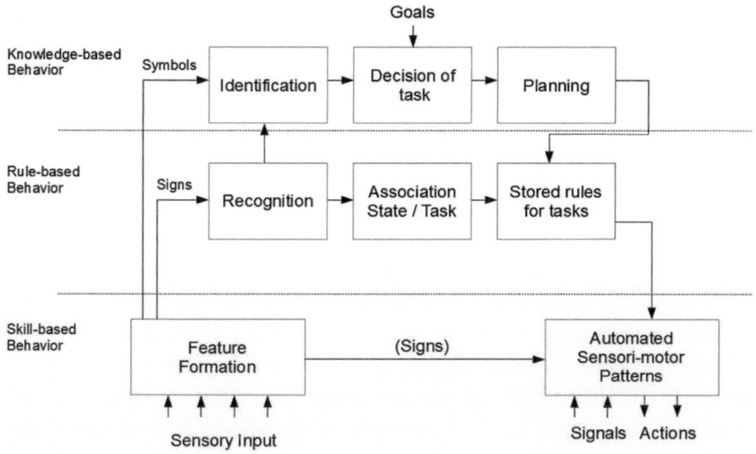


**Fig. 1.1** Skill-Rule-Knowledge based model of human information processing (Rasmussen 1983, 1986, 1987)

human behaviors in industrial settings, e.g., factory and machine operation, have been studied for the purposes of accident and error prevention.

In the SRK model, human information processing depends on three levels of cognitive control: the skill-based level, the rule-based level, and the knowledge-based level. Several stages of information processing are assumed at each level.

At the *skill-based* level, a behavior is based on sensori-motor control. Through experience and extensive training, the link between specific stimulus and action is learned. Once this linkage is established, perception of a stimulus immediately activates the action linked to it. It is automatic and not demanding. For example, the performance of an expert sports player is regarded to be processed at this level. An action conducted this way is a pattern of behavior which has been acquainted through extensive experience. Our daily life consists of innumerable acts developed through repeated experience. Basically, a behavior performed at this level does not include intentional control or monitoring, and so modification of ongoing behavior is quite difficult.

At the *rule-based* level, a behavior is controlled by stored rules learned through previous experience. Most of our daily routines are at the rule-based level. After sensory input is recognized, what task is to be performed in the current situation is identified by reference to the stored rules that are appropriate for the current situation. Sometimes the rules are so familiar that the detail of them cannot be consciously specified. At this level, recognized sensory information is used as a sign not for consciously controlling and monitoring the processes but for selecting and modifying the rules to execute the familiar processes.

In the novel, unfamiliar situation in which no rules are available, a behavior is controlled at the *knowledge-based* level. Information processing at this level is consciously controlled, and information is processed as a symbol that refers to a concept to identify the situation. What task is required to perform in the current situation is decided by reference to the explicitly recognized goal, and a sequence of actions is planned accordingly. Rasmussen (1986, 1987) elaborates this decision process at the knowledge-based level and suggests the step ladder model containing eight stages of decision making: activation, observation, identification, interpretation, evaluation, task definition, procedure formulation, and execution. In the step ladder model, data processing in a stage produces states of knowledge, and they are used for data processing in the next stage. Stereotyped mental processes sometimes take a shortcut to the state that is otherwise the result gained by several steps further, and this shortcut may result in an error in decision making. For example, when we observe the situation to obtain information for decision making, we are likely to determine what to do by improperly adopting some habitual rule without checking its pertinence by carefully analyzing the information.

Norman and Shallice (1986) propose a general account of the control of action. In this model, human behavior is controlled in two ways: automatic schema-based control and conscious control by the supervisory attentional system. A schema refers to organized packets of information about the situation, task, events, etc., which are acquired through experience and instructions. When a behavior is based on well-established skills or habits, perceived stimulus automatically activates certain schemas linked to it, and the schemas lead to appropriate actions. When activities come into conflict, the contention scheduling intervenes to resolve the conflict by adopting simple rules and setting priorities on the activities. The contention scheduling is regarded to work relatively automatically. When the automatic control of behavior is unable to sufficiently respond to the situation because of its novelty or complexity, the supervisory attention system (SAS) intervenes in the streams of the automatic behavior routines. The concept of the supervisory control has been used for envisaging the central executive of working memory system (Baddeley 1986, 1997).

The SRK model and the Norman and Shallice model have similar structures with regard to the explanation of human behavior. The skill-based level is comparable to the control based on automatic schema activation; the rule-based level is comparable to the control intervened by the contention scheduling; and the knowledge-based level is comparable to the control by the SAS. These two models are useful to describe the internal processes of human behavior and especially of human behavior in the real world.

### *1.2.2 Attention in the Human Information Processing*

Attention is an important concept for describing human behavior. It is obvious that we can do many things without attention, but we need to use full attention when we face an unfamiliar or difficult task. The three levels of the SRK model have different relations to attention. Attention is not required at the skill-based level, because the direct link from perceived stimulus to the activation of sensori-motor pattern is completely automated. Attention is only minimally required at the rule-based level. Attention is needed to perform each stage of processing at the knowledge-based level. In any case, attention should be taken account of in order to model human behavior in terms of human information processing.

There are three forms of attention: selective attention, divided attention, and sustained attention (Schmeichel and Baumeister 2010). 'Selective attention' refers to the function of selecting information, i.e., focusing attention on one object and ignoring other aspects in the environment. *The cocktail party phenomenon* (Cherry 1953) is such that we can select one voice to attend in the noisy circumstance, such as a cocktail party. The phenomenon is well known as an effect relating to the auditory selective attention. In the early studies of attention, the locus of selection of auditory message has been extensively studied by adopting a procedure called '*dichotic listening*.' In the dichotic listening procedure, different messages are presented to each ear by headphones. Participants are asked to attend to one of the two messages, and to repeat aloud (shadow) it while ignoring the other. After doing this task, participants are asked to answer questions about the unattended message. The typical result is that participants cannot answer what the unattended message says, but can identify the physical features of it, such as the gender of the voice. Interestingly, some subjects can detect their own name from the unattended messages, suggesting that semantic processing is not totally impossible for the unattended information. To explain these phenomena, several models of auditory selective attention have been developed, such as the filter theory (Broadbent 1958), the attenuation theory (Treisman 1964), and the late selection model (Deutsch and Deutsch 1963). These models are concerned with the problem of when the attended auditory information is selected.

'Divided attention' refers to the function of allocating attention to several sources or streams of information simultaneously. It is necessary for post-selection information processing to invest attention as a mental 'resource' or 'fuel'. Attentional resources are divided and allocated to several streams of information processing.

Wickens and McCarley (2008) propose a simple model of attention (Fig. 1.2). In this model, information from external and internal events is filtered and selected at the first stage of processing. Thus, selective attention is regarded as the process of information filtering to avoid the capacity-limited cognitive processes being overloaded. Filtering is affected by such factors, as expectancy, value, salience, and effort. For example, if information is expected to be important and valuable, it is likely to be selected. These factors pertain to the top-down processing, the process consciously controlled to achieve the task goal. Any stimulus with salient features, such as high luminance and prominent color, is easy to be detected, because it automatically captures attention.

Selected information is stored and processed in the working memory. Attentional resources are needed if information processing is to be intentionally executed. Usually, there are several ongoing cognitive processes working at the same time, and so attentional resources are divided and separately supplied to them.

The role of attention can be formulated in the SRK framework. Selective attention is mainly related to the feature formation stage and the recognition stage. Divided attention contributes to the stages of knowledge-based level

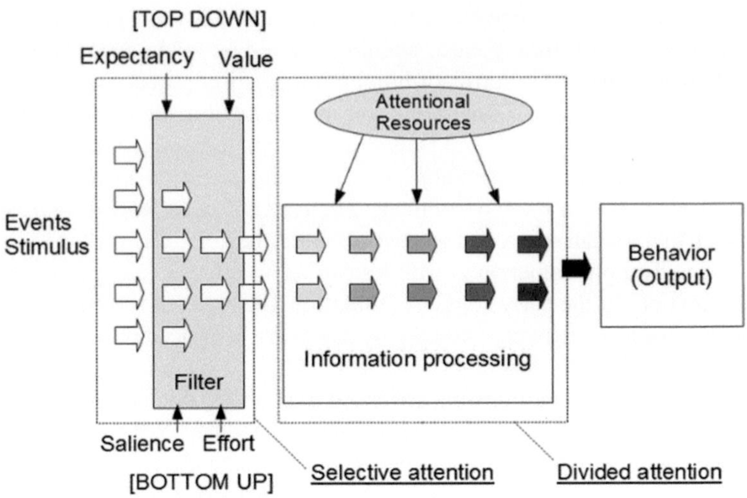


**Fig. 1.2** A simple model of attention (Wickens and McCarley 2008)

processes, and to those of rule-based level processes to a lesser extent. The skill-based level processing is usually unrelated to attention control; but if the automatic activation of action is to be inhibited, divided attention is used to interrupt the link between sensory input and action (Shinohara 2011).

## 1.3 Perceptual Processes

Perception is "the acquisition and processing of sensory information in order to see, hear, taste, or feel objects in the world and it also guides an organism's actions with respect to those objects" (Sekuler and Blake 2001). There are numerous important findings and topics on human perception, such as illusion, depth and size perception, perceptual organization, and so on. For lack of space, this section only reviews the acquisition and selection of visual information as an important perceptual process. The concept of *visual focused attention* is introduced in order to describe this process.

### *1.3.1 Orientation of Visual Attention*

The popular metaphor that visual focused attention is like a *spotlight* has been used for describing the character of visual attention (Posner et al. 1980). This metaphor assumes that information in the spatial area on which the spotlight is casted is rapidly selected and efficiently processed, and that attention can move as the

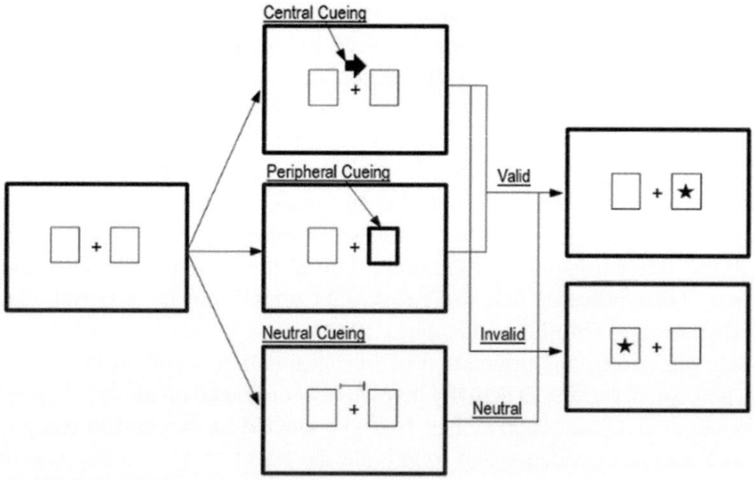


**Fig. 1.3** Event sequence in the spatial cueing paradigm

direction of spotlight changes. To study the movement of attentional spotlight, or in other words, to study the characteristics of the orientation of visual attention, the spatial cueing paradigm has been adopted (Posner 1980; Posner et al. 1980).

In the typical spatial cuing paradigm (Fig. 1.3), participants are required to fix their eyes on a fixation point, and to make a response when they detect the onset of the target stimulus. Before presenting the target, a cue pointing to the position of the target is provided. The validity of the cue is systematically manipulated; a *valid* cue actually indicates the position of the target and an *invalid* cue does not. The cue does not always indicate a specific location, in which case the cue is called *neutral*. Typically, the valid cue is presented in the majority of trials. Participants usually move their attention to the position that the cue indicates. When the cue is invalid, participants have to shift their attention to the position where the target is actually presented. Since the presentation of the cue and target is only for a short time and participants are instructed not to move their eye during the trial, the orientation of visual attention in the spatial cueing paradigm is covert.

There are two types of cues: central cues and peripheral cues. A central cue is presented near the fixation point in the form of a symbol or character, and semantically indicates the position at which the target is supposed to appear. Participants are expected to actively move their attention according to the central cue. A peripheral cue is presented at the position of the target. A peripheral cue indicates the position of the target physically and directly. For example, a brief illumination of a possible location of the target or a changing color of the placeholder in which the target is subsequently presented has been used as a peripheral cue. It is expected that visual attention is automatically captured by the onset of a peripheral cue. The shifts of attention induced by a central cue and a peripheral cue are called the ‘endogenous orientation’ and the ‘exogenous orientation,’ respectively (Jonides 1981).

Cost-benefit analysis is used to analyze the result of a spatial cuing paradigm: the cost is the difference in reaction time between the invalid condition and the neutral condition, and the benefit is the difference in reaction time between the valid condition and the neutral condition. Thus, the cost reflects how long it takes to move attention from the cued position to the un-cued position, and the benefit reflects how much time is saved in attending to the cued position. Not surprisingly, the typical result is that reaction is faster and more accurate in valid trials than in invalid trials. When the cue-target interval is short, the benefit of valid cue is obtained for the exogenous orientation, but not obtained for the endogenous orientation. This indicates that the exogenous orientation by a peripheral cue is rapid and automatic (Müller and Rabbitt 1989).

Though the exogenous orientation of visual attention seems to be a simple and transient perceptual process, it actually involves a coordinated operation of several brain areas. Posner et al. (Posner and Cohen 1984) conducted an experiment using a spatial cuing task for neglect patients, and found that the effect of cue validity was different among participants with different areas of lesion. By analyzing the pattern of the results, Posner et al. suggested that there were several processes of attentional control, and that the activities of the posterior parietal lobe, the superior colliculus, and the pulvinar nucleus are respectively responsible for disengaging attention from the previously attended position, shifting attention from the disengaged position to the newly oriented stimulus, and engaging attention to stimulus (Posner and Peterson 1990).

### *1.3.2 Spatial Width of Visual Attention*

The spotlight metaphor of visual attention mentioned above assumes that visual attention is distributed around the fixation point, and that it can be moved to the location where detailed visual information can be acquired. Some researchers have emphasized that the spatial area over which visual attention is distributed is variable according to the task requirement.

While the spotlight metaphor is useful to describe the characteristics of the orientation of visual attention, the zoom lens metaphor (Eriksen and St. James 1986) is proposed to describe the variable size of the area of visual attention. For example, LaBerge (1983) presents a five-letter word and asks participants one of the two questions: whether or not the middle letter in a word is a target and whether or not the word is a name (Fig. 1.4). The former instruction leads participants to focus their attention on the middle letter, and the latter instruction does to focus their attention on the whole word. Additionally, some trials present a row of #s with a probe letter at each letter of a word, and asks participants to respond to the probe. When participants focus on the middle letter in a word, the response to the probe appeared in the position of the middle letter is fastest, and the response to the probe farthest from the center is slowest. When attention is directed to the whole word, the speed of response to any position is constant. These results suggest that the area of visual attention varies with the task requirement. Eriksen and St. James (1986) examine how the deployment of visual

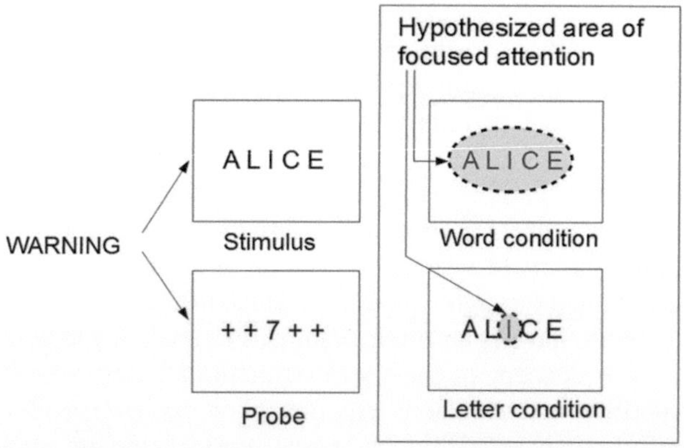


**Fig. 1.4** Hypothesized area of focused attention in LaBerge (1983)

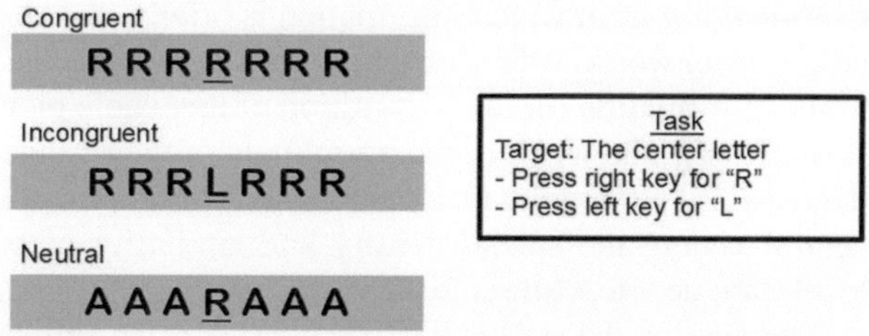


**Fig. 1.5** The flanker compatibility paradigm

attention is controlled. In their study, what position is to be attended is indicated by providing cues immediately before presenting stimulus. They have found that the reaction time to stimulus depends on the number of cues and the time interval between cue presentation and stimulus onset. This result suggests that it takes time to manipulate the size of visual attention.

The flanker compatibility paradigm (Eriksen and Eriksen 1974) is useful to measure the size of visual attention distributed around the fixation point. In this paradigm (Fig. 1.5), several letters are presented in line, and each participant is asked to respond to the target letter, the one located in the middle, by pressing a certain key, and to ignore the flanker letters, the ones flanking the target letter. Target and flanker letters may be the same, and only distinguished by location. Responses in the congruent trial in which both the target and the flanker are assigned to the same key are usually much faster than in the incongruent trial in which the response to the target is different from that to the flanker. This phenomenon is called the 'flanker-compatibility effect.' This effect is thought to be caused by the conflict of response information. Both the target and the flanker automatically activate a response based on the pre-determined response mapping. When the target and the flanker activate different responses, the response to the target interferes with the response to the flanker, and a response conflict arises. It is known that the spatial separation between the target and the flanker has a strong influence on the flanker compatibility effect; as the distance between the target and the flanker increases, the flanker compatibility effect decreases (Eriksen and Eriksen 1974; Miller 1991). This is evidence that information in the area of visual attention distributed around the target is inevitably selected and processed. Interestingly, a slight spatial separation, 1 degree of visual angle, can reduce the flanker compatibility effect, suggesting that it can reflect the minimum size of the area of visual attention.

It is important for behavior in the real world to coordinately control the movements of fixation points and the varying field of visual attention around each fixation point for optimal information acquisition from the surrounding situation. Optimized visual information acquisition is a kind of cognitive skill which is obtained through experience. Thus, the pattern of fixation and the width of visual attention have been often examined to analyse the behavior in the real world.

The concept of the useful (functional) field of view, which is similar to the spotlight and the zoom lens metaphor, has been widely used for analyzing visual information acquisition in a real world behavior, such as car driving (e.g., Clay et al. 2005). The definition of the useful field of view is that "the area around the fixation point from which information is briefly stored and read out during a visual task" (Mackworth 1965), or that "the area around the fixation point from which usable information for the recognition of the whole picture is extracted" (Saida and Ikeda 1979). The width of the useful field of view ranges from approximately 4–20 degrees, and varies with many mental and environmental factors, such as the load of central vision, the spatial density and similarity of background objects, arousal level, fatigue, etc. (Miura 2012).

The size of the useful field of view is often critical for real-world tasks. For example, driving a car depends on visual information, and requires the driver to obtain visual information by continuously moving the fixation point around the scene. In essence, it is desirable that the driver can obtain as much information as possible from the area around the fixation point. In other words, if the useful field of view is wide, it is better for safe driving. Miura (1986, 2012) measured the useful field of view of drivers in the real-world setting by using the dual task paradigm. He required participants to drive a car while detecting and responding by pressing a key to the visual stimulus which was occasional emission of LEDs attached on the front window. The size of the useful field of view was estimated on the basis of the reaction time to the visual stimulus. He found that drivers adaptively modulated the size of the useful field of view as the need arose in the driving situation. The movement of visual fixation is minimum, and the size of the useful field of view is large in the low demand conditions, such as highway driving; whereas drivers frequently move their fixation point, and the size of useful field of view is small, in the high demand conditions, such as driving in a congested downtown area.

### *1.3.3 Visual Search and Feature Integration Theory*

In the real world, we often have to find a particular object among many objects by vision. We move our eyes and sequentially check each visual object. This is a visual search task. It has been frequently used as an experimental task to investigate visual attention. A spatial cuing task is used to examine the covert orientation, the movement of attention without eye movement. A visual search task is used to examine the overt orientation, the movement of attention with eye movement.

Typically, a visual search task requires searching for a target defined by physical features (e.g., white circle) in a display with distractors. Participants judge whether the target is in a display or not, and respond by pressing a response key as immediately as possible. How long participants take to detect the target is analyzed. When the target is defined by simple features, e.g., the target is a red circle and the distractor is a blue circle, search time does not depend on the number of visual objects in the display. This indicates that the feature of the target automatically stands out and no search is needed. On the contrary, when the target is defined by a conjunction of several features, e.g., the target is a red circle and the distractor is either a circle or a rectangle and colored either red or blue, search time depends on the number of visual objects. Thus, search time increases as a function of the number of visual objects. The slope of the search time function indicates the efficiency of visual search.

Treisman and Gelade (1980) propose the feature integration theory (Fig. 1.6) of perception of objects. In this model, when visual information is in a particular location of the retina at which the spotlight of attention is directed, basic features of visual information are automatically processed on the basis of separate feature maps, such as color map or orientation map, and then these features of information are combined to create an object. Attention is required to combine features into an object. An object is a temporal representation of visual target, and it is further processed by a higher-order cognitive mechanism, such as recognition network. The target can be found in the feature search without any attentional process; the defining features of the target are processed in a parallel manner. Each visual object must be sequentially searched in the conjunction search; the target can be perceived after its features are combined by the attention demanding process. This process of combining features of a target is called 'feature binding'.

As the primary support for the feature integration theory, Treisman and Schmidt (1982) reports the phenomenon of illusory conjunctions which is a binding error occurred when attention is diverted from the visual display containing several objects. Several studies (e.g. Nakayama and Silverman 1986; Duncan and Humphreys 1989) have reported the findings that cannot be explained by the original feature integration theory, and the feature integration theory has been revised (Treisman and Sato 1990; Treisman 1993).

### *1.3.4 Scene Perception and Coherence Theory*

When an original image A and a partially modified image B repeatedly alternate with a brief blank field between the successive images, it is surprisingly difficult to identify where the modified part is. This phenomenon occurs even when the modification is quite large or when viewers expect that something is different between the images.

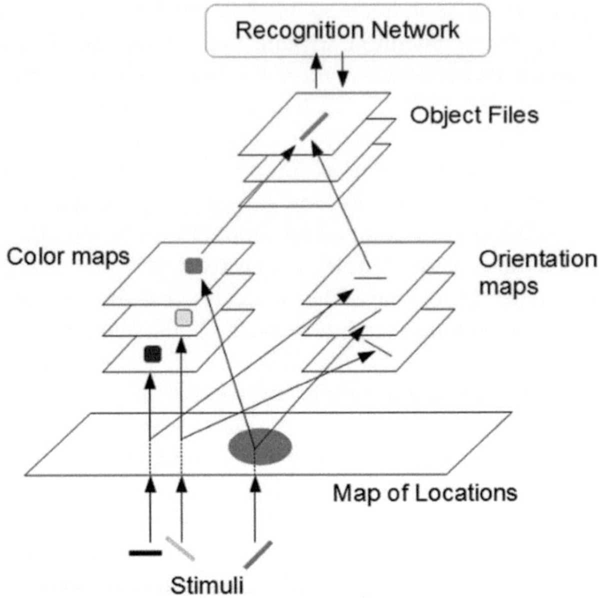


**Fig. 1.6** Feature integration theory

This phenomenon is called 'change blindness' (Simons and Levin 1997; Simons and Ambinder 2005). O'Reagan et al. (1999) report that change blindness occurs even without a blank field; when a brief visual disruption, e.g., an image looking like a "mudsplash," is presented at the moments of switching images, observers fail to notice differences between them. This suggests that change blindness is related not to masking or occlusion but to selection and representation of visual information. Lavie (2006) conducts a neuroimaging experiment to examine the brain activity during the change blindness task. He has found that the fusiform gyrus, the bilateral parietal lobe, and the prefrontal cortex are active. This finding suggests that frontoparietal activity has a role in visual awareness.

Coherence theory (Rensink 2000, 2002) is able to explain the perception of change (Fig. 1.7a). In this model, incoming visual stimuli are continuously processed, and proto-objects are created. Proto-objects are volatile and last only for a short time. They are easily replaced by new stimuli at their locations. Focused attention selects several proto-objects, and they set up a coherence field, i.e., a reciprocal connection between proto-objects and a single higher level nexus. A higher level nexus pools information contained in an object by summing all inputs.

As for change blindness, to judge which part (object) in the image changes or differs requires object perception. It is difficult to notice a change in any proto-object. When attention is released from a particular location in the image due to a brief blank field or a visual disruption between the images, a coherence field disappears and an object divides into volatile proto-objects.

Rensink (2000) proposes the concept of a virtual representation, according to which a coherent representation of items required for performing the current task is

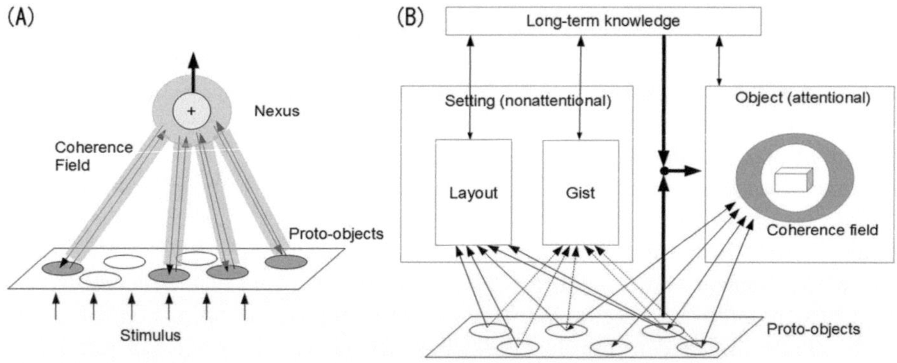


**Fig. 1.7** Coherence theory (**a**) and Triadic architecture (**b**) (Rensink 2000, 2011)

created in seeing a scene. It is important that a representation of all items in a scene is not created and coherent representations are created in a just-in-time manner; in other words, they are created when they are needed. Rensink (2000, 2002, 2011) proposes the triadic architecture (Fig. 1.7b) comprising three independent systems: the early visual system, the object system, and the setting system. The early visual system involves an automatic process of creating proto-objects; the object system involves an attentional process to construct coherent fields by linking proto-objects to a nexus; and the setting system is a non-attentional system responsible for the process of extraction of the abstract meaning (gist) and the spatial arrangement of objects in the scene (layout information). Both the attentional "object" system and the non-attentional setting system are linked to the long-term memory, i.e., the memory which activates stored knowledge of the relevant objects and scenes. The former system is executed in a top-down fashion: the access to the long-term memory is intentionally controlled on the basis of the meaning and the importance of each object and scene. The latter system is executed in a bottom-up fashion: it is automatically and compulsory triggered by salient stimuli in the scene.

## 1.4 Cognitive Processes

The information selected by the perceptual processes is further processed in the working memory system. It is necessary for information to be consciously processed that attention is allocated to the process at work. The function of attention in the perceptual processes is to select the necessary information to perform the task out of a large amount of information coming from the surrounding environment. The function of attention in the cognitive processes is to maintain the processes for the selected information and to inhibit the processes automatically activated by external stimuli or by internal signals retrieved from the long term memory or schema.

### 1.4.1 Attentional Resources and Multitasking

When we have multiple tasks to do, it is necessary to divide our attention among them. Especially when tasks are difficult and/or at high stakes, it is crucially important how to allocate attention. Some people may believe that they have an excellent capacity to multitask. It has been reported that some people, who are called "super taskers," show excellent multitasking performances (Watson and Strayer 2010). However, the ability to multitask is in principle limited. For example, when Charron and Koechilin (2010) examined the role of the medial frontal cortex (MFC), the brain region involved in motivating and selecting behavior, both in the single task and in the dual task, they found that the right and left MFCs processed two separate task goals concurrently, and the anterior prefrontal cortex (APC) coordinated these two processes. This suggests that the multitasking ability is limited and can pursue only two concurrent goals. Therefore, even if one seems to simultaneously perform many tasks, it is actually based on the management of two cognitive processes in a time sharing manner. As for divided attention in multitasking, allocation of attention has two purposes: to maintain the processes for each task and to control the time sharing process.

Kahneman (1973) proposes the *unitary-resource theory* according to which common attentional resource, which is closely related to physiological arousal, can be used to sustain a wide variety of task performances (Fig. 1.8). Attentional resources, albeit limited, are allocated to several tasks as they are necessary; allocation of attentional resources depends on the allocation policy. When several tasks are performed simultaneously, the performance of each task depends on the amount of resources each task demands and the amount of supplies from the general resource pool. Capacity demands are evaluated after tasks are performed, and then the allocation policy is modified. The amount of resources can vary as the arousal level changes; the amount of mental resources needed for a performance may be

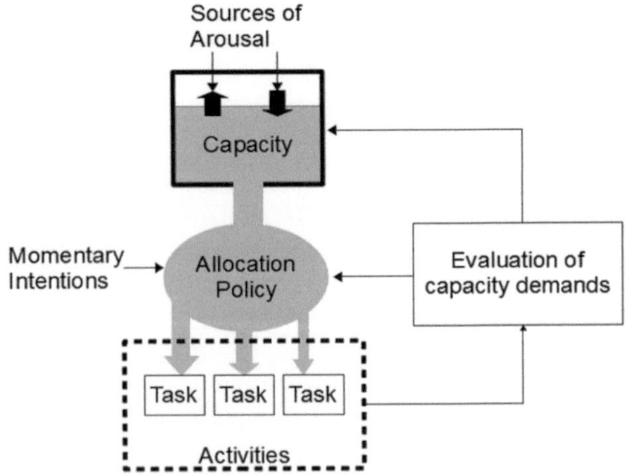


**Fig. 1.8** Unitary-resource model (Kahneman 1973)

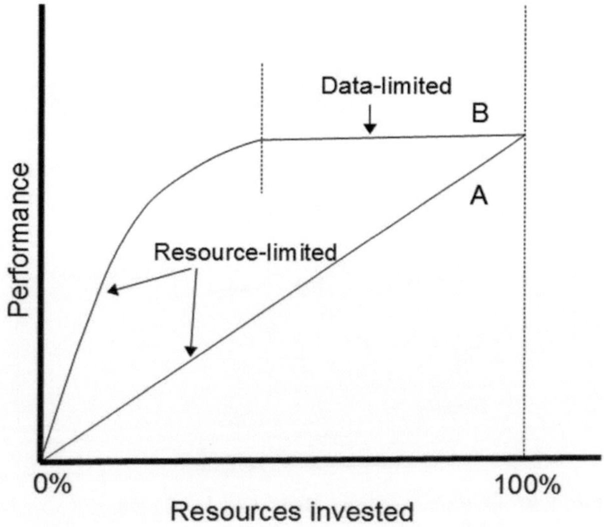


**Fig. 1.9** The performance resource function

explained by the difficulty of the performed task (Wickens and McCarley 2008). As the Yerkes Dodson Law (Yerkes and Dodson 1908) indicates, task performance is not a simply-increasing linear function of the amount of available attentional resources. The optimal amount of attentional resources for a task is presumed to be determined by the characteristics of the task, such as its difficulty and complexity.

Task performance usually increases with the amount of attentional resources allocated to the task. However, task performance is affected by the characteristics of information processing during the task, as well as attentional resources. To describe the relationship between task performance and attentional resources, the performance resource function (Norman and Bobrow 1975) has been used (Fig. 1.9). If the task is difficult and must be performed intentionally, the task performance linearly improves as the amount of attentional resource increases (curve A). The task with this feature is *resource-limited*. On the contrary, if the task is easy, well automated to some extent, or if perceptual and/or memorized information for the task is sufficient, the task performance can be maximal with a little amount of attentional resources; there is no improvement in performance if more attention is allocated (curve B). The task with this feature is *data-limited*.

Though the general resource theory is useful to explain many phenomena concerning attention and performance, it cannot explain certain aspects of task performance. When we drive a car while listening to music from the radio, it causes no problem in most cases. This example suggests that different kinds of attention are involved in visual task and auditory task performances.

Wickens proposes the *multiple resource theory* (Fig. 1.10) (Wickens 1984, Wickens and Hollands 1999, Wickens and McCarley 2008). In this model, there are five dimensions of attention resources: the stages of processing, modalities, processing codes, responses, and visual processing. An important assumption of the

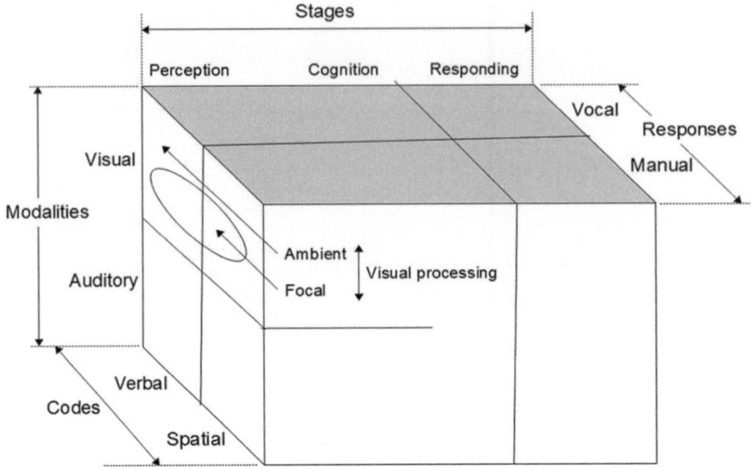


**Fig. 1.10** Multiple resources model (Wickens 1984, Wickens and Hollands 1999, Wickens and McCarley 2008)

multiple resource view is that task performance is not disrupted when different attention-demanding tasks depend on different dimensions of attention.

The "stages of processing" dimension involves perceptual and cognitive processing, and selection and execution of response. It has been found in dual task studies that manipulating the difficulty of response does not affect the concurrent performance of perceptual and cognitive tasks (e.g., Wickens and Kessel 1980).

The "modalities" dimension is intuitively easy to understand with an example. It is easy in most situations to read a book while listening to music; this is so especially when the book is readable and the music is relaxing. Studies of human-machine interface have shown that cross-modal displays are better than intramodal displays (e.g., Wickens et al. 1983).

In addition, focal and ambient vision are distinguished with regard to visual processing channels. Wickens and Hollands (1999) suggests that focal vision is used for fine detail and pattern recognition, and ambient vision involving peripheral vision is used for sensing orientation and ego motion. These two aspects of visual processing have different roles in acquiring visual information, and contribute to efficient time-sharing among concurrent visual tasks. For example, when we drive a car, we obtain information about other cars, road signs and signals, by focal vision. At the same time, we use ambient vision to obtain information about the location, speed, and direction of the car we drive.

The "processing codes" dimension is related to the form of information coded in the working memory. As is previously discussed, there are storages for verbal, categorical, or symbolic information, and for analog and spatial information in the working memory system. The separation of spatial and verbal resources is demonstrated by the classic study of Brooks (1968). In his study, participants were asked to visually imagine a block capital letter in their mind and navigate through it. Then, they were asked to fixate their "mind's eye" at the bottom left of the letter and examine whether each corner involved the bottom or top line of the letter in a clockwise, one-by-one manner. Participants answered "Yes" if the corner was on the bottom or top line and "No" if it was not, either by verbal response or by pointing to the word on the paper. The result showed that it is far more difficult to respond by pointing than by verbal response, suggesting that the verbal codes for verbal response, not the spatial codes for pointing, interfere with the visuospatial codes dealing with the letter.

The processing codes dimension is closely related to the "responses" dimension. Wickens et al. have found that a manual tracking task and a discrete verbal task are performed simultaneously with minimal interference. This finding suggests that manual responses, including tracking and pressing, are usually spatial (Wickens et al. 1983, Wickens and Liu 1988).

The dual-task paradigm has been extensively used in the psychological study of attention. In the dual task paradigm, participants are required to perform two tasks simultaneously or each task singly. Manipulating experimental factors, such as combination of tasks, task difficulty, and task priority, task performance in the dual task condition is compared with task performance in the single task (baseline) condition. In many cases, and particularly in the cases of real world task, primary and secondary tasks are distinguished: the former has high priority and is more important than the latter. Assuming with the general resource view that common attentional resources are shared between tasks and the primary task is given priority over the secondary task, the performance of the secondary task is likely to be affected by experimental manipulation. Thus, the performance of the secondary task is used as an index of residual attention which is not allocated to any cognitive processes and is still available for additional allocation. This unused attentional resource is called 'spare' attentional resources. This type of dual task paradigm is called the 'subsidiary task paradigm.'

Given the multiple resource view, the amount of interference in performance of combined two tasks can be examined. If no interference is observed, it is inferred that two tasks do not share the common dimension of attentional resources.

### *1.4.2 Working Memory and Attention*

The working memory is responsible both for holding information for a short term and for actively processing information by interacting with the long term memory. According to Baddeley (2000), the working memory system has a central executive system and sub-systems: the latter include a phonological loop for holding speech-based information, a visuospatial sketchpad for holding spatial and visual coded information, and an episodic buffer for holding and integrating a variety of information. There are interactions between each sub-system and its corresponding function in the long term memory. The function of the working memory has been studied in psychological experiments, and the brain activities underlying each component of the working memory have been extensively investigated in the neuroimaging and neurological studies of patients with brain lesions (see Chap. 3).

The concept of the working memory is closely related to that of attention. Baddeley (1993) states that the general term 'attention' is used to refer to the control processes operating throughout the working memory system, and that the label "working attention" would have been used if the working memory system had been studied with a focus on its control mechanisms. Furthermore, Baddeley and Logie (1999) describes the role of the working memory as a mediator of conscious awareness that maintains and coordinates "information from a number of sources including the present, specific episodes from the past, and projections to the future". This function seems to include attentional processes.

The central executive is regarded as an attentional system, which is a modality-free component operating in the whole cognitive system (Baddeley and Logie 1999). It has endogenous functions of receiving and processing information from the external world to adjust internal task goals, and of selectively generating actions. Baddeley (1996) proposes four functions of the central executive: focusing attention, dividing attention across different sources, switching attention between tasks, and using attention to link the working memory with the long term memory. As is mentioned before, the supervisory attention system (SAS) proposed by Norman and Shallice (1986) is a possible model for the central executive.

The capacity of the working memory is different from person to person, and typically assessed by the span task. The capacity often affects the performance of attentional tasks. For example, when a dichotic listening task requires attending to the message from one ear and ignoring the message from the other ear, participants with low working memory span notice their name contained in the unattended message more often than participants with high working memory capacity (Conway et al. 2001). This finding seems counter-intuitive at first glance, but it can be reasonably interpreted based on the function of working memory. While participants with high capacity efficiently focus on the message from one ear and ignore the message from the other ear, participants with low capacity inevitably pay attention to the message from the other ear. This is because the latter subjects cannot control attention well, compared to the former subjects.

Neuroimaging studies (e.g. Bunge et al. 2001; D'Esposito et al. 1995) have revealed that the brain areas responsible for the functions of the central executive are the dorsolateral prefrontal cortex (DLPFC) and the anterior cingulate cortex (ACC). The DLPFC is responsible for control attention in the dual task performance, and the ACC is involved in conflict monitoring and response inhibition. Osaka et al. (2004) examine focused attention in the performance of the reading span test with fMRI, and find that the left DLPFC, the ACC, and the left superior parietal lobule (SPL) are activated. They propose that the SPL has a role of visual attention controller, and fixate and shift attention in accordance with the ACC and the DLPFC (Osaka and Osaka 2007).

Baddeley's model of the working memory can be regarded as a kind of multiple resource model. For example, phonological information and visuospatial information are separately stored in the phonological loop and the visuospatial sketchpad, respectively. This separation between phonological process and visuospatial process is similar to the "processing codes" dimension of Wickens' multiple resource theory (see Sect 1.3.1).

In the study of the working memory, as well as in the neuropsychological approaches, the dual task paradigm has been used to characterize the usage of the working memory in performing a particular task. A *loading task* imposes a processing load exclusively on a particular component of the working memory to suppress its function. When the component of the working memory essential for the task in question is suppressed by the loading task, the performance of the primary task is expected to be significantly impaired. The articulatory suppression suppresses the phonological loop by requiring reading aloud simple words repeatedly; the spatial tapping suppresses the visuospatial sketchpad by requiring pressing spatially arranged keys in an instructed order; and the random number generation suppresses the central executive by requiring continuously producing digits or characters in a random order. For example, Robbins et al. (1996) examine how chess players use their working memory to select a chess moves in a dual task experiment. Chess players are asked to select moves while performing loading tasks. The experiment reveals that the secondary task supressing the central executive and the visuospatial sketchpad impairs the quality of selected moves; the loading task has little effect on the phonological loop. These results suggest that the cognitive skills of chess depend on the processes involved in the central executive and the spatial sketchpad.

## 1.5 Summary

In this chapter, several models and concepts of human information processing pertaining to perceptual and cognitive processes have been reviewed. A wide variety of models for human information processing in perception and cognition have been proposed and examined by researchers. New findings have accumulated in many ways: psychological experiments with well-developed tasks and paradigms, neuropsychological studies with functional brain imaging and physiological techniques, and neurological studies of patients with brain damages or decreased brain functions. On the basis of the proposed models, we can understand how we perceive and recognize surrounding environments and how we decide to behave in response to them.

It should be noted that it is important not only to elaborate a model by examining the relationships between the functions of perceptual and cognitive processes and the brain activities, but also to consider human behavior in the real world. Human behavior in the real world depends on real-time, simultaneous processing of sensory, perceptual, and cognitive information. Obviously, it is very difficult to establish a model of human behavior which can comprehensively explain behavior in all sorts of situation. At the first stage of explanation, it is necessary to accumulate observational and case data of human behavior in particular situation, and to interpret the underlying structure of psychological processes by referring to the basic models of perception and cognition. An integrative model of human cognition, which is necessary for robots in the future behaving as well as human, will be attained through accumulating these local models of human behavior. If the model is perfect in the true sense, robots operated by the model may make various mistakes as if humans do.

## Exercise

Distracted driving happens when one drives a car while engaging in other activities, such as mobile phone use and texting. It can take driver's attention away from driving and induce traffic accidents. Describe the degradation of driver's abilities while distracted driving on the basis of the psychological models of human perception and cognition discussed in this chapter.

## References

Baddeley, A.D.: Working Memory. Oxford University Press, New York (1986)

Baddeley, A.D.: Working memory or working attention? In: Baddeley, A., Weiskrantz, L. (eds.) Attention: Selection, Awareness and Control. A Tribute to Donald Broadbent, pp. 152–170. Clarendon Press University, Oxford (1993)

Baddeley, A.D.: Exploring the central executive. Q. J. Exp. Psychol. **49**, 5–28 (1996)

Baddeley, A.D.: Human Memory: Theory and Practice Revised Edition. Psychology Press, Hove (1997)

Baddeley, A.D.: The episodic buffer: a new component of working memory? Trends Cogn. Sci. **4**, 417–423 (2000)

Baddeley, A.D., Logie, R.H.: Working memory: the multiple-component model. In: Miyake, A., Shah, P. (eds.) Models of Working Memory: Mechanisms of Active Maintenance and Executive Control. Cambridge University Press, New York (1999)

Broadbent, D.: Perception and Communication. Penguin Press, London (1958)

Brooks, L.: Spatial and verbal components of the act of recall. Can. J. Psychol. **22**, 349–368 (1968)

Bunge, S.A., Ochsner, K.N., Desmond, J.E., Glover, G.H., Gabriel, J.D.E.: Prefrontal regions involved in keeping information in and out of mind. Brain **124**, 2072–2086 (2001)

Charron, S., Koechlin, E.: Divided representation of concurrent goals in the human frontal lobes. Science **16**, 360–363 (2010)

Cherry, E.C.: Some experiments on the recognition of speech, with one and with two ears. J. Acoust. Soc. Am. **25**, 975–979 (1953)

Clay, O.J., Wadley, V.G., Edwards, J.D., Roth, D.L., Roenker, D.L., Ball, K.K.: Cumulative meta-analysis of the relationship between useful field of view and driving performance in older adults: current and future implications. Optom. Vis. Sci. **82**, 724–731 (2005)

Conway, A.R.A., Cowan, N., Bunting, M.F.: The cocktail party phenomenon revisited: the importance of working memory capacity. Psychon. Bull. Rev. **8**, 331–335 (2001)

D'Esposito, M., Detre, J.A., Alsop, D.C., Atlas, R.K., Grossman, M.: The neural basis of the central executive system of working memory. Nature **378**, 279–281 (1995)

Deutsch, J.A., Deutsch, D.: Attention: some theoretical considerations. Psychol. Res. **70**, 80–90 (1963)

Duncan, J., Humphreys, G.W.: Visual search and visual similarity. Psychol. Rev. **96**, 433–458 (1989)

Eriksen, B.A., Eriksen, C.W.: Effects of noise letters upon the identification of a target letter in nonsearch task. Percept. Psychophys. **16**, 143–149 (1974)
Eriksen, C.W., St. James, J.D.: Visual attention within and around the field of focal attention: a zoom lens model. Percept. Psychophys. **40**, 225–240 (1986)
Jonides, J.: Voluntary versus automatic control over the mind's eye's movement. In: Long, J.B., Baddeley, A.D. (eds.) Attention and Performance, IX, pp. 187–203. Lawrence Erlbaum, Hillsdale (1981)
Kahneman, D.: Attention and Effort. Prentice-Hall, Englewood Cliffs (1973)
Kahneman, D.: Thinking, Fast and Slow. Penguin, London (2011)
Kahneman, D.: Maps of Bounded Rationality: A Perspective on Intuitive Judgment and Choice. Nobel Prize Lecture. http://www.nobelprize.org/nobel_prizes/economics/laureates/2002/kahnemann-lecture.pdf (2002)
LaBerge, D.: Spatial extent of attention to letters and words. J. Exp. Psychol. Hum. Percept. Perform. **9**, 371–379 (1983)
Lavie, N.: The role of perceptual load in visual awareness. Brain Res. **1080**, 91–100 (2006)
Mackworth, N.: Visual noise causes tunnel vision. Psychon. Sci. **3**, 67–68 (1965)
Miller, J.: The flanker compatibility effect as a function of visual angle, attentional focus, visual transients and perceptual load: a search for boundary conditions. Percept. Psychophys. **49**, 270–288 (1991)
Miura, T.: Coping with situational demands: a study of eye movements and peripheral vision. In: Gale, A.G., Freeman, M.H., Haslegrave, C.M., Smith, P., Taylor, S.P. (eds.) Vision in Vehicles, pp. 205–216. Elsevier Science Publishers, B.V, Amsterdam (1986)
Miura, T.: Visual behavior: research on eye movement, useful field of view, and depth attention relating to this study. In: Miura, T. (ed.) Visual Attention and Behavior: Bridging the Gap Between Basic and Practical Research. Kazama Shobo, Tokyo (2012)
Müller, H.J., Rabbitt, P.M.A.: Reflexive orienting of visual attention: time course of activation and resistance to interruption. J. Exp. Psychol. Hum. Percept. Perform. **15**, 315–330 (1989)
Nakayama, K., Silverman, G.H.: Serial and parallel processing of visual feature conjunctions. Nature **320**, 264–265 (1986)
Norman, D.A., Bobrow, D.G.: On data-limited and resource-limited processes. Cogn. Psychol. **7**, 44–64 (1975)
Norman, D.A., Shallice, T.: Attention to action: willed and automatic control of behavior. In: Davidson, R.J., Schwarts, G.E., Shapiro, D. (eds.) Consciousness and Self-Regulation. Advances in Research and Theory, vol. 4, pp. 1–18. Plenum Press, New York (1986)
O'Reagan, J.K., Rensink, R.A., Clark, J.J.: Change-blindness as a result of 'mudsplashes'. Nature **398**, 34 (1999)
Osaka, M., Osaka, N.: Neural bases of focusing attention in working memory. In: Osaka, N., Logie, R.H., E'sposito, M. (eds.) The Cognitive Neuroscience of Working Memory, pp. 99–118. Oxford University Press, Oxford (2007)
Osaka, N., Osaka, M., Kondo, H., Morishita, M., Fukuyama, H., Shibasaki, H.: The neural basis of executive function in working memory: an fMRI study based on individual differences. Neuroimage **21**, 623–631 (2004)
Posner, M.I.: Orienting of attention. Q. J. Exp. Psychol. **32**, 3–25 (1980)
Posner, M.I., Cohen, Y.: Components of visual orienting. In: Bouma, H., Bouwhuis, D.G. (eds.) Attention and Performance X, pp. 531–556. Erlbaum, Hillsdale (1984)
Posner, M.I., Peterson, S.E.: The attention system of the human brain. Annu. Rev. Neurosci. **13**, 25–42 (1990)
Posner, M.I., Snyder, C.R.R., Davidson, B.J.: Attention and the detection of signals. J. Exp. Psychol. Gen. **109**, 160–174 (1980)
Rasmussen, J.: Skills, rules, and knowledge; signals, signs and simbols, and other distinctions in human performance model. IEEE Transactions on Systems, Man, and Cybernetics, SMC-13, 257–266 (1983)

Rasmussen, J.: Information Processing and Human-Machine Interaction: An Approach to Cognitive Engineering. North Holland, New York (1986)
Rasmussen, J.: The definition of human error and a taxonomy for technical system design. In: Rasmussen, J., Duncan, K., Leplat, J. (eds.) New Technology and Human Error, pp. 23–30. Wiley, Chichester (1987)
Rensink, R.A.: The dynamic representation of scenes. Vis. Cogn. **7**, 17–42 (2000)
Rensink, R.A.: Change detection. Annu. Rev. Psychol. **53**, 245–277 (2002)
Rensink, R.A.: The management of visual attention in graphic displays. In: Roda, C. (ed.) Human Attention in Digital Environment, pp. 63–92. Cambridge University Press, Cambridge (2011)
Robbins, T., Anderson, E., Barker, D., Bradley, A., Fearneyhough, C., Henson, R. . . ., Baddeley, A.: Working memory in chess. Mem. Cogn. **24**, 83–93 (1996)
Saida, S., Ikeda, M.: Useful visual field size for pattern perception. Percept. Psychophys. **25**, 119–125 (1979)
Schmeichel, B., Baumeister, R.F.: Effortful attention control. In: Bruya, B. (ed.) Effortless Attention: A new Perspective in the Cognitive Science of Attention and Action, pp. 29–49. MIT Press, Cambridge (2010)
Sekuler, R.W., Blake, R.: Perception, 4th edn. McGraw-Hill, New York (2001)
Shinohara, K.: Attention and human error. In: Harada, E., Shinohara, K. (eds.) Theories and Applications of Cognitive Psychology 4: Attention and Safety, pp. 186–208. Kitaoji Syobo, Kyoto (in Japanese) (2011)
Simons, D., Ambinder, M.: Change blindness: theory and consequences. Curr. Dir. Psychol. Sci. **14**, 44–48 (2005)
Simons, D., Levin, D.: Change blindness. Trends Cogn. Sci. **1**, 261–267 (1997)
Stanovich, K.E., West, R.F.: Individual differences in reasoning: implications for the rationality debate. Behav. Brain Sci. **23**, 645–665 (2000)
Treisman, A.: Verbal cues, language, and meaning in selective attention. Am. J. Psychol. **77**, 206–219 (1964)
Treisman, A.: The perception of features and objects. In: Baddeley, A., Weiskrantz, L. (eds.) Attention: Selection, Awareness and Control. A Tribute to Donald Broadbent, pp. 5–35. Clarendon Press University, Oxford (1993)
Treisman, A., Gelade, G.: A feature-integration theory of attention. Cogn. Psychol. **12**, 97–136 (1980)
Treisman, A., Sato, S.: Conjunction search revisited. J. Exp. Psychol. Hum. Percept. Perform. **16**, 459–478 (1990)
Treisman, A., Schmidt, H.: Illusory conjunctions in the perception of objects. Cogn. Psychol. **14**, 107–141 (1982)
Watson, J.M., Strayer, D.L.: Supertaskers: profiles in extraordinary multitasking ability. Psychon. Bull. Rev. **17**, 479–485 (2010)
Wickens, C.D.: Processing resources in attention. In: Parasuraman, R., Davies, D.R. (eds.) Varieties of Attention, pp. 63–102. Academic, New York (1984)
Wickens, C.D., Hollands, J.G.: Engineering Psychology and Human Performance, 3rd edn. Prentice Hall, Upper Saddle River (1999)
Wickens, C.D., Kessel, C.: Processing resource demands of failure detection in dynamic systems. J. Exp. Psychol. Hum. Percept. Perform. **6**, 564–577 (1980)
Wickens, C.D., Liu, Y.: Codes and modalities in multiple resources: a success and a qualification. Hum. Factors **30**, 599–616 (1988)
Wickens, C.D., McCarley, J.S.: Applied Attention Theory. CRC Press, Boca Raton (2008)
Wickens, C.D., Sandry, D., Vidulich, M.: Compatibility and resource competition between modalities of input, output, and central processing. Hum. Factors **25**, 227–248 (1983)
Wickes, C.D.: Multiple resources and mental workload. Hum. Factors **50**, 449–455 (2008)
Yerkes, R.M., Dodson, J.D.: The relation of strength of stimulus to rapidity of habit-formation. J. Comp. Neurol. Psychol. **18**, 459–482 (1908)

# Chapter 2
# Emotion

**Yasunobu Yasoshima**

**Abstract** Emotions have impacts on a variety of behaviors in animals and human. But they are not easy to define, and there exist multiple definitions of emotions. In this chapter, the basic concepts, functions, and features of emotions are summarized, and the roles and natures of emotional facial expressions are introduced. Two main roles of emotion are distinguished: intra-individual (intra-personal) roles and inter-individual (inter-personal or social) roles. Emotion is related to primary drives that arise from innate needs, and hence have intra-individual roles. In addition, an emotional output of a person may influence the other's emotion and behavior, and hence emotion has inter-individual roles.

**Keywords** James-Lange theory • Drive-reduction theory • Homeostasis • Fight-or-flight • Amygdala • Facial expressions • Facial feedback theory • Flat affect • Conditioned taste aversion (CTA) • Generalization • Palatability shift • communication box • Ultrasonic vocalization • Duchenne smile • Orbitofrontal cortex

## 2.1 Introduction

Emotions have impacts on a variety of behaviors in animals and human. But they are not easy to define, and there exist multiple definitions of emotions. Historically, William James (1884) raised a question, "What is an emotion?" Carl Lange proposed a theory of emotion (Lange 1885). Their proposals accelerate the research of emotion. Many investigators contribute to understand and clarify a variety of types, functions, roles and mechanisms of emotions. Scientific studies in different fields, such as biology, psychology, neuroscience, psychiatry, anthropology, economics and other social sciences, propose distinctive interpretations and perspectives about emotion. Other fields, such as behavioral economics (Camerer 1999) and neuroeconomics, recently provide additional viewpoints of emotions and their impacts on economic behavior of human. These fields have a growing influence on economics, politics, and engineering. The wide range of viewpoints produce a

Y. Yasoshima (✉)
Graduate School of Human Sciences, Osaka University, Suita, Osaka, Japan
e-mail: yasosima@hus.osaka-u.ac.jp



M. Kasaki et al. (eds.), *Cognitive Neuroscience Robotics B*,
DOI 10.1007/978-4-431-54598-9_2

complex picture of emotions, and its complexity leads to difficulties in studying emotions.

In this chapter, the basic concepts, functions, and features of emotions are summarized. In addition, the roles and natures of emotional facial expressions are introduced.

Emotions are subjectively felt experiences; that is, a sensory stimulus does not necessarily produce the same emotional experience in different persons. Subjective experience is a complex process consisting of consciousness, bodily sensation and behavioral output. Charles Darwin (1872) has proposed the view that emotional reactions are innate. The similarity of emotional expressions between many mammals and humans has led him to speculate that some facial expressions of emotions are universal across many human cultures. He speculates that emotions, or at least 'basic' emotions, are rooted in biological architectures and genetic determinants. Ekman and Friesen (1971) have found that several emotional facial expressions in photographs of the Western people, such as happiness, sadness, anger, fear and surprise, are recognized by the Eastern people. On the basis of this finding, Ekman (1999) proposes the concept of 'basic emotions.' Although not all emotions correspond to specific facial expressions, Ekman (1999) postulates that a small subset of emotions is universal in human beings. The facial expression patterns of the 'basic emotions' seem to be innately hardwired in all humans, and elicited by purely biological processes. For this reason, the recognition of them is universal across cultures. A number of recent studies investigate neural mechanisms and brain regions underlying facial expression recognition, using neuroimaging techniques.

Instead of trying to unify emotional studies and other disciplines, we first review the basic concepts and roles of emotions. In particular, we focus on the roles of emotional control and cognition of facial expression. At least, two aspects of emotion should be accounted for: intra-individual (intra-personal) and inter-individual (inter-personal or social) roles of emotion.

## 2.2 Intra-individual Roles of Emotions

### *2.2.1 Basics*

#### 2.2.1.1 Drive and Emotion

First, emotion is related to primary drives that are not necessarily learned and arise from innate needs. Emotion produces a driving force of behavior and plays a role in motivating animals and humans to initiate and control behaviors. For example, fear usually evokes escape behavior when it is caused by a stimulus indicative of danger. In this case, we consciously or unconsciously perceive that the stimulus is fearful at the time of and/or after encountering it. Usually, a fearful event establishes fear memory. When psychological experience of fear comes to our mind even after the fearful stimulus disappears, it is a recalling (retrieval) of fear memory.

Another example is that emotional experience triggers adaptive behavior in response to internal signals. If the internal homeostatic system in the human body detects a decrease in blood glucose level, the interoceptive sensory system generates signals alerting about low energetic status and they are sent to the brain to regulate energetic balance. The brain may produce a subjective psychological experience that corresponds to the lower energetic status of being 'hungry.' The feeling and sensation of 'hungry' may initiate food-seeking (appetitive) behavior. Emotions associated with 'hungry' and related statuses reinforce seeking behavior ("foraging") to satisfy the energetic needs. When animals and humans find food or capture a prey, eating behavior may start to achieve the satisfaction of the specific drive 'hunger.' According to the drive-reduction theory (Hull 1943), eating behavior is initiated to reduce the drive 'hunger.' The theory has proposed that animals and humans are motivated by primary (hunger, thirst, sex, etc.) drives to maintain a certain state and equilibrium of body. The drive-reduction theory is based on the physiological concept of *homeostasis*. Some motivational behaviors can be accounted for by this theory; that is, such behaviors are motivated by primary drives, such as hunger, thirst, pain avoidance, and sex drive. When the excitatory state associated with a primary drive is reduced or satisfied by performing motivational behavior, animals and humans usually experience positive emotions, such as satisfaction, satiety and pleasure. However, there are some criticisms of the drive reduction theory. Most prominently, some human behaviors seem not to be motivated by primary drives: many people challenge to climb a high mountain in the winter, and enjoy risky things like bungee jumping and riding of a roller coaster, without having any direct biological benefits and survival advantages. Why do they perform such dangerous and detrimental behaviors? Why do fear experiences fail to lead them to avoid those dangerous activities? In these situations, the emotions elicited by dangerous experiences are powerless and fail to lead to the avoidance of risky activities. Human risk-taking behaviors seem not to be necessarily adequate to fulfill biological needs or increase evolutionary 'fitness' (Cosmides and Tooby 2000). The cost-benefit maximizing model and the marginal model (Krebs and Davis 1993) seem to have no explanatory power for these behaviors. Some reports suggest that the risk-taking behavior of humans is related to sensation seeking and its individual variations based on differential biological traits (Bell 2009).

Emotional experiences are related to physiological status and its changes. For example, healthy subjects terminate eating behavior when they feel full or their interoceptive system produces satiating signals. They may feel satiety and/or satiation, as well as happiness, after consumption of sufficient food. Recent studies suggest that many physiological factors, such as blood glucose elevation and gut hormone releases, change after consumption of sufficient food, and these changes contribute to produce subjective experiences of satiety. However, other reports reveal that some bulimic persons do not terminate eating behavior even after they consume sufficient food. For such bulimic persons, consumption of a great amount of food may fail to reduce a drive to eat food, resulting in overconsumption. They are persistently motivated to continue the consummatory eating behavior. Ingestion of a test meal by bulimic patients does not increase anorexinergic gut peptide hormone, while ingestion of the same meal by normal subjects raises the secretion of the peptide hormones (Hannon-Engel 2012; Monteleone et al. 2005; Kojima et al. 2005). In the case of Prader-Willi syndrome (PWS) patients, their satiety dysfunction caused by genetic deficiency does not stop them from eating (Shapira et al. 2005). Their persistent and enhanced appetite may be induced by the alteration of the motivational function. It has been suggested that bulimic and PWS patients fail to have emotional experiences concerning satiety and satiation, resulting in lack of the sense of self-control. A stomach-derived orexigenic peptide, ghrelin, is higher in PWS children than in obese and lean children (Bizzarri et al. 2010). Hypoactivation in cortical regions (dorsolateral prefrontal cortex and orbitofrontal cortex) and inhibitory control have been suggested to be associated with the abnormal eating motivation in PWS patients (Holsen et al. 2012). Neural changes in PWS patients may disrupt the sense of self-control and satiation. Patients with eating disorders suffer from lack of pleasantness and satiation associated with eating behavior. Other psychiatric disorders involve dysfunction in certain brain regions, such as the amygdala and prefrontal cortex, and they are associated with negative emotions (Blair 2003) Negative emotions concerning aggression are regarded to be caused by neural changes in these brain regions (Davidson et al. 2000). These evidence suggest that emotional statuses and emotional experiences are generated through interaction among brain functions, homeostatic physiological processes, and endocrine signals.

#### 2.2.1.2 Behavioral Priority and Emotions

Emotions play a role in response to emergencies. When animals and ancient humans suddenly encountered a predator, they had to escape or fight to survive. Fight-or-flight behavioral responses to a threatening stimulus, such as predator, fire, or disaster, are produced by a specific emotional experience 'fear' or 'flight.' Fear can enable animals and humans to terminate their behavior, such as eating and sleeping, and to initiate escaping and/or fighting behavior. From the evolutional viewpoint, fear response might be specialized in decreasing the risk of encountering or increasing the chance of escaping from dangerous stimuli during the exploration of a new area, foraging, or predator vigilance. In some cases, the mechanisms of emotion respond to a variety of cues from the external environments, and activate various neural and psychological systems simultaneously. The simultaneous activation of different systems may induce conflicting behaviors, e.g., sleeping and fighting. If small diurnal animals have a nap, as they usually do, without caution, after they smell and see a predator at night, they are likely to fall into crisis. In the circumstances like this, their emotional systems usually make them frightened, and their sleep system is deactivated. Smells of predators evoke fear responses and cause predator avoidance behavior in many animals. The fear system elicits both general arousal and preparatory changes in physiological status, such as an increase in heart rate, a reduction in blood flow to the surface of the body, and an increase in blood flow to the arms, legs, brain, and various sensory organs. These preparatory changes in the arousal and autonomic nervous systems help organism to prepare for performing defensive behaviors. For example, if blood flow in the skin decreases through vasoconstriction, blood loss in case of injury may be reduced. Many adaptive behaviors normally override other behaviors that are simultaneously activated. Emotions play a role in selecting and overriding behaviors. From the evolutionary perspective, emotions are the superordinate programs of the mind that consist of simultaneous activation of many independent subprograms for a variety of behaviors (Cosmides and Tooby 2000). Each emotion contributes to select a behavior among many options by selectively activating a set of neural and psychological systems and deactivating and modifying a different set of physiological psychological systems. In other words, emotions may help to determine behavioral priority in cases of emergency.

#### 2.2.1.3 Neural Mechanisms of Fear Emotion

A review by LeDoux and Phelps (2008) summarizes the importance of the amygdala for fear processing. The amygdala is located at the deep brain area. They suggest that the brain uses dual pathways to generate fear experiences. External information from the environment stimulates the sensory systems and is sent to the sensory relay station in the brain, i.e., the thalamus. The flow of sensory information branches into two different pathways at the thalamus. One pathway (the what LeDoux calls the 'low road') goes directly to the amygdala and bypasses the cerebral cortex which is the higher center of the brain to process and integrate complicated sensory information than the amygdala. This thalamo-amygdaloid pathway contributes to trigger rapid emotional responses (i.e., immediate fear) to a threatening stimulus. On the contrary, the other pathway goes to the sensory area (s) in the cerebral cortex and then reaches the amygdala. Sensory information can be processed more carefully in the sensory cortex. The thalamo-cortico-amygdaloid pathway (what LeDoux calls the 'high road') contributes to slower but more elaborate emotional responses. The "low road" pathway helps to initiate behavioral reactions emotionally (e.g., fearfully) to stimuli very quickly. Suppose that you are hiking in the bushes and encounter something moving in a rustling bush. The visual and auditory information of this thing are sent to the amygdala via the sensory thalamus along the "low road," and these sensory information triggers emotional and preparatory bodily responses to this, possible threat, such as an increase in heart rate and muscle contracts. These immediate fear reactions will be exerted by the 'low road' function. Monkeys fear a lunging snake and show stereotypic aversive behaviors, such as escaping and shouting. Some humans show similar reactions to fearful entities; their reactions resemble primate's immediate reactions to the threat. But when they find out that only leaves and twigs are rustling, their fear reactions are reduced and adjusted through the "high road" pathway. The dual pathways for emotional processing enable animals and humans to respond rapidly to very threatening stimuli and to reduce unnecessary fear responses to non-harmful stimuli.

## 2.2.2 *Emotion and Facial Expressions*

### 2.2.2.1 Mutual Interaction of Facial Expressions and Emotions

In some species of primates, including humans, emotional outputs are associated with facial expressions; emotional expressions are observed in other animals, e.g., rodents and dogs. Facial expressions may be an important factor for humans to change emotions. Ekman et al. (1983) have found that the emotions of participants instructed to adopt a certain facial expression using their hands are affected by that artificial facial expression. This finding was scientific evidence for the facial feedback theory (Buck 1980). Roughly speaking, the facial feedback theory implies that emotion is produced by ascending sensory experience of configurational changes in our facial muscular activity. According to this theory, when we smile, the subjective experience of happiness and positive emotion are generated. Kleinke et al. (1998) also argue that facial emotional expressions affect emotional status.

Facial expressions elicit motor (efferent/descending) and sensory (afferent/ascending) neural information in animals and humans. When we move our body, the brain sends descending motor command signals to a set of muscles and receives back their internal copy and feedback sensory information derived from actual movements. The internal copy remodels the pattern and configuration of muscular activities, i.e., the information on how to move and how to control each muscle prior to actual movements. The remodeling information derived from the motor command (feedforward signal) is called the 'efferent copy.' Using an "efferent copy," the brain could predict how each muscle moves and how to coordinate each muscular activity prior to receiving actual sensory feedback signals generated from movements and actions. When an efferent copy matches the actual sensory feedback signals derived from actual muscular movements, a movement will be exerted without any modification of the original motor command. When the actual feedback signals from the peripheral motor system (e.g., muscles for facial expressions) are different from the efferent copy, the motor command will be modified to minimize the difference. According to the reafferent principle (von Holst 1954), the efferent copy of a given facial expression is produced by and associated with the motor commands of certain facial movements. A facial efferent copy and actual sensory feedback information may affect emotional processing. The dissociation between facial feedback information and real emotion may have an influence on emotional processing. Recent studies with Botox (botulinum toxin) have revealed a strong correlation between facial expressions and emotions. Botox selectively inhibits facial motor movements by paralyzing facial muscles. It blocks acetylcholine receptors at neuromuscular junctions, resulting in no muscular feedback information. When Botox is effective, the motor command and/or its efferent copy are dissociated from the ascending information from muscular movements. Botox injection leads to a lack of facial expression (face expressionless). The use of Botox blunts the magnitude of self-reported emotional experience (Davis et al. 2010) because of lack of actual facial feedback impact. Psychological and mood disorders, such as depression and autistic disorder, could be the causes of "flat affect" (Winograd-Gurvich et al. 2006). Patients in these disorders tend to have difficulties in showing emotional facial expressions and in facilitating emotional changes. These studies indicate that facial expressions, even if not required for real emotions, partly contribute to them.

#### 2.2.2.2 Learning-Dependent Emotional Facial Outputs

The other aspects and roles of facial expression are connected to the basic behavioral outputs of emotion via a reflex-like process. External sensory cues, such as taste and smell, also produce orofacial expressions. The taste-elicited (oro)facial expressions provide a good behavioral measure for animal's emotional outputs and neural underpinnings. For example, taste reactivity has been reported for various animal species including rodents, primates and humans (Berridge 2000; Steiner et al. 2001). When a sweet taste substance is put in an oral cavity of a rodent, it shows specific behavioral outputs called 'positive (hedonic) taste reactions,' examples of which are tongue protrusion, lateral tongue protrusion, and month movement. When a bitter, sour, or very salty food or fluid is presented to an oral cavity of a rodent, the animal shows negative behavioral patterns called 'aversive taste reactions,' examples of which are gaping, chin rubbing, and forelimb flailing (Fig. 2.1) (Berridge 2000). These behaviors consist of stereotyped and reflective patterns that are controlled by the neural systems in the brainstem (Grill and Norgren 1978a, b). The aversive taste reactive orofacial expressions are seen in infants of humans, as well as those of rodents, when they receive bitter or sour solutions (Berridge 2000; Steiner 1973; Steiner et al. 2001). This suggests that these aversive and positive facial expressions are evolutionary conserved. Moreover, hydrocephalic infants show positive (hedonic) and aversive taste reactivity in the same ways as normal infants do (Berridge 2000; Steiner 1973; Steiner et al. 2001). According to these findings, the orofacial reactions to a particular taste stimulus are basically innate (*unconditioned*) traits; however, the reaction patterns to a taste stimulus may not be immutable even though it is sweet or palatable.

A unique taste-related emotional learning changes affective and behavioral reaction patterns to a taste. When animals ingest a novel sweet taste substance (e.g., sucrose) and suffer from subsequent visceral malaise, they will change their taste reactivity patterns concerning the sweet taste substance from positive (preferred) to negative (aversive). The change in orofacial taste reaction is a kind of experience-dependent learning, and called 'conditioned taste aversion (CTA)' or 'taste aversion learning' (Bures et al. 1998; Yamamoto et al. 1994). CTA, also known as Garcia effect, is a form of classical (or Pavlovian) conditioned avoidance/rejection that is established through association of a taste conditioned stimulus

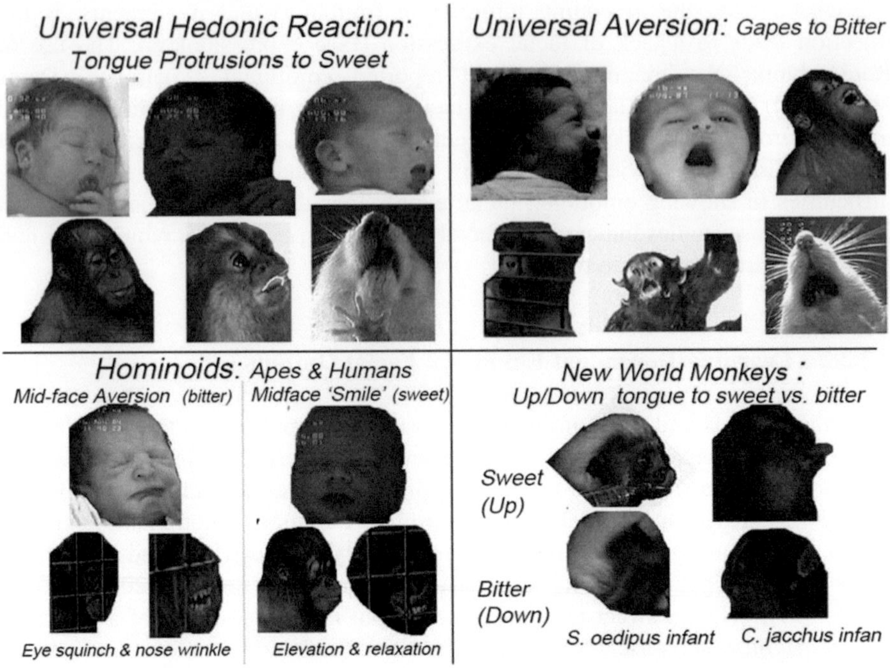


**Fig. 2.1** Evolutionary conservation of hedonic and aversive taste reactions to taste stimuli in a variety of species (Adapted from Berridge (2000) with permission from Elsevier)

(CS) with visceral malaise information (unconditioned stimulus, US). CTA is one of the one-trial learning behaviors that can be established by a single pairing of a CS and a US. Animals conditioned to sucrose (original CS) show aversive taste reactions to other sweet taste substances, such as saccharin and fructose, without any pairings of ingestion of a sweetener and malaise. In other words, conditioned animals usually show generalized aversion to substances that have a similar taste to the original CS. This phenomenon is called 'generalization.' Generalization of CTA enables conditioned animals to avoid substances with a similar taste to the CS, and result in the avoidance of potentially toxic substances that are not yet known to be safe. A CTA could be established not only for a sweet taste substance but also for a salty, sour, bitter, or umami substance. When the taste of a CS is novel for animals, the CTA to the novel CS is strong; however, a familiar taste substance is less likely to establish a strong CTA, suggesting that taste familiarity influences CTA formation.

CTA provides a good experimental model to study neural mechanisms of palatability shift (hedonic/affective evaluation) regarding a given taste. Forebrain structures are necessary for CTA (Grill and Norgren 1978c). Many studies suggest that the amygdala, which belongs to the limbic systems, contributes to affective evaluation processing of innately positive or aversive taste stimuli. The amygdala is also recruited in acquisition, retention and expression of CTA learning (Bures et al. 1998; Yamamoto et al. 1994).

Taste-responsive neurons activated by innately aversive tastes, such as bitter, sour, and concentrated sodium chloride, are categorized as aversive taste-hedonic neurons, and they are related to encoding of taste unpleasantness. Neuronal reactions to a taste CS in the amygdala and the insular cortex are changed after CTA acquisition (Yamamoto et al. 1989; Yamamoto et al.; Yasoshima et al. 1995; Yasoshima and Yamamoto 1998). After a sweet taste CS is paired with visceral malaise, aversive taste-hedonic neurons in these brain regions are activated by the oral presentation of the sweet CS (Yamamoto et al. 1989, 1994; Yasoshima et al. 1995; Yasoshima and Yamamoto 1998).

Human infants show aversive orofacial expressions to bitter or sour substances (Steiner 1973; Steiner et al. 2001). These aversive reactions are considered to be innate and unconditioned behaviors. When they become adults, however, they tend to show positive (hedonic) taste reactions to aversive bitter substances, e.g., coffee and beer. This type of behavioral change is called 'conditioned taste preference' or 'taste preference learning.' Conditioned taste preference (CTP) is usually mediated by pairing a taste of a substance with postingestive rewarding properties (e.g., caloric load, satiation and taste palatability). It has been suggested that CTP is mediated by the brain reward system including the mesolimbic dopaminergic system and/or nutrient sensing (Sclafani and Ackroff 2012). For example, the flavor (taste and smell) of beer increases dopaminergic release in the ventral striatum in male persons without having pharmacological effects of the alcohol (Oberlin et al. 2013). Flavor cues of an alcoholic beverage may be repeatedly associated with the rewarding effect of alcohol by alcohol drinking experiences, and the association may develop a CTP for its flavor. Due to this CTP process, the flavor of the alcoholic beverage alone may have the potential to activate the brain reward system. In the processes of CTA or CTP learning, sensory (taste or flavor) cue(s) of a CS become an elicitor to evoke learned emotional outputs.

Conditioned changes in behavioral reactions to a taste CS in CTA and CTP are mediated by the learning process that is associated with taste memory formation. A different type of affective/emotional and behavioral change concerning a taste has been reported. Cabanac (1971) reported a learning-independent change in taste reactivity, that is, a change of reaction from aversive to preferable in accordance with physiological changes in homeostatic states. When animals and humans are acutely depleted of internal sodium, they prefer and accept ingesting a very salty substance containing a high sodium chloride concentration. The craving and preference for salt is called 'salt appetite,' and it is an example of *taste alliesthesia* (Cabanac 1971). Salt-depleted animals demonstrate hedonic (positive) taste reactions to a concentrated sodium solution, but non-depleted and normal animals show aversive taste reactions to the same solution (Daniels and Fluharty 2004). When the repletion of internal sodium concentration reaches the normal physiological level in previously depleted animals, their taste reactions become aversive to a very salty solution and they avoid ingesting it. The behavioral changes are independent of learning process. It has been suggested that changes in neural activity in one of the ventral basal ganglia nuclei (the 'the ventral pallidum') are associated with this type of change in hedonic reaction to very salty taste (Tindell et al. 2006).

In summary, the affective/emotional experiences associated with conditioned and/or unconditioned processes lead to the change and modification of orofacial emotional outputs through innate and learning-related neural mechanisms.

## 2.3 Inter-Individual Roles of Emotions

### *2.3.1 Emotional Outputs as an Elicitor of Other's Emotional Responses*

In a social situation, an emotional output of a person may influence the other's emotion and behavior. For example, sudden loud vocalization and violent behaviors of an angry person astonish surrounding others and result in their negative emotionality.

Sensory stimuli associated with one's emotional outputs are potential mediators, and they evoke other's emotional experiences. An animal model for mimicking inter-individual affective communication has been reported. In this model, two groups of mice are placed in an experimental apparatus called a '*communication box*' (CB) (Ogawa and Kuwahara 1966; Ogawa et al. 1990). Mice in the shocked group receive painful electric footshocks and show emotional outputs, such as evacuation, urination and/or (ultrasonic) vocalization, in the footshock compartments of the CB. Mice in the non-shocked group are placed in the non-footshocked compartments of the CB, and exposed to the sensory information associated with the emotional outputs of the shocked mice. The non-shocked mice suffer from stomach lesions (Ogawa et al. 1990) and show changes in autonomic and endocrine response (Ishikawa et al. 1992). This result suggests that emotional outputs of shocked animals produce affective socio-psychological stress in non-shocked animals. In particular, chemosensory signals (i.e., body odors and alarm pheromones) are important for rodents to communicate danger with surrounding conspecifics. A recent study (Mujica-Parodi et al. 2009) shows that emotional stress is evoked in non-feared persons by persons who have suffered from a fearful event, when they have human body 'odor' communication. Chemosensory signals from feared persons activate the amygdala of non-feared persons. Other body odors of humans affect emotional states as well (Pause 2012).

Vocalization transmits negative emotions from one to other surrounding conspecifics. Rodents elicit ultrasonic vocalization when they receive negative stimuli, such as pain and stressor (Sanchez and Meier 1997). Human voice is a crucial tool for communication, and humans have high sensitivity to other's vocal cues. Adult human listeners can recognize the affective status of a speaker by identifying the affective contents of the speaker's voice (Bryant and Barrett 2007; Sauter and Scott 2007). It is reported that the human brain is equipped with the mechanisms for distinguishing between affective and non-affective auditory signals of human vocalization (Sauter and Eimer 2010).

In sum, the visual, auditory and olfactory information associated with one's emotional outputs have power to change the other's emotional experiences. In evolutionary processes, many social animals develop and conserve capacities to detect the emotional sensory stimuli elicited by one's emotional experiences, and thus humans inherit the capacity for social communication.

### 2.3.2 Emotional Facial Expression as a Communication Tool

Emotion has a function in communication. Imagine the following situation: if you were talking with a person who did not show any emotional facial expression or verbal/non-verbal response, you would be disappointed, and your motivation to talk with him/her might be deteriorated. By contrast, positive emotional facial expressions (e.g., smiling and nodding) of a listener often encourage a speaker to continue to talk with the listener. Talk with a negative listener is likely to induce a cautious and descriptive thinking style for a speaker (Beukeboom 2009). Facial emotional expressions of listeners have a high influence on interpersonal communication and mutual information exchange.

Humans recognize facial expressions rapidly by various biopsychological processes. Facial expression recognition plays an important role in social communication; facial expressions convey one's intentions and emotional states to others. Socially interacting animals transmit their emotional states to others by auditory and visual signals, such as vocalization, songs, speech, body displaying, and facial expressions. Charles Darwin, in his (1872) book *the Expression of Emotions in Man and Animals*, points out that two opposite behavioral attitudes, such as hostility and affection, may be produced by opposite subjective emotions, such as angry and affection. He holds that facial expression is a kind of bodily behavioral expression of emotions. From this point of view, it seems that human beings have acquired the cognitive capacity to perceive subtle differences in facial expressions and to speculate emotional states. This capacity seems to be very important for social animas to make communication among members of the same species.

Human females have superiority in recognizing emotional facial expressions than men are. Hampson et al. (2006) report that male subjects take significantly longer to discriminate six facial expressions than female subjects do. In addition, they report, male subjects take 30 % longer to respond to negative facial expressions than they do to positive ones; whereas female subjects take only 9 % longer. These findings seem to be consistent with the hypothesis proposed by the fitness threat hypothesis according to which it is predicted that the superiority of facial recognition processing in females is larger for negative emotions than for positive ones.

You can easily imagine a dissociation between facial expression and (real) emotion: for example, a 'fake smile.' A smile usually expresses a delightful and pleased emotional state. One, however, can make a 'fake smile' without crinkling around one's eyes (this is called the 'Duchenne smile') (Ekman et al. 1990). Damasio (1994) suggests that 'real' smiles with emotional experience are controlled by neural circuits in the limbic cortical areas and the basal ganglia in the human brain.

It has been suggested that perception of fearful facial expressions depends on the amygdala. A patient with bilateral amygdala damage has a difficulty in recognizing others' fearful expressions (Adolphs et al. 1994; Adolphs 2008). Interestingly, facial expressions of disgust are mediated by different neural substrates than those for other facial expressions (Chapman and Anderson 2012). Pleasant and unpleasant stimuli, regardless of sensory modality, activate the core neural network consisting of the orbitofrontal cortex, the temporal pole, and the superior frontal gyrus in the left hemisphere (Royet et al. 2000). However, only emotional olfactory stimuli activate the bilateral amygdala (Zald and Pardo 1997). The amygdala response to emotional stimuli changes with age (Mather et al. 2004). The reactivity to positive and negative emotional visual information in the amygdala is compared between young (the age of 18–29 years) and old subjects (the age of 70–90 years), with event-related functional magnetic resonance imaging (fMRI). The amygdala reactivity to negative information in young subjects is greater than that in old subjects; whereas no significant age difference is found in the amygdala reactivity to positive information. These results suggest that the amygdala reactivity to negative information is diminished selectively when people become old.

Identification and recognition of emotional facial expressions are different among patients with schizophrenia, patients with Parkinson's disease, and healthy persons. The sensitivity of Parkinson's disease patients to the facial expression for the emotion of disgust is lower than that of healthy subjects (Assogna et al. 2008; Suzuki et al. 2006). Subjects without facial feedback information due to Botox-induced facial muscle paralysis show slower comprehension of emotional sentences (Havas and Matheson 2013). Loss of facial expression (hypomimia) in the Parkinson' disease patients (Jankovic 2008) might cause a "flat affect," leading to loss of understanding of the other's emotions. These findings suggest that facial feedback provides an effective simulation of the other's emotion (Havas and Matheson 2013).

## 2.4 Summary and Conclusion

This chapter has provided a brief overview of certain roles of emotion. Emotions are constituted by complex psychological and neural processes; to understand emotions, we should pay attention to perception and learning. Emotion plays a crucial role in homeostasis of the body and mind and social communication. Future research of the roles and psychobiological basis of emotion may help to improve the quality of our lives. It is worthwhile to study how a sign of a favorable stimulus provides pleasure to humans and animals, and how a sign of danger elicits displeasure (aversiveness). Emotional experiences, such as pleasure and displeasure, help to motivate useful behaviors and prohibit dangerous behaviors. Social lives of animals and non-human primates may have promoted evolutionary development of brain so as to differentiate affective from non-affective signals of other conspecifics, and the human brain may have been benefitted from this evolutionary progress and adapted for social behaviors. However, unfortunately in the present era, the capacity for affective social communication produces socio-psychological stress and mental disorders in many persons. The research of emotion helps to understand our emotions and provides clues to treat affective disorders derived from socio-psychological stress.

## Exercises

1. Describe the influence of facial expressions on emotions.
2. Imagine an example of inter-individual emotional transmission (communication) and discuss its behavioral and sensory mechanisms.
3. Is it possible that a person does not always elicit identical emotional responses to an identical sensory stimulus? If it is, then state why he/she shows different responses.
4. Describe an example of sensory-evoked emotion and consider a strategy for implementing its emotional outputs in a robot. Note that emotion is subjective experience.

## References


Adolphs, R.: Fear, faces, and the human amygdala. Curr. Opin. Neurobiol. **18**, 166–172 (2008)

Adolphs, R., Tranel, D., Damasio, H., Damasio, A.: Impaired recognition of emotion in facial expressions following bilateral damage to the human amygdala. Nature **372**, 669–672 (1994)

Assogna, F., Pontieri, F.E., Caltagirone, C., Spalletta, G.: The recognition of facial emotion expressions in Parkinson's disease. Eur. Neuropsychopharmacol. **18**, 835–848 (2008)

Bell, A.M.: Approaching the genomics of risk-taking behavior. Adv. Genet. **68**, 83–104 (2009)

Berridge, K.C.: Measuring hedonic impact in animals and infants: microstructure of affective taste reactivity patterns. Neurosci. Biobehav. Rev. **24**, 173–198 (2000)

Beukeboom, C.J.: When words feel right: how affective expressions of listeners change a speaker's language use. Eur. J. Soc. Psychol. **39**, 747–756 (2009)

Bizzarri, C., Rigamonti, A.E., Luce, A., Cappa, M., Cella, S.G., Berini, J., Sartorio, A., Müller, E. E., Salvatoni, A.: Children with Prader-Willi syndrome exhibit more evident meal-induced responses in plasma ghrelin and peptide YY levels than obese and lean children. Eur. J. Endocrinol. **162**, 499–505 (2010)

Blair, R.J.R.: Neurobiological basis of psychopathy. Br. J. Psychiatry **182**, 5–7 (2003)

Bryant, G., Barrett, H.C.: Recognizing intentions in infant-directed speech: evidence for universals. Psychol. Sci. **18**, 746–751 (2007)

Buck, R.: Nonverbal behavior and the theory of emotion: the facial feedback hypothesis. J. Pers. Soc. Psychol. **38**, 811–824 (1980)

Bures, J., Bermúdez-Rattoni, F., Yamamoto, T.: Conditioned Taste Aversion: Memory of a Special Kind. Oxford University Press, Oxford (1998)

Cabanac, M.: Physiological role of pleasure. Science **173**, 1103–1107 (1971)
Camerer, C.: Behavioral economics: reunifying psychology and economics. Proc. Natl. Acad. Sci. U. S. A. **96**, 10575–10577 (1999)
Chapman, H.A., Anderson, A.K.: Understanding disgust. Ann. N. Y. Acad. Sci. **1251**, 62–76 (2012)
Cosmides, L., Tooby, J.: Evolutionary psychology and the emotion. In: Lewis, M., Haviland-Jones, J.M. (eds.) Handbook of Emotions, 2nd edn. The Guilford Press, New York (2000)
Damasio, A.: Descartes' Error: Emotion, Reason, and the Human Brain. G. P. Putnam's Sons, New York (1994)
Daniels, D., Fluharty, S.J.: Salt appetite: a neurohormonal viewpoint. Physiol. Behav. **81**, 319–337 (2004)
Darwin, C.: The Expression of Emotion in Man and Animals. The Project Gutenberg EBook (EBook #1227). http://www.gutenberg.org/files/1227/1227-h/1227-h.htm. (01/17/2014). (Original work published 1872) (1998)
Davidson, R.J., Putnam, K.M., Larson, C.L.: Dysfunction in the neural circuitry of emotion regulation-a possible prelude to violence. Science **289**, 591–594 (2000)
Davis, J.I., Senghas, A., Brandt, F., Ochsner, K.N.: The effects of Botox injections on emotional experience. Emotion **10**, 433–440 (2010)
Ekman, P.: Basic emotions. In: Dalgleish, T., Power, M. (eds.) Handbook of Cognition and Emotion. Wiley, Sussex (1999)
Ekman, P., Friesen, W.V.: Constants across cultures in the face and emotion. J. Pers. Soc. Psychol. **17**, 124–129 (1971)
Ekman, P., Levenson, R.W., Friesen, W.V.: Autonomic nervous system activity distinguishes among emotions. Science **221**, 1208–1210 (1983)
Ekman, P., Davidson, R.J., Friesen, W.V.: The Duchenne smile: emotional expression and brain physiology. II. J. Pers. Soc. Psychol. **58**, 342–353 (1990)
Grill, H.J., Norgren, R.: The taste reactivity test. I. Mimetic responses to gustatory stimuli in neurologically normal rats. Brain Res. **143**, 263–279 (1978a)
Grill, H.J., Norgren, R.: The taste reactivity test. II. Mimetic responses to gustatory stimuli in chronic thalamic and chronic decerebrate rats. Brain Res. **143**, 281–297 (1978b)
Grill, H.J., Norgren, R.: Chronically decerebrate rats demonstrate satiation but not bait shyness. Science **201**, 267–269 (1978c)
Hampson, E., van Anders, S.M., Mullin, L.I.: A female advantage in the recognition of emotional facial expressions: test of an evolutionary hypothesis. Evol. Hum. Behav. **27**, 401–416 (2006)
Hannon-Engel, S.: Regulating satiety in bulimia nervosa: the role of cholecystokinin. Perspect. Psychiatr. Care **48**, 34–40 (2012)
Havas, D.A., Matheson, J.: The functional role of the periphery in emotional language comprehension. Front. Psychol. **4**, 294 (2013). doi:10.3389/fpsyg.2013.00294
Holsen, L.M., Savage, C.R., Martin, L.E., Bruce, A.S., Lepping, R.J., Ko, E., Brooks, W.M., Butler, M.G., Zarcone, J.R., Goldstein, J.M.: Importance of reward and prefrontal circuitry in hunger and satiety: Prader-Willi syndrome vs simple obesity. Int. J. Obes. (Lond) **36**, 638–647 (2012)
Hull, C.: Principles of Behavior. Appleton, New York (1943)
Ishikawa, M., Hara, C., Ohdo, S., Ogawa, N.: Plasma corticosterone response of rats with sociopsychological stress in the communication box. Physiol. Behav. **52**, 475–480 (1992)
James, W.: What is an emotion? Mind **9**, 188–205 (1884)
Jankovic, J.: Parkinson's disease: clinical features and diagnosis. J. Neurol. Neurosurg. Psychiatry **79**, 368–376 (2008)
Kleinke, C.L., Peterson, T.R., Rutledge, T.R.: Effects of self-generated facial expressions affect emotional status. J. Pers. Soc. Psychol. **74**, 272–279 (1998)
Kojima, S., Nakahara, T., Nagai, N., Muranaga, T., Tanaka, M., Yasuhara, D., Masuda, A., Date, Y., Ueno, H., Nakazato, M., Naruo, T.: Altered ghrelin and peptide YY responses to meals in bulimia nervosa. Clin. Endocrinol. (Oxf) **62**, 74–78 (2005)

Krebs, J.R., Davies, N.B.: An Introduction to Behavioural Ecology, pp. 48–76. Wiley-Blackwell, Oxford (1993)
Lange, C. G.: The Mechanism of the Emotions (1885) (translated by Rand, B.) Reprinted from Rand, B. The Classical Psychologists, pp. 672–684. Houghton Mifflin, Boston (1912)
LeDoux, J., Phelps, E.: Emotional networks in the brain. In: Lewis, M., Haviland-Jones, J., Barrett, L. (eds.) Handbook of Emotions, pp. 159–179. Guilford Press, New York (2008)
Mather, M., Canli, T., English, T., Whitfield, S., Wais, P., Ochsner, K., Gabrieli, J.D.E., Carstensen, L.L.: Amygdala responses to emotionally valenced stimuli in older and younger adults. Psychol. Sci. **15**, 259–263 (2004)
Monteleone, P., Martiadis, V., Rigamonti, A.E., Fabrazzo, M., Giordani, C., Muller, E.E., Mai, M.: Investigation of peptide YY and ghrelin responses to a test meal in bulimia nervosa. Biol. Psychiatry **57**, 926–931 (2005)
Mujica-Parodi, L.R., Strey, H.H., Frederick, B., Savoy, R., Cox, D., Botanov, Y., Tolkunov, D., Rubin, D., Weber, J.: Chemosensory cues to conspecific emotional stress activate amygdala in humans. PLoS One **4**, e6415 (2009). doi:10.1371/journal.pone.0006415
Oberlin, B.G., Dzemidzic, M., Tran, S.M., Soeurt, C.M., Albrecht, D.S., Yoder, K.K., Kareken, D. A.: Beer flavor provokes striatal dopamine release in male drinkers: mediation by family history of alcoholism. Neuropsychopharmacology **38**, 1617–1624 (2013)
Ogawa, N., Kuwahara, K.: Psychophysiology of emotion-communication of emotion. Jpn. J. Psychosom. Med. **6**, 352–357 (1966)
Ogawa, N., Hara, C., Ishikawa, M.: Characteristic of socio-psychological stress induced by the communication box method in mice and rats. In: Mannine, O. (ed.) Environmental Stress, pp. 417–427. ACES Publishing Ltd, Tampere (1990)
Pause, B.M.: Processing of body odor signals by the human brain. Chemosens. Percept. **5**, 55–63 (2012)
Royet, J.P., Zald, D., Versace, R., Costes, N., Lavenne, F., Koenig, O., Gervais, R.: Emotional responses to pleasant and unpleasant olfactory, visual, and auditory stimuli: a positron emission tomography study. J. Neurosci. **20**, 7752–7759 (2000)
Sánchez, C., Meier, E.: Behavioral profiles of SSRIs in animal models of depression, anxiety and aggression. Are they all alike? Psychopharmacology (Berl) **129**, 197–205 (1997)
Sauter, D.A., Eimer, M.: Rapid detection of emotion from human vocalizations. J. Cogn. Neurosci. **22**, 474–481 (2010)
Sauter, D.A., Scott, S.K.: More than one kind of happiness: can we recognize vocal expressions of different positive states? Motiv. Emot. **31**, 192–199 (2007)
Sclafani, A., Ackroff, K.: Role of gut nutrient sensing in stimulating appetite and conditioning food preferences. Am. J. Physiol. Regul. Integr. Comp. Physiol. **302**, R1119–R1133 (2012)
Shapira, N.A., Lessig, M.C., He, A.G., James, G.A., Driscoll, D.J., Liu, Y.: Satiety dysfunction in Prader-Willi syndrome demonstrated by fMRI. J. Neurol. Neurosurg. Psychiatry **76**, 260–262 (2005)
Steiner, J.E.: The gustofacial response: observation on normal and anencephalic newborn infants. In: Bosma, J.F. (ed.) Fourth Symposium on Oral Sensation and Perception. Development in the Fetus and Infant, pp. 254–278. US Department Health, Education, and Welfare, National Institutes of Health, Bethesda (1973)
Steiner, J.E., Glaser, D., Hawilo, M.E., Berridge, K.C.: Comparative expression of hedonic impact: affective reactions to taste by human infants and other primates. Neurosci. Biobehav. Rev. **25**, 53–74 (2001)
Suzuki, A., Hoshino, T., Shigemasu, K., Kawamura, M.: Disgust-specific impairment of facial expression recognition in Parkinson's disease. Brain **129**(Pt 3), 707–717 (2006)
Tindell, A.J., Smith, K.S., Pecina, S., Berridge, K.C., Aldridge, J.W.: Ventral pallidum firing codes hedonic reward: when a bad taste turns good. J. Neurophysiol. **96**, 2399–2409 (2006)
von Holst, E.: Relations between the central nervous system and the peripheral organs. Br. J. Anim. Behav. **2**, 89–94 (1954)

Winograd-Gurvich, C., Fitzgerald, P.B., Georgiou-Karistianis, N., Bradshaw, J.L., White, O.B.: Negative symptoms: a review of schizophrenia, melancholic depression and Parkinson's disease. Brain Res. Bull. **70**, 312–321 (2006)

Yamamoto, T., Matsuo, R., Kiyomitsu, Y., Kitamura, R.: Taste responses of cortical neurons in freely ingesting rats. J. Neurophysiol. **61**, 1244–1258 (1989)

Yamamoto, T., Shimura, T., Sako, N., Yasoshima, Y., Sakai, N.: Neural substrates for conditioned taste aversion in the rat. Behav. Brain Res. **65**, 123–137 (1994)

Yasoshima, Y., Yamamoto, T.: Short-term and long-term excitability changes of the insular cortical neurons after the acquisition of taste aversion learning in behaving rats. Neuroscience **84**, 1–5 (1998)

Yasoshima, Y., Shimura, T., Yamamoto, T.: Single unit responses of the amygdala after conditioned taste aversion in conscious rats. Neuroreport **6**, 2424–2428 (1995)

Zald, D.H., Pardo, J.V.: Emotion, olfaction, and the human amygdala: amygdala activation during aversive olfactory stimulation. Proc. Natl. Acad. Sci. USA **94**, 4119–4124 (1997)

# Chapter 3
# Working Memory as a Basis of Consciousness

**Mariko Osaka**

**Abstract** 'Working memory' refers to the capacity-constrained active memory in which information is temporarily maintained and concurrently processed for the use in an ongoing goal-directed activity. The neural mechanisms responsible for consciousness are located in certain brain regions, such as the DLPFC, PPC TPJ and ACC, and these brain regions are coupled with a network that includes the central executive of working memory. In this chapter, we explore the nature of the neural basis of working memory and try to explain the mechanisms of working memory. In order to understand the neural basis of active consciousness, we also investigate how information is controlled by the neural basis of working memory. We use reading span test (RST), which measures the working memory capacity to memorize the target words of sentences during reading, to measure individual differences in working memory capacity.

**Keywords** Working memory • Consciousness • Capacity • Awareness • Dual process • Central executive system • Phonological loop • Visuospatial sketchpad • Episodic buffer • Reading span test (RST) • Language comprehension • Inhibitory control • Focusing attention • DLPFC • ACC • Individual differences • Superior parietal lobule • Posterior parietal cortex (PPC) • Recursive-consciousness

## 3.1 Neural Basis of Consciousness and Working Memory

Consciousness plays an essential role in human cognitive functions, such as language comprehension, self- and other-recognition, complex reasoning, and problem solving. Neuroimaging techniques like functional magnetic resonance imaging (fMRI) are recently coupled with psychological methods for studying consciousness and its neural basis. Neuroimaging study has promoted the understanding of consciousness by gradually revealing the complex neural networks that dynamically connect the areas of the brain that are involved in high-level cognition.

M. Osaka (✉)
Division of Cognitive Neuroscience Robotics, Institute for Academic Initiatives, Osaka University, Suita, Osaka, Japan
e-mail: mosaka@hus.osaka-u.ac.jp



M. Kasaki et al. (eds.), *Cognitive Neuroscience Robotics B*,
DOI 10.1007/978-4-431-54598-9_3

Current evidence from cognitive neuroscience and computational neurobiology indicates that the neural mechanisms responsible for consciousness are located in different regions, including the dorsolateral and ventrolateral prefrontal cortex (DLPFC and VLPFC), the posterior parietal cortex (PPC), the medial PFC coupled with the superior temporal sulcus (STS) and temporo-parieal junction area (TPJ), the anterior cingulate cortex (ACC), and the orbitofrontal cortex. Consciousness has two aspects by its nature: one is the passive aspect which perceives the surrounding world and the other is the active aspect which plans to act to adjust to the surrounding current world.

Recently, it has been shown that working memory is essential for understanding consciousness (Osaka 1997, 1998). ‘Working memory’ refers to the capacity-constrained active memory in which information is temporarily maintained and concurrently processed for the use in an ongoing goal-directed activity. The active aspect of working memory involves conscious tasks, such as rehearsal, inner speech, visual imagery, and verbal report. These tasks are similar to the tasks performed by active consciousness, which people use in order to adjust themselves to the current world.

The executive function of working memory is regarded to be essential for the relationship between consciousness and working memory. It depends on the central system of working memory and a flexible attention control system for performing cognitive tasks. Awareness arises when the resources of working memory are divided to perform dual tasks. A possible hypothesis is that active consciousness is a portion of working memory that is activated by the cognitive control of the executive function of working memory. The brain regions mentioned above in which consciousness is located are coupled with a network that includes the central executive of working memory. In the present chapter, we explore the nature of the neural basis of working memory and try to explain the mechanisms of working memory. In order to understand the neural basis of active consciousness, we also investigate how information is controlled by the neural basis of working memory.

## 3.2 Working Memory

Our daily activities often require the dual process of storing and processing information over a short time. When we are driving to the store, for example, we must observe the traffic signals on the street, while keeping in mind what we have to buy. Working memory serves to store and process information simultaneously (Baddeley 1986). Higher cognitive brain functions require the dual process like this. This means that working memory supports a wide range of functions that are needed for complex cognitive activities, such as reading texts or talking with people.

### 3.2.1 Baddeley's Model of Working Memory

Traditionally, human memory was regarded to be composed of two main storage components: STM (short term memory) and LTM (long term memory) (Atkinson and Shiffrin 1968). Patient studies later revealed that the two memory types were dissociated; a lesion in the left temporo-parietal lobe was found to impair only STM (Shallice and Warrington 1970), and pure amnesic patients were observed to perform well STM-associated tasks, despite their grossly impaired LTM (Baddeley and Warrington 1970). On the basis of these findings, Baddeley and Hitch (1974) replaced the concept of a single short-term store with that of a three-component system. The three-component system, as they conceptualized it, is comprised of an attentional controller (also known as the central executive system) and two subsidiary systems: the phonological loop and the visuospatial sketchpad. The phonological loop processes and articulates vocal and subvocal information, and the visuospatial sketchpad stores visual and spatial information.

Norman and Shallice (1986) proposed the Supervisory Attentional System (SAS), and Baddeley (1986) adopted it as a model for the central executive component of working memory. Baddeley proposed that the central executive had four roles: to focus attention, to divide attention across different subsystems, to switch attention between tasks, and to use attention to link working memory with LTM (Baddeley 1996).

However, the three-component system fails to explain the crucial phenomenon of chunking in which one's existing knowledge is used to increase the STM span (Miller 1956). Baddeley, then, added a fourth component: the episodic buffer. It was assumed to be a limited capacity attentional storage system based on multidimensional information, and is controlled by the central executive (Baddeley 2000). The episodic buffer binds together different sources of information into chunks, a process that is assumed to be central to conscious awareness in learning and performing complex tasks. A recent model Baddeley offers for the flow of information from perception to working memory is shown in Fig. 3.1 (Baddeley 2012).

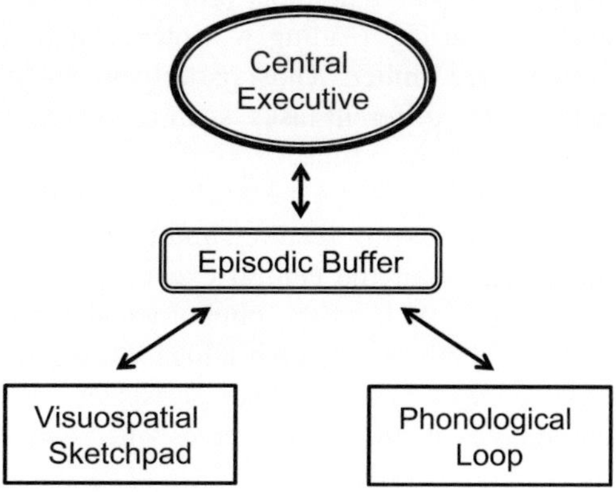


**Fig. 3.1** Baddeley's model (Baddeley 2012)

## 3.3 Individual Differences in Working Memory

Working memory plays an important role in language comprehension, learning, and reasoning (Baddeley 1986; Just and Carpenter 1992). It also plays a particularly critical role in text reading. While reading text, incoming information is decoded perceptually, reorganized, and integrated with a contextual interpretation, and the constituent products of each of these processes are stored for a short period of time (Kintsch and Van Dijk 1978; Daneman and Carpenter 1980). Working memory is important for storing the intermediate and final products of successive data, allowing for integrating text contents and putting words into context.

Because the resources of working memory are limited, an individual must selectively maintain representations that are most needed for current task goals. At the same time, the individual must allocate the resources appropriately to perform current tasks. The central executive serves as an attention controller, and assigns and coordinates the limited resources for storage and processing (Baddeley 1996; Baddeley and Logie 1999; Engle et al. 1999).

The resources available for working memory to maintain and process information are finite. Individuals show differences in working memory, and their differences consist in how they allocate the resources to task goals (Daneman and Carpenter 1980; La Pointe and Engle 1990; Turner and Engle 1989). The differences can account for different performances in cognitive functions, such as language comprehension (Just and Carpenter 1992).

### *3.3.1 Reading Span Test (RST)*

The reading span test (RST) was developed to measure behavioral differences between individuals in verbal working memory capacity during reading sentences (Daneman and Carpenter 1980). In the RST, participants read a few sentences aloud and memorize the last word of each sentence. According to the resource sharing model for working memory proposed by Daneman and Carpenter, mental resources available during reading a sentence and memorizing the target word of each sentence are limited. Thus, participants must allocate portions of working memory resources to different tasks, such as processing information and memorizing information for a short time.

The RST measures the working memory capacity to memorize the target words of sentences during reading, and the contents measured by the RST are similar to the functions of the central executive control processes and not to those of the subsystems, such as the phonological loop (Baddeley 1992; Just and Carpenter 1992). Therefore, resource allocation in the RST must be controlled by the executive control system which serves as an attention controller. It allocates and coordinates attentional resources when one reads and maintains the representations of the target words.

### 3.3.2 RST and Language Comprehension

By measuring the processing and storage during reading, the RST can account for various aspects of language comprehension (Daneman and Carpenter 1980; Just and Carpenter 1992; Daneman and Merikle 1996). Although the correlation between short term memory and reading comprehension is low, (Perfetti and Goldman 1976), the RST estimates show a higher correlation with reading comprehension (Masson and Miller 1983; Baddeley et al. 1985). In fact, Daneman and Carpenter (1980) have found that participants with high working memory capacity (high-span participants in the RST) are more successful at remembering target words than are participants with low working memory capacity (low-span participants in the RST). Furthermore, high-span participants are more successful at interpreting the meaning of an ambiguous word when it appears separately from the words necessary for clarifying its meaning. Other span tasks, such as the listening span test (LST) and operation span test (OST), also show that there is a high correlation between working memory capacity and language comprehension scores (Turner and Engle 1989). These results indicate that the correlation is independent of the stimulus modality (reading or listening) or the task (reading or arithmetic).

### 3.3.3 Japanese RST

There are different versions of the RST that correspond to different languages. For example, in the Japanese version of the RST, a target word can be selected from one word in a sentence (Osaka and Osaka 1992, 1994, see Table 3.1), whereas in the English RST, a target word is always the last word of a sentence (Daneman and Carpenter 1980). In Japanese, the last word of a sentence is usually a verb, and it is rarely a noun. In addition, as a feature of Japanese syntactic mechanisms, the last word of a Japanese sentence is rarely a focus word (Kuno 1978), whereas in English it is (Bolinger 1986). The Japanese RST requires the variability of target words; because the target word occupies different positions in different sentences, it needs to be underlined in each sentence (Osaka et al. 2002). Despite the differences between Japanese and English, the scores of the Japanese RST and English RST show a strong correlation (Osaka and Osaka 1992). Just as in the English RST, span scores are strongly correlated with reading comprehension cores in the Japanese RST (Osaka and Osaka 1994).

**Table 3.1** Japanese reading span test. Three sentence condition

| |
|---|
| • 水泳教室に通っているためか、母は最近とても元気である。<br>(Lately, my mother is well because she goes to swimming school.) |
| • 熱帯の植物なので、その花は寒さには弱い。<br>(Tropical plants are vulnerable to cold weather.) |
| • 雷のため、電車の切符販売機が故障した。<br>(Thunder caused the ticket machine at the train station to break down.) |

### 3.3.4 What Does RST Measure?

In order to explain what kinds of processes are involved in sentence comprehension, several hypotheses have been proposed. The resource-sharing model provides an interpretation of different performances on the RST in terms of capacity differences. Daneman and Carpenter (1983) suggest that the semantic processing of sentence comprehension is attributable to differences in capacity. High-span participants devote fewer resources to the semantic processing of a sentence, and therefore they retain sufficient resources to remember words. Another suggestion is that high-span participants make greater use of various strategies (Carpenter and Just 1989). Meta-analysis of studies on working memory span tasks show that passive storage measures, such as STM measures, correlate less with reading skills than working memory span measures do (Daneman and Merikle 1996).

The inhibitory control is also important. When one performs poorly on working memory tasks, one often has a deficit in inhibiting irrelevant information and performs poorly on the RST as well. For this reason, it has been proposed that successful performance on the RST requires good inhibitory mechanism (Conway and Engle 1994; Engle et al. 1995; De Beni et al. 1998; May et al. 1999).

Another interpretation of the RST is based on a task switching difference. Towse et al. (1998) report that sentence processing and word storage do not compete for working memory resources during the RST. They conclude that high span participants utilize a task-switching strategy, and they alternate easily reading a sentence and holding the last word of it (Towse et al. 2000).

Osaka (2002) reports that RST participants often use strategies like a rehearsal using the phonological loop. Participants who show good performance employ several strategies. They change strategies during the RST. This suggests that they can tell whether the adopted strategy is effective or not. Self-monitoring influences performance on the RST. Consistent with this observation, an earlier study shows that high-span participants are likely to use more strategies than low-span participants do (Osaka and Nishizaki 2000). Thus, high-span participants can monitor their performance and change strategies more effectively than low span participants.

## 3.4 Focusing Attention

Focusing attention is important for language comprehension in reading and listening (Carpenter and Just 1977; Blutner and Sommer 1988; Osaka et al. 2002). When one reads a sentence, one initially directs one's attention to the focus word in a sentence (Carpenter and Just 1977). The focus word in a sentence is a critical word for text integration, and it is considered to play an important role in comprehending text reading. It has been reported that focusing on a word enhances memory (Birch and Garnsey 1995). They proposed that the focus word in a sentence facilitates the process of integrating information in sentences, and it is critical for creating a coherent understanding.

Focusing attention is important for attentional control systems in the central executive (Cowan 2001). The central executive is responsible for the control and selection of the currently relevant parts of long term memory representations. The activated parts of long term memory are regarded to work under the surveillance of attention (Cowan 1999, 2001). Cowan (2001) suggests that the focus of attention represents a capacity-limited part of working memory that constitutes approximately four independent units and holds a restricted set of items. As was stated in the last section, low-span-participants have a deficit in inhibiting irrelevant information for the task (Conway and Engle 1994; Engle et al. 1995; May et al. 1999). In addition, it is difficult for low-span participants to inhibit information on which they have previously focused attention (De Beni et al. 1998).

### *3.4.1 Focused RST vs Non-focused RST*

In order to confirm the importance of focusing attention in the span task, Osaka et al. (2002) have developed two versions of the RST: the focused-RST (F-RST) and the non-focused RST (NF-RST). While the NF-RST does not, the F-RST uses the focus word in each sentence as the target word, i.e., the word to be remembered.

The focus word in a sentence is defined as the word most critical for understanding the sentence (Birch and Garnsey 1995). To identify the focus word, a preliminary survey has been conducted among students who do not participate in the experiments. They are asked to identify which word of a sentence is most important and critical for understanding it. When a word is chosen by more than 70 % of the students, it is selected as the focus word in that sentence. (Osaka et al. 2002).

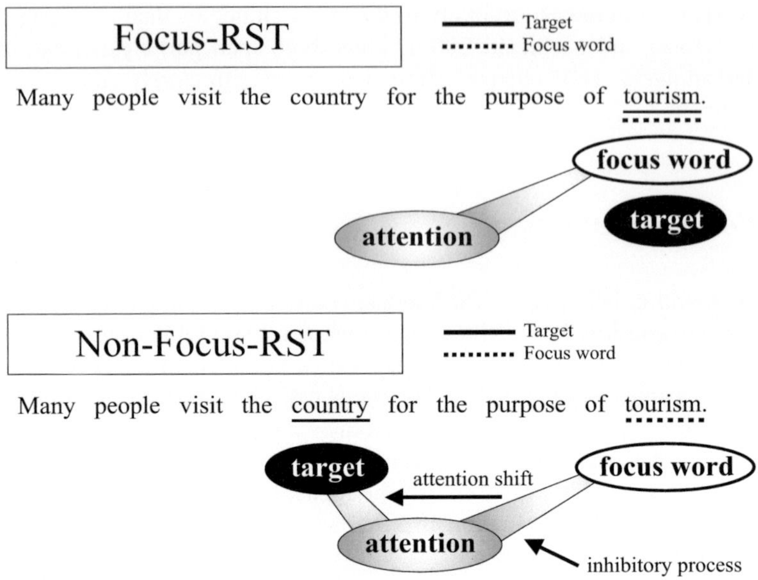


**Fig. 3.2** Sample sentences of the F-RST and NF- RST

Figure 3.2 shows sample sentences used in both the F- RST and the NF- RST. In the sentence "Many people visit the country for the purpose of tourism," the word "tourism" was chosen as the focus word by a preliminary estimate. The F-RST then uses "tourism" as the target word, and the NF-RST uses other words like "country."

### *3.4.2 Focusing Attention and Inhibitory Processes in the NF-RST*

When the target word coincides with the focus word in a sentence, it should be easier to memorize the target word because attention is easily focused on the target word. When the focus word in a sentence is not the target word, participants have to shift their attention from the focus word to the target word.

In the F-RST, the focus word is the target word, but not in the NF-RST. In the NF-RST, it is required for participants to inhibit attention on the focus word in a sentence, because their task goal is to remember the target word.

Not surprisingly, it has been reported that participant's performance is significantly higher in the F-RST than in the NF-RST (Osaka et al. 2002). In particular, the number of intrusion errors is significantly higher in the NF-RST than in the F-RST. Most intrusion errors in the NF-RST are focus intrusion errors: participants recall the focus word rather than the target word. The prevalence of focus-intrusion errors in the NF-RST suggests that it is difficult for participants to inhibit the word on which they have previously focused their attention.

Moreover, the frequency of intrusion errors, including focus intrusion errors, is higher in low-span participants than in high-span participants in the NF-RST. These findings indicate that low-span participants have deficits in their abilities to shift attention to the target and to inhibit the irrelevant words. They have more difficulty in inhibiting irrelevant information after they pay attention to it.

## 3.5 Neural Basis of Working Memory

Recent brain-imaging studies have attempted to identify the brain anatomy underlying the working memory systems. On the basis of Baddeley's original model (Baddeley 1986), two types of working memory processes are distinguished: the central executive system and modality-specific buffers, such as the phonological loop and the visuo-spatial sketchpad. The phonological loop is responsible for the retention of verbal information. Verbal information activates the left ventrolateral prefrontal cortex (VLPFC), while visuo-spatial information activates the right homologues (Jonides et al. 1993; Paulesu et al. 1993; Awh et al. 1996; Smith et al. 1996; Courtney et al. 1998; Owen et al. 1998).

Positron emission tomography (PET) has revealed the frontal lobe activities during episodic memory encoding and retrieval. The right PFC during episodic retrieval typically shows that the VLPFC (Broadman Area; BA 45) and the anterior extent of the PFC (BA 10) are involved (the latter is also known as the frontopolar cortex) (Shallice et al. 1994; Tulving et al. 1994). The ACC (BA 24/32) is found to be activated during episodic retrieval tasks (Nyberg 1998).

The central executive system serves as an attention controller, and allocates and coordinates attentional resources in performing cognitive tasks (Baddeley 1996; Baddeley and Logie 1999; Engle et al. 1999). Neuroimaging studies have explored the neural basis of this executive attention control system, and suggested that the system is located in the prefrontal cortex, and mainly in the DLPFC (BA9/46) and ACC (D'Esposito et al. 1995, 1998, 1999; Owen et al. 1996; Cohen et al. 1997; Smith and Jonides 1999; Bunge et al. 2000; Smith et al. 2001; Kane and Engle 2003; Osaka et al. 2003, 2004; Linden 2007).

Brain activities in the DLPFC increase as working memory task demands increase (Braver et al. 1997; Rypma et al. 1999; Bunge et al. 2000). D'Esposito et al. (1995) have found that DLPFC activation increases only during a dual task; it does not increase during a single task, regardless of how difficult it is. Rypma et al. (1999) report on the relationship between activation in the DLPFC and remembering digits. Although the DLPFC is not activated when participants remember one to three digits, activities in the DLPFC increase when participants remember six digits. It is within the capacity limitation to remember three digits, whereas it exceeds the capacity to remember six digits (Cowan 2001). In order to remember six digits, participants need the aid of the executive attention control, which increases activities in the DLPFC.

### *3.5.1 Attention Control in DLPFC vs ACC*

It is important for working memory performance to dissociate the DLPFC and ACC. MacDonald et al. (2000) dissociate them using the Stroop paradigm (Stroop 1935): activation in the DLPFC is observed in congruent color-word trials, and activation in the ACC occurs when participants engage in incongruent color-naming trials (but not in congruent trials). MacDonald et al. conclude that the DLPFC plays a role in providing top-down support for attention maintenance in task-appropriate behaviors. On the other hand, they regard the ACC to be subserved by the attention control system when attention is strongly controlled in incongruent color-naming trials. Smith and Jonides (1999) propose that both the DLPFC and ACC play an executive role in working memory tasks. The ACC mediates the inhibition of a preprogrammed response, such as word reading, and inhibition occurs automatically in incongruent color-naming trials so as to release any conflict.

It has been reported that the ACC has an executive function and the posterior cingulate cortex has an evaluative function (Vogt et al. 1992). More specifically, the dorsal site and ventral side of the ACC are regarded to be involved in cognitive activity and emotional division, respectively (Bush et al. 1998, 2000). An increase in activation of the ACC is reported to occur in error trials or high-conflict trials, such as go/no-go trials, oddball trials, and two-alternative forced-choice selections (Barch et al. 1997; Bush et al. 1998; Carter et al. 1998; Braver et al. 2001).

### *3.5.2 Neural Correlates of Span Tasks*

The functions or processes measured in span tasks are considered similar to those for which the executive control of the working memory system is responsible (Baddeley 1992; Just and Carpenter 1992). Given this, it is plausible that resource allocation is controlled by the executive control system when one takes the RST. fMRI studies give evidence for this claim by showing that increases in activation in the frontal regions are associated with task demands in the RST. For example, Just et al. (1996) have found that activation in the left frontal and temporal language areas increases during the RST, but not during the single reading task. Bunge et al. (2000) have found that activation in the PFC increases during the RST. This suggests that the increase in activation in the frontal region is affected by dual task demands. Activation in the left DLPFC increases during the OST as well (Smith et al. 2001). This increase occurs only in poor performers, and also depends on the level of the task demand.

These studies lead to further questions concerning the neural bases of working memory that explain the differences between high-span and low-span participants in span tasks, and in particular the activation differences in the frontal region, such as the DLPFC.

### 3.5.3 Neural Basis of Individual Differences

Osaka et al. (2003) investigated the neural substrates to which the differences between high-span and low-span participants are attributable. An fMRI study showed that a significant increase in activation of the left DLPFC and ACC occurred during the LST, but not in single task conditions. While a significant increase in activation of the DLPFC has been found in both high-span and low-span participants, an increase in the ACC is significant only in high-span participants.

Figure 3.3 (left) shows fMRI images of the brain areas in high-span and low-span participants that are activated in the LST condition (Osaka et al. 2003). Figure 3.3 (right) compares activations in the ACC regions between the high and low span participants. In both, the DLPFC shows an increase in activation. An increase in activation of the ACC, however, is confirmed only in the group of high-span participants.

In Fig. 3.4, fMRI shows that there are significant increases in activation mainly in three regions in the RST: the ACC, left PFC, and superior parietal lobule (SPL) (Osaka et al. 2004). A group difference in activation is observed in these three regions: the increases are higher in high-span participants than in low-span participants.

### 3.5.4 Functional Connectivity Between DLPFC and ACC

The possible functional connectivity between the DLPFC and ACC was compared between high-span and low-span groups by computing the average time courses of the activated voxels of fMRI data (Osaka et al. 2003, 2004). The correlation coefficient (a measure of the similarity in voxel activation between the DLPFC and ACC) is higher in the high-span group for both the LST and RST (Osaka et al. 2003, 2004). Higher correlations between different cortical areas throughout the activation time course are taken to indicate an increase in functional connectivity (Diwadkar et al. 2000).

Structural equation modeling (SEM) has been used to investigate the network connectivity between the DLPFC and ACC in the OST (Kondo et al. 2004a). As with the RST and LST, signal changes in the ACC are greater in the high-span group during the OST. The SEM result indicates that the effective connectivity from the ACC to the left DLPFC is positive and high in the high-span group, and it is negative and low in the low-span group. A significant positive correlation

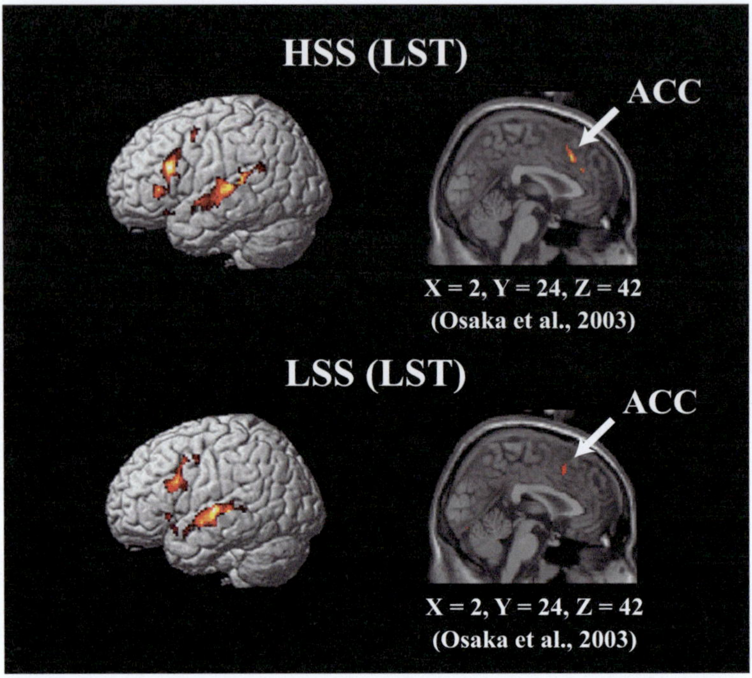


**Fig. 3.3** (*Left*) Rendered fMRIimages of activated brain areas of high-span and low-span participants in the LST condition. (*Right*) Activated brain areas in the ACC of high-span and low-span participants in the LST condition. *HSS* high-span participants, *LSS* low-span participants

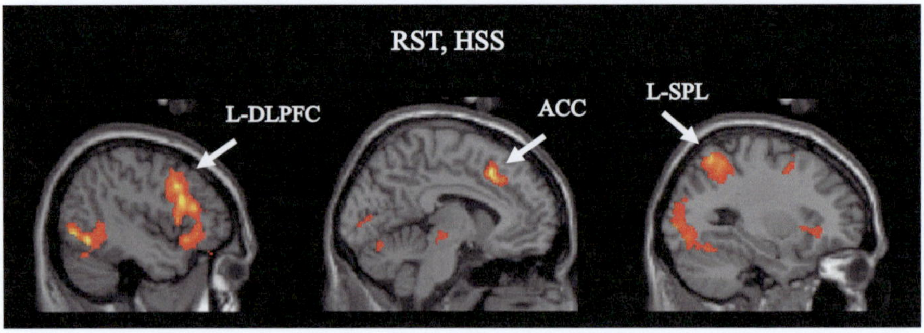


**Fig. 3.4** Rendered fMRI images of activated brain areas in the DLPFC, ACC, and SPL of high span participants in the RST condition

between signal changes in the right DLPFC and right ACC is confirmed by using a spatial span task (SST) in which five letters and five arrows were alternately presented and participants verify whether the letters were normal or mirror-imaged while concurrently retaining the orientations of arrows (Kondo et al. 2004b). These results suggest that an effective connection exists between the DLPFC and ACC only in the high-span group.

Osaka et al. (2003, 2004) conclude on the basis of these findings that the ACC subserves the attention control system of working memory; it inhibits irrelevant information and monitors attention control processes in accordance with the DLPFC. Furthermore, Osaka et al. propose that when one performs span tasks, such as the LST and RST, the attention controller of the central executive is regulated by the DLPFC and ACC; the DLPFC maintains attention processes and the ACC inhibits them. As for capacity differences, it has been found that activation differences between low-span and high-span participants can be explained in terms of differences between the DLPFC and ACC. A higher functional connectivity between the DLPFC and ACC is always observed in the high-span group.

### *3.5.5 Role of the Parietal Cortex*

A recent study demonstrates that activity in the inferior parietal lobule (IPL) is associated with the disengagement and reorientation of attention to the relevant target presented outside the current focus of attention (Corbetta et al. 2008). It is consistent with this report that the IPL has a role in the basic attentional process of the central executive. In addition, the SPL involves the lateral intraparietal area and is generally related to attention and saccade-related eye movements (Culham and Kanwisher 2001). RST participants show activation in the left SPL, as well as in the DLPFC and ACC, suggesting that the SPL may contribute to shifting and focusing attention (Osaka et al. 2004). Attention shift may explain the performance difference between high-span and low-span participants. When they perform the F- or NF-RST, activation in the left SPL (BA 7) is found to be enhanced in both groups. However, only the high-span group shows a greater increase in left SPL activation during the NF-RST than during the F-RST (Osaka et al. 2007).

Because low-span participants receive less aid from the SPL in N F-RST, it is difficult for them to shift attention from the focus word to the target word. This difficulty results in strong conflicts and confusions regarding the goals of the task. Thus, the differences between the low-span and high-span groups in their efficiency of shifting and focusing attention may depend on the SPL and the aid from the DLPFC and ACC network.

## 3.6 Neural Model of Executive Function of Working Memory

It may be supposed on the basis of the span task results that conflict perception may strengthen the executive control system that is mediated by the ACC, DLPFC and posterior parietal cortex (PPC) including the SPL or IPL. Once a conflict is detected, the executive control mechanism strengthens focusing of attention onto the task relevant stimulus and filters out information from irrelevant stimuli. The attentional system, if thus enhanced, seems to coordinate attention in working memory better.

The model in Fig. 3.5 proposes a distributed network that is structured around the central executive systems in the PFC, ACC and PPC. They contribute to control attention by maintaining, inhibiting, and focusing attention with the aids of the occipital, temporal, and parietal cortices. When a participant performs working memory tasks like the RST and LST, she must focus attention on a certain word and establish a mental representation of it. Her performance depends on how well she can allocate attention or shift the mental focus.

In the network proposed here, it is easy to control attention for self-monitoring both consciously and unconsciously in working memory tasks.

### 3.6.1 *Working Memory and Consciousness Revisited*

It has gradually become clear that active consciousness and working memory share some common neural representations. Active consciousness functions on the basis

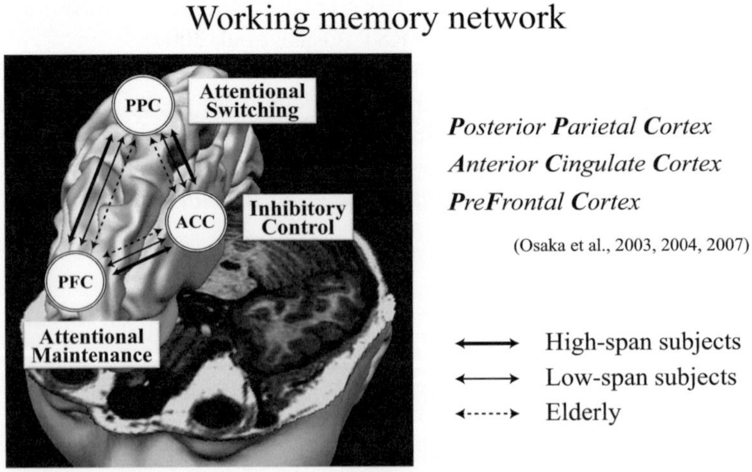


**Fig. 3.5** Working memory network model of high-span and low-span participants

**Fig. 3.6** Layered model of active consciousness and working memory

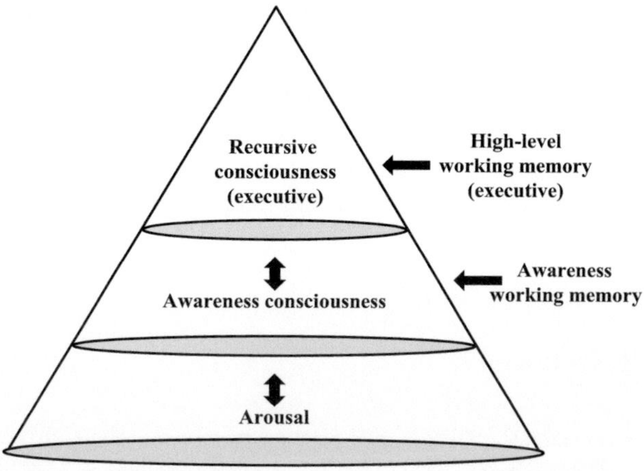


of a capacity-constrained and goal-directed neural system, and it is very similar to the central executive system of working memory. Osaka (2000) proposes a layered model that connects the two systems. Figure 3.6 shows a three-layered model of consciousness: it includes arousal-, awareness-, and recursive-consciousness, and recursive-consciousness is top-down controlled.

These layers correspond to arousal-, awareness-, and recursive-consciousness, and interact with each other in a bidirectional manner. The first level of consciousness is shaped by the arousal system generated by reticular formation of the brain. A biologically-driven arousal system is boosted by the brain stem system that is regulated by neurotransmitters. The second level of consciousness is called 'awareness driven by the awareness system at bottom' and closely related with the attentionalsystem. Perception of environments and attentional motor control for performing goal-directed behavior are led by awareness. Finally, the third level of consciousness is called 'recursive-consciousness' and goes with those cognitive processes like thinking which requires recursive function of information processing. This high-level recursive consciousness includes a top-down executive control that is responsible for, e.g., social interactions with others. When consciousness layers are compared with working memory, awareness working memory corresponds to the awareness consciousness, and high-level working memory corresponds to the recursive consciousness with executive function. High-level working memory has an executive function that effectively controls attention and monitors one's own performance so as to achieve task goals effectively. High-level working memory shifts the focus of attention and self-monitors in order to improve performance in tasks that require higher cognitive brain function. Cooperative activation of different brain areas is important for effective working memory performance and active consciousness.

## Exercises

1. What are differences between working memory and short term memory (STM)?
2. How working memory capacity is measured by the reading span test (RST)?
3. What regions of the brain compose the neural bases of working memory?
4. What kind of executive function corresponds to the highest level of active consciousness?

## References

Atkinson, R.C., Shiffrin, R.M.: Human memory: a proposed system and its control processes. In: Spence, K.W., Spence, J.T. (eds.) The Psychology of Learning and Motivation: Advances in Research and Theory, vol. 2, pp. 89–195. Academic, New York (1968)

Awh, E., Jonides, J., Smith, E.E., Schumacher, E.H., Koeppe, R.A., Katz, S.: Dissociation of storage and rehearsal in verbal working memory: evidence from positron emission tomography. Psychol. Sci. **7**(1), 25–31 (1996)

Baddeley, A.: Working Memory. Oxford University Press, Oxford (1986)

Baddeley, A.: Working memory. Science **255**(5044), 556–559 (1992)

Baddeley, A.: Exploring the central executive. Q. J. Exp. Psychol. **49A**(1), 5–28 (1996)

Baddeley, A.: The episodic buffer: a new component of working memory? Trends Cogn. Sci. **4** (11), 417–423 (2000)

Baddeley, A.: Working memory: theories, models, and controversies. Annu. Rev. Psychol. **63**, 1–29 (2012)

Baddeley, A.D., Hitch, G.J.: Working memory. In: Bower, G.H. (ed.) The Psychology of Learning and Motivation, vol. 8, pp. 47–89. Academic, New York (1974)

Baddeley, A.D., Logie, R.H.: Working memory: the multiple component model. In: Miyake, A., Shah, P. (eds.) Models of Working Memory: Mechanisms of Active Maintenance and Executive Control, pp. 28–61. Cambridge University Press, Cambridge (1999)

Baddeley, A.D., Warrington, E.K.: Amnesia and the distinction between long- and short-term memory. J. Verbal Learn. Verbal Behav. **9**(2), 176–189 (1970)

Baddeley, A., Logie, R., Nimmo-Smith, I., Brereton, N.: Components of fluent reading. J. Mem. Lang. **24**(1), 119–131 (1985)

Barch, D.M., Braver, T.S., Nystrom, L.E., Forman, S.D., Noll, D.C., Cohen, J.D.: Dissociating working memory from task difficulty in human prefrontal cortex. Neuropsychologia **35**(10), 1373–1380 (1997)

Birch, S.L., Garnsey, S.M.: The effect of focus on memory for words in sentence. J. Mem. Lang. **34**(2), 232–267 (1995)

Blutner, R., Sommer, R.: Sentence processing and lexical access: the influence of the focus-identifying task. J. Mem. Lang. **27**(4), 359–367 (1988)

Bolinger, D.: Intonation and Its Parts. Stanford University Press, Stanford (1986)

Braver, T.S., Cohen, J.D., Nystrom, L.E., Jonides, J., Smith, E.E., Noll, D.C.: A parametric study of prefrontal cortex involvement in human working memory. Neuroimage **5**(1), 49–62 (1997)

Braver, T.S., Barch, D.M., Gray, J.R., Molfese, D.L., Snyder, A.: Anterior cingulate cortex and response conflict: effects of frequency, inhibition and errors. Cereb. Cortex **11**(9), 825–836 (2001)

Bunge, S.A., Klinberg, T., Jacobsen, R.B., Gabrieli, J.D.E.: A resource model of the neural basis of executive working memory. Proc. Natl. Acad. Sci. U. S. A. **97**(7), 3573–3578 (2000)

Bush, G., Whalen, P.J., Rosen, B.R., Jenike, M.A., Mclnerney, S.C., Rauch, S.L.: The counting stroop: an interference task specialized for functional neuroimaging -validation study with functional MRI. Hum. Brain Mapp. **6**(4), 270–282 (1998)

Bush, G., Luu, P., Posner, M.I.: Cognitive and emotional influences in anterior cingulate cortex. Trends Cogn. Sci. **4**(6), 215–222 (2000)

Carpenter, P.A., Just, M.A.: Integrative processes in comprehension. In: LaBerge, D., Samuels, S.J. (eds.) Basic Processes in Reading: Perception and Comprehension, pp. 217–241. Erlbaum, Hillsdale (1977)

Carpenter, P.A., Just, M.A.: The role of working memory in language comprehension. In: Klahr, D., Kotovsky, K. (eds.) Complex Information Processing: The Impact of Herbert A. Simon, pp. 31–68. Erlbaum, Hillsdale (1989)

Carter, C.S., Braver, T.S., Barch, D., Botvinick, M.M., Noll, D., Cohen, J.D.: Anterior cingulate cortex, error detection, and the online monitoring of performance. Science **280**(5364), 747–749 (1998)

Cohen, J.D., Peristein, W.M., Braver, T.S., Nystrom, L.E., Noll, D.C., Jonides, J., Smith, E.E.: Temporal dynamics of brain activation during a working memory task. Nature **386**(6625), 604–608 (1997)

Conway, A.R.A., Engle, R.W.: Working memory and retrieval: a resource-dependent inhibition model. J. Exp. Psychol. Gen. **123**(4), 354–373 (1994)

Corbetta, M., Patel, G., Shulman, G.L.: The reorienting system of the human brain: from environment to theory of mind. Neuron **58**(3), 306–324 (2008)

Courtney, S.M., Petit, L., Maisog, J.M., Ungerleider, L.G., Haxby, J.V.: An area specialized for spatial working memory in human frontal cortex. Science **279**(5355), 1347–1351 (1998)

Cowan, N.: An embedded-processes model of working memory. In: Miyake, A., Shah, P. (eds.) Models of Working Memory: Mechanisms of Active Maintenance and Executive Control, pp. 62–101. Cambridge University Press, Cambridge (1999)

Cowan, N.: The magical number 4 in short-term memory: a reconsideration of mental storage capacity. Behav. Brain Sci. **24**(1), 87–185 (2001)

Culham, J.C., Kanwisher, N.G.: Neuroimaging of cognitive functions in human parietal cortex. Curr. Opin. Neurobiol. **11**(2), 157–163 (2001)

D'Esposito, M., Detre, J.A., Alsop, D.C., Atlas, R.K., Grossman, M.: The neural basis of the central executive system of working memory. Nature **378**(6554), 279–281 (1995)

D'Esposito, M., Aguirre, G.K., Zarahn, E., Ballard, D., Shin, R.K., Lease, J.: Functional MRI studies of spatial and nonspatial working memory. Cogn. Brain Res. **7**(1), 1–13 (1998)

D'Esposito, M., Postle, B.R., Ballard, D., Lease, J.: Maintenance versus manipulation of information held in working memory: an event-related fMRI study. Brain Cogn. **41**(1), 66–86 (1999)

Daneman, M., Carpenter, P.A.: Individual differences in working memory and reading. J. Verbal Learn. Verbal Behav. **19**(4), 450–466 (1980)

Daneman, M., Carpenter, P.A.: Individual differences in integrating information between and within sentences. J. Exp. Psychol. Learn. Mem. Cogn. **9**(4), 561–583 (1983)

Daneman, M., Merikle, P.M.: Working memory and language comprehension: a meta-analysis. Psychon. Bull. Rev. **3**(4), 422–433 (1996)

De Beni, R., Palladino, P., Pazzaglia, F., Cornoldi, C.: Increases in intrusion errors and working memory deficit of poor comprehenders. Q. J. Exp. Psychol. **51A**(2), 305–320 (1998)

Diwadkar, V.A., Carpenter, P.A., Just, M.A.: Collaborative activity between parietal and dorsolateral prefrontal cortex in dynamic spatial working memory revealed by fMRI. Neuroimage **12**(1), 85–99 (2000)

Engle, R.W., Conway, A.R.A., Tuholski, S.W., Shisler, R.J.: A resource account of inhibition. Psychol. Sci. **6**(2), 122–125 (1995)

Engle, R.W., Tuholski, S., Laughlin, J.E., Conway, A.R.A.: Working memory, short-term memory, and general fluid intelligence: a latent-variable approach. J. Exp. Psychol. Gen. **128**(3), 309–331 (1999)

Jonides, J., Smith, E.E., Koeppe, R.A., Awh, E., Minoshima, S., Mintun, M.A.: Spatial working memory in humans as revealed by PET. Nature **363**(6430), 623–625 (1993)

Just, M.A., Carpenter, P.A.: A capacity theory of comprehension: individual differences in working memory. Psychol. Rev. **99**(1), 122–149 (1992)

Just, M.A., Carpenter, P.A., Keller, T.A.: The capacity theory of comprehension: new frontiers of evidence and arguments. Psychol. Rev. **103**(4), 773–780 (1996)

Kane, M.J., Engle, R.W.: Working-memory capacity and the control of attention: the contributions of goal neglect, response competition, and task set to stroop interference. J. Exp. Psychol. Gen. **132**(1), 47–70 (2003)

Kintsch, W., Van Dijk, T.A.: Toward a model of text comprehension and production. Psychol. Rev. **85**(5), 363–394 (1978)

Kondo, H., Morishita, M., Osaka, N., Osaka, M., Fukuyama, H., Shibasaki, H.: Functional roles of the cingulo-frontal network in performance on working memory. Neuroimage **21**(1), 2–14 (2004a)

Kondo, H., Osaka, N., Osaka, M.: Cooperation of the anterior cingulated cortex and dorsolateral prefrontal cortex for attention shifting. Neuroimage **23**(2), 670–679 (2004b)

Kuno, S.: Discourse and Grammar. Taishukan-Shoten, Tokyo (1978) (in Japanese)

La Pointe, L.B., Engle, R.W.: Simple and complex word spans as measures of working memory capacity. J. Exp. Psychol. Learn. Mem. Cogn. **16**(6), 1118–1133 (1990)

Linden, D.E.: The working memory networks of the human brain. Neuroscientist **13**(3), 257–267 (2007)

MacDonald III, A.W., Cohen, J.D., Stenger, V.A., Carter, C.S.: Dissociating the role of the dorsolateral prefrontal and anterior cingulate cortex in cognitive control. Science **288**(5472), 1835–1838 (2000)

Masson, M.E., Miller, J.A.: Working memory and individual differences in comprehension and memory of text. J. Educ. Psychol. **75**(2), 314–318 (1983)

May, C.P., Hasher, L., Kane, M.J.: The role of interference in memory span. Mem. Cognit. **27**(5), 759–767 (1999)

Miller, G.A.: The magical number seven, plus or minus two: some limits on our capacity for processing information. Psychol. Rev. **63**(2), 81–97 (1956)

Norman, D.A., Shallice, T.: Attention to action: willed and automatic control of behavior. In: Davidson, R.J., Schwartz, G.E., Shapiro, D. (eds.) Consciousness and Self-Regulation: Advances in Research and Theory, vol. 4, pp. 1–18. Plenum Press, New York (1986)

Nyberg, L.: Mapping episodic memory. Behav. Brain Res. **90**(2), 107–114 (1998)

Osaka, N.: In the theatre of working memory of the brain. J. Conscious. Stud. **4**(4), 332–334 (1997)

Osaka, N.: Working memory and consciousness. In: Osaka, N. (ed.) Brain and Working Memory, pp. 1–18. Kyoto University Press, Kyoto (2000) (in Japanese)

Osaka, M.: Working Memory: The Sketch Pad of the Brain. Shinyosha, Tokyo (2002). (in Japanese)

Osaka, N.: Understanding Consciousness in the Brain. Iwanami Publishers, Tokyo, (1998) (in Japanese)

Osaka, M., Nishizaki, Y.: How working memory works in the central executive. In: Osaka, N. (ed.) Brain and Working Memory, pp. 203–223. Kyoto University Press, Kyoto (2000) (in Japanese)

Osaka, M., Osaka, N.: Language-independent working memory as measured by Japanese and English reading span tests. Bull. Psychon. Soc. **30**(4), 287–289 (1992)

Osaka, M., Osaka, N.: Working memory capacity related to reading: measurement with the Japanese version of reading span test. Jpn. J. Psychol. **65**(5), 339–345 (1994) (in Japanese with an English summary)

Osaka, M., Nishizaki, Y., Komori, M., Osaka, N.: Effect of focus on verbal working memory: critical role of the focus word in reading. Mem. Cognit. **30**(4), 562–571 (2002)

Osaka, M., Osaka, N., Kondo, H., Morishita, M., Fukuyama, H., Aso, T., Shibasaki, H.: The neural basis of individual differences in working memory capacity: an fMRI study. Neuroimage **18**(3), 789–797 (2003)

Osaka, N., Osaka, M., Kondo, H., Morishita, M., Fukuyama, H., Shibasaki, H.: The neural basis of executive function in working memory: an fMRI study based on individual differences. Neuroimage **21**(2), 623–631 (2004)

Osaka, M., Komori, M., Morishita, M., Osaka, N.: Neural bases of focusing attention in working memory. Cogn. Affect. Behav. Neurosci. **7**(2), 130–139 (2007)

Owen, A.M., Evans, A.C., Petrides, M.: Evidence for a two-stage model of spatial working memory processing within the lateral frontal cortex: a positron emission tomography study. Cereb. Cortex **6**(1), 31–38 (1996)

Owen, A.M., Stern, C.E., Look, R.B., Tracey, I., Rosen, B.R., Petrides, M.: Functional organization of spatial and non-spatial working memory processing within the human lateral frontal cortex. Proc. Natl. Acad. Sci. U. S. A. **95**(13), 7721–7726 (1998)

Paulesu, E., Frith, C.D., Frackowiak, R.S.: The neural correlates of the verbal component of working memory. Nature **362**(6418), 342–345 (1993)

Perfetti, C.A., Goldman, S.R.: Discource memory and reading comprehension skill. J. Verbal Learn. Verbal Behav. **15**(1), 33–42 (1976)

Rypma, B., Prabhakaran, V., Desmond, J.E., Glover, G.H., Gablieli, J.D.E.: Load-dependent roles of frontal brain regions in the maintenance of working memory. Neuroimage **9**(2), 216–226 (1999)

Shallice, T., Warrington, E.K.: Independent functioning of verbal memory stores: A neuropsychological study. Q. J. Exp. Psychol. **22**(2), 261–273 (1970)

Shallice, T., Fletcher, P., Frith, C.D., Grasby, P., Frackowiak, R.S., Dolan, R.J.: Brain regions associated with acquisition and retrieval of verbal episodic memory. Nature **368**(6472), 633–635 (1994)

Smith, E.E., Jonides, J.: Storage and executive processes in the frontal lobes. Science **283**(5408), 1657–1661 (1999)

Smith, E.E., Jonides, J., Koeppe, R.A.: Dissociating verbal and nonverbal working memory using PET. Cereb. Cortex **6**(1), 11–20 (1996)

Smith, E.E., Geva, A., Jonides, J., Miller, A., Reuter-Lorenz, P., Koeppe, R.A.: The neural basis of task-switching in working memory: effects of performance and aging. Proc. Natl. Acad. Sci. U. S. A. **98**(4), 2095–2100 (2001)

Stroop, J.R.: Studies of interference in serial verval reactions. J. Exp. Psychol. **18**(6), 643–662 (1935)

Towse, J.N., Hitch, G.J., Hutton, U.: A reevaluation of working memory capacity in children. J. Mem. Lang. **39**(2), 195–217 (1998)

Towse, J.N., Hitch, G.J., Hutton, U.: On the interpretation of working memory span in adults. Mem. Cognit. **28**(3), 341–348 (2000)

Tulving, E., Kapur, S., Markowitsch, H.J., Craik, F.I., Habib, R., Houle, S.: Neuroanatomical correlates of retrieval in episodic memory: auditory sentence recognition. Proc. Natl. Acad. Sci. U. S. A. **91**(6), 2012–2015 (1994)

Turner, M.L., Engle, R.W.: Is working memory capacity task dependent? J. Mem. Lang. **28**(2), 127–154 (1989)

Vogt, B.A., Finch, D.M., Olson, C.R.: Functional heterogeneity in cingulate cortex: the anterior executive and posterior evaluative regions. Cereb. Cortex **2**(6), 435–443 (1992)

# Chapter 4
# Primate Social Behavior: Understanding the Social Relationships of Japanese Macaques

**Masayuki Nakamichi**

**Abstract** For many years, a variety of primate species have been studied in the wild or in captivity in various academic fields, such as biology, ecology, anthropology, psychology, and medical and veterinary science. Japanese primatologists began a thorough investigation of wild Japanese macaques after World War II, when research on other wild primate species was rarely conducted. Thus, Japanese macaques are among the primate species whose social organizations and life histories in the wild can be clearly described. This chapter provides an overview of the research on social relationships of Japanese macaques. Firstly, the dominance relationships and matrilineal blood relationships of adult females in a group, the relationships of adult males in a group, and the social events in which both females and males take part, are described. Secondly, the mother-young offspring relationships and social development of immature individuals are discussed. Lastly, the behavioral characteristics of old individuals are summarized.

**Keywords** Dominance relationships • Mother-offspring relationships • Social development • Kawamura's rules • Alpha male • Social grooming • Alliance • Mating • Group division • Malformed infant • Dead infant • Grandmother hypothesis • Aging

## 4.1 Introduction

The common ancestor of all primates first split into two branches: nonhuman primates and humans. Is this correct or not?

The answer is negative. The common ancestor of primates including humans appeared on the earth approximately 55 million years ago (Martin 1990). The common ancestor first split into two branches: prosimians and simians (Fig. 4.1). The common ancestor of the New World monkeys, now inhabiting Central and

M. Nakamichi (✉)
Graduate School of Human Sciences, Osaka University, Suita, Osaka, Japan
e-mail: naka@hus.osaka-u.ac.jp



M. Kasaki et al. (eds.), *Cognitive Neuroscience Robotics B*,
DOI 10.1007/978-4-431-54598-9_4

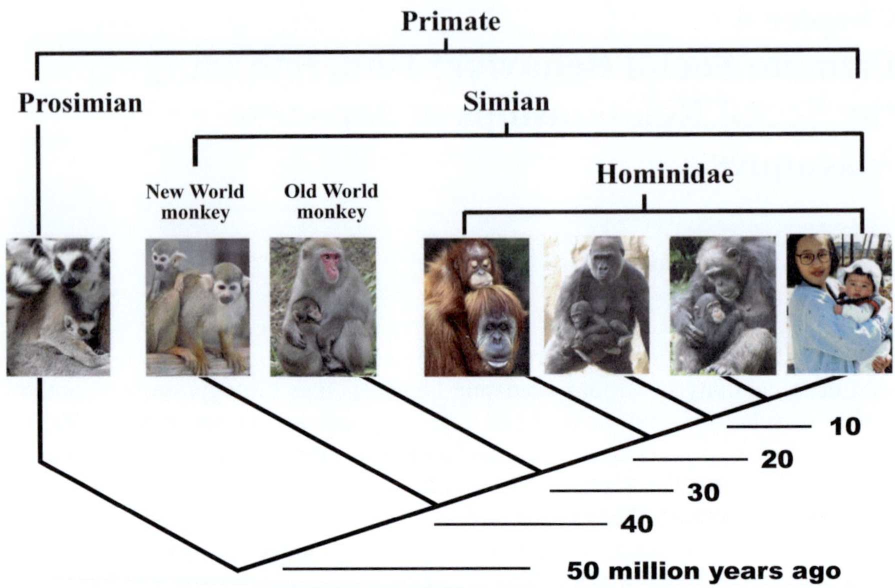


**Fig. 4.1** A phylogenetic tree of primates including humans. Species are from *left* to *right*: ring-tailed lemur (*Lemur catta*), squirrel monkey (*Saimiri sciureus*), Japanese macaque (*Macaca fuscata*), orangutan (*Pongo pygmaeus*), gorilla (*Gorilla gorilla*), chimpanzee (*Pan troglodytes*), and human (*Homo sapiens*)

South America, diverged from the common ancestor of simians, and then the common ancestor of the Old World monkeys, now inhabiting Africa and Asia, diverged. The common ancestors of lesser apes and great apes, such as orangutans, gorillas, and chimpanzees, diverged from the common ancestor of the Old World monkeys. The common ancestor of chimpanzees and humans diverged about 7 million years ago, and they share 98 % of DNA (Dunbar and Barrett 2000).

At present, more than 300 nonhuman primate species live on the earth. The smallest species, rufous (or brown) mouse lemur (*Microcebus rufus*), weighs as little as less than 100 g, and the largest species, gorilla (*Gorilla gorilla*), can weigh as much as more than 150 kg (Dunbar and Barrett 2000; Rowe 1996). While some prosimian species and orangutans live a solitary life, most primates form social groups, varying in size and composition. Major forms of primate social groups are the following: a monogamous group (*pair-bond*) comprises one adult male and one adult female; a *polyandrous* group comprises one adult female and two or more adult males with whom she is mated; a *one-male* group comprises one adult male, several adult females, and their immature offspring; and a *multi-male, multi-female* group comprises two or more adult males, adult females, and their immature offspring (Boyd and Silk 2000).

While most primates inhabit the tropical or semi-tropical regions of Africa, Asia, Central America, and South America, Japanese macaques (*Macaca fuscata*) primarily inhabit the temperate regions. There are also some Japanese macaques at the southern limit of their range, and they are distributed in subtropical evergreen broad-leaved forests, such as in Yakushima Island, Kagoshima Prefecture. Other Japanese macaques are found in the northern Japan, where the temperature drops below −20 °C and the snow is several meters deep in the winter (Hanya 2010). It is because of these northern-dwelling macaques that Japanese macaques are widely referred to as 'snow monkeys.' The northern limit of Japanese macaque distribution, Shimokita Peninsula in Aomori Prefecture, is in fact the northern limit of all living nonhuman primate species.

Japanese macaques form multi-male, multi-female groups. The number of adult males in each group is much smaller than that of adult females, because most males leave their natal group upon maturity, while females usually remain in their natal group throughout their life. Most females experience their first birth at 5–7 years of age, give birth every 2–3 years thereafter, and die in the early 20s. Only a small number of females survive to be older than 25 years of age (Itoigawa et al. 1992).

For many years, a variety of primate species have been studied in the wild or in captivity in various academic fields, such as biology, ecology, anthropology, psychology, and medical and veterinary science. Japanese primatologists began a thorough investigation of wild Japanese macaques after World War II, when research on other wild primate species was rarely conducted. Thus, Japanese macaques are among the primate species whose social organizations and life histories in the wild can be clearly described. The descriptions depended greatly on the two methods, provisioning and individual identification, which Japanese primatologists used to observe wild groups of Japanese macaques for the first time.

When researchers began observing wild macaques in the early 1950s, the macaques were not easy to observe, because they were very shy, easily got frightened, and quickly fled, whenever they encountered people. Past hunting was responsible for their shyness and tendency to avoid humans (Yamagiwa 2010). To promote research (and tourism), then, researchers, with the help from local people, attempted to habituate monkeys. The attempt consisted in provisioning monkeys with artificial food; they received sweet potato, wheat, and soybeans in the foothills of a mountain in their home range. Monkeys started approaching the feeding site to eat the provisioned food and spending more time near the feeding site during the day (Fig. 4.2). Researchers spent long time with monkeys at close range after successful provisioning. As a result, they became able to identify each individual monkey by its facial and body characteristics, with no need for any artificial markings like hair dyeing. Individual monkeys were then named (not numbered). Using the methods of provisioning and individual identification by name, various aspects of the social lives of Japanese macaque groups, such as group organization, social relationships among group members, mother-infant relationships, and social development, were revealed (see Yamagiwa 2010 for the research history of Japanese macaques by Japanese primatologists). Not only researchers but also tourists visit artificial feeding sites referred to as 'wild monkey parks' and watch monkeys at close range, although the number of wild monkey parks and tourists visiting them have recently decreased.

**Fig. 4.2** Japanese macaques of the Katsuyama free-ranging, provisioned group (Maniwa City, Okayama Prefecture) picking up and eating wheat scattered around the feeding area (*left*). All the individuals in the group have been identified since 1958, the beginning of provisioning. A mother (formal name, *Be71'79*) is 27 years old and the oldest animal in the group at the time of the photograph. *Be71'79* is pictured sitting with her 8- and 10-year-old daughters (formal names, *Be71'79'98* and *Be71'79'96*, respectively), both holding their newborn infants (*left* and *right*, respectively)

Provisioning improved the nutritional condition of Japanese macaques by decreasing travel time and distance. It also raised their birth rate and longevity, and lowered their mortality rate (Yamagiwa 2010). As a result, the usual group size of provisioned Japanese macaques increased to more than 100 individuals. By contrast, the usual group size of non-provisioned wild Japanese macaques is less than 100. On the face of this contrast, some researchers after the mid-1970s suspected that non-provisioned wild groups might behave differently than monkeys at artificial feeding sites do. These researchers began to habituate Japanese macaques without provisioning. With considerable efforts, they succeeded at some field sites. Studies of Japanese macaques in their natural habitats provided interesting and important data which were indispensable for understanding Japanese macaques from a socioecological perspective. Differences between provisioned and non-provisioned wild groups were found in demography, foraging, nomadic activities, and social behaviors. However, the observed differences should not be considered to be completely separate but to be different points along a continuum. For this reason, it is useful for promoting understanding of the whole picture of the social lives of Japanese macaques to investigate both free-ranging non-provisioned and provisioned groups.

In this chapter, I describe social relationships of Japanese macaques. Firstly, I focus on the dominance relationships and matrilineal blood relationships of adult females in a group, the relationships of adult males in a group, and the social events in which both females and males take part. Secondly, I discuss the mother-young offspring relationships and social development of immature individuals. Lastly, I survey the behavioral characteristics of old individuals.

## 4.2 Social Relationships Among Adult Females

### *4.2.1 Dominance Relationships Among Adult Females*

#### 4.2.1.1 Kawamura's Rules

If you were to observe a social group of Japanese macaques for a while, you might witness the following episode: adult female A approaches adult female B who is just sitting on the ground, and B walks away from her seat. Which female do you think is dominant? In this case, female A is dominant to B. As in this example, if an individual successfully forces another individual to change its position just by approaching it, i.e., without displaying any obvious dominant behavior, such as gazing or opening the mouth, the approaching one is the dominant animal of the two. This type of social interaction is called 'supplanting.' If the approached one does not avoid the approaching one but shows a grimace at or turns its head away from the approaching one, the approaching one is the dominant one. Dominance relationships between two animals are also observed through unidirectional agonistic interactions in which an animal's aggressive behavior, e.g., threatening, is followed by another animal's subordinate behavior, e.g., bared-teeth grimace (Fig. 4.3). Dominance relationships can be observed in the food-dominance test for provisioned groups. For example, when a soybean is tossed between two animals, one animal takes and eats it, but the other remains sitting or moves away without trying to take the soybean. The soybean-taking animal is dominant.

Figure 4.4 shows the dominance matrix for 16 adult females, their blood-relationships through maternal lines, and ages in years when the data were recorded in a provisioned group at Katsuyama, Okayama Prefecture. For example, *Be71* is a female who was born to *Be* in 1971. *Be71'79* and *Be71'83* are *Be71*'s daughters born in 1979 and 1983, respectively, and thus are sisters 4 years apart in age. *Be71'79* is a niece of *Be67*, who is *Be71*'s older sister. The number within each cell indicates the total number of dominant behaviors that the individual in each row

**Fig. 4.3** Facial expressions made by adult female Japanese macaques: open mouth threat face (*left*) and bared-teeth grimace (*right*)

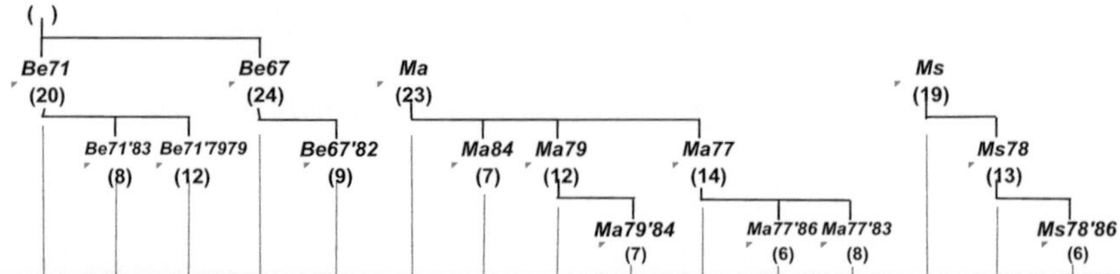


| | Be71 | Be71'83 | Be71'79 | Be67 | Be67'82 | Ma | Ma84 | Ma79 | Ma79'84 | Ma77 | Ma77'86 | Ma77'83 | Ms | Ms78 | Ms78'86 |
|---|---|---|---|---|---|---|---|---|---|---|---|---|---|---|---|
| Be71 | | 4 | 4 | 7 | 6 | 1 | 3 | 3 | 4 | 5 | — | — | 1 | 1 | — |
| Be71'83 | 0 | | 8 | 3 | 5 | 5 | 3 | 2 | 3 | 3 | — | 2 | 1 | 2 | 1 |
| Be71'79 | 0 | 0 | | 9 | 7 | — | 3 | 5 | 3 | 5 | — | — | 1 | 1 | 1 |
| Be67 | 0 | 0 | 0 | | 10 | 3 | 4 | 3 | 2 | 2 | — | 1 | 2 | 3 | — |
| Be67'82 | 0 | 0 | 0 | 0 | | 6 | 4 | 2 | 8 | 3 | — | — | — | 2 | 1 |
| Ma | 0 | 0 | — | 0 | 0 | | 1 | 6 | — | 6 | — | 3 | 4 | 3 | — |
| Ma84 | 0 | 0 | 0 | 0 | 0 | 0 | | 3 | 5 | 8 | 1 | 7 | 3 | 1 | 2 |
| Ma79 | 0 | 0 | 0 | 0 | 0 | 0 | 0 | | 9 | 7 | — | 2 | 3 | 1 | 1 |
| Ma79'84 | 0 | 0 | 0 | 0 | 0 | — | 0 | 0 | | 1 | 3 | 3 | 2 | 2 | 1 |
| Ma77 | 0 | 0 | 0 | 0 | 0 | 0 | 0 | 0 | 1 | | 1 | 10 | 4 | 1 | 1 |
| Ma77'86 | — | — | — | — | — | — | 0 | — | 0 | 0 | | 2 | — | — | 1 |
| Ma77'83 | — | 0 | — | 0 | — | 0 | 0 | 0 | 0 | 0 | 0 | | 3 | 2 | 4 |
| Ms | 0 | 0 | 0 | 0 | — | 0 | 0 | 0 | 0 | 0 | — | 0 | | 6 | 3 |
| Ms78 | 0 | 0 | 0 | 0 | 0 | 0 | 0 | 0 | 0 | 0 | — | 0 | 0 | | 2 |
| Ms78'86 | — | 0 | 0 | — | 0 | — | 0 | 0 | 0 | 0 | 0 | 0 | 0 | 0 | |

**Fig. 4.4** Dominant-subordinate matrix for 16 adult female Japanese macaques of three kin groups in the Katsuyama group and their blood-relationships traced through maternal lines. Rows and columns represent winners and losers of agonistic interactions, respectively. The *number* in *parentheses* indicates the age of each individual in years. The *number* in a cell indicates the observed number of dominant behaviors in each dyad, and the *dashes* indicate that no dominant-subordinate interactions are recorded. The *gray* areas indicate the results of agonistic conflicts between related females. For example, *Be71* (20 years old) was recorded to be dominant to her daughter *Be71'83* (8 years old) 4 times, but *Be71'83* was never recorded to be dominant to *Be71*. No dominant-subordinate interactions were recorded between *Be71* and the unrelated female *Ma77'86* (6 years old)

displayed toward the individual in each column. These behaviors include supplanting, agonistic interactions, and food-taking behaviors in food dominance tests. For example, the first three numbers in the first row are 4, 4, and 7. This means that *Be71* was recorded to be dominant to *Be71'83* four times, to *Be71'79* four times, and to *Be67* seven times. The first three numbers in the first column are 0. These numbers indicate that *Be71* was never recorded to be subordinate to any of these three females.

Before proceeding, it may be helpful to use the dominance interactions in Fig. 4.4 to find some of the rules or characteristics that govern the dominance relationships among these females.

If you examine the dominance relationships between mothers and daughters, you may find that mothers are dominant to daughters in all of the 11 mother-daughter dyads. If you examine the sister-sister dyads, you can see that younger siblings are dominant to older siblings in all of the 6 dyads. In this figure, all numbers greater than or equal to 2 are above the diagonal line, while all the numbers in the corresponding cells below the diagonal line are 0. This indicates that in the most dyads for which multiple episodes are recorded, one animal is consistently dominant to the other animal. Moreover, all the females in the *Be* kin group are dominant to the females of the *Ma* and *Ms* kin groups, and all the females in the *Ma* kin group are dominant to the females of *Ms* kin group. You might then surmise that the females of one kin group are collectively ranked above or below the females of other kin groups. You might also presume that the dominance rank order of the 16 females is almost linear, as there are no circular dominance relationships (i.e., it is not the case that individual A is dominant to B, B is dominant to C, and C is dominant to A).

Early in the history of Japanese primatology, Kawamura (1958) investigated the dominance relationships among 11 adult and young adult females in a provisioned group of 29 Japanese macaques. On the basis of his observations, Kawamura described two characteristics of the dominance relationships among adult females: (1) females are ranked just below their mothers, and (2) sisters are ranked in reverse age order, which is called 'the youngest ascendancy.' These two rules are called 'Kawamura's rules.' Although he did not state other findings clearly, Kawamura's data showed that one female was consistently dominant to another, that dominance rank order was almost linear, and that all females in one kin group were ranked collectively.

Kawamura's rules have been confirmed not only in other groups of Japanese macaques (as in Fig. 4.4), but also in groups of other macaque species and baboons similar to Japanese macaques. These species form multi-male, multi-female groups. Females stay in their natal groups throughout their life, while males usually leave their natal groups upon maturity. Note that the phenomenon of youngest ascendancy among sisters is controversial, because it has been confirmed in free-ranging, provisioned groups and in captive groups of Japanese macaques, but not in wild groups of Japanese macaques. This implies that youngest ascendancy might be a by-product of provisioning (see Hill and Okayasu 1995 or Kutsukake 2000 for the discussion).

#### 4.2.1.2 Dominance Rank and Alliances

Alliance plays a determining role in female's acquisition and maintenance of a rank. Alliance is manifested when a third individual supports one of the two individuals in an agonistic interaction. Mothers tend to support their younger offspring in agonistic interactions between daughters. With support from the mother, a younger sibling, even if immature, behaves in a dominant manner toward her older sister when her mother is nearby. However, it may be difficult for a younger sibling to behave dominantly toward an older sibling when her mother is not nearby. Therefore, the dominance ranks of younger siblings relative to older siblings depend greatly on the mother. In other words, the ranks of younger siblings are regarded as *dependent ranks*, i.e., ranks dependent on their mother's support. However, younger siblings usually become consistently dominant toward older siblings before reaching young adulthood, regardless of their mother's presence in the vicinity. This means that before attaining adulthood, younger siblings can gain a *basic rank* relative to their older siblings (Kawai 1958).

Because mothers support daughters when they have agonistic interactions with others, young daughters can first gain a dependent rank and then gain a basic rank before they become mature. Females are more likely to give agonistic aid to their close relatives than to distant relatives, and they more often aid relatives than they do non-relatives. Therefore, an immature sister often becomes dominant to younger siblings after their mother's death, because they no longer have support from their mother. Younger siblings can gain a rank just below their older sibling, due to support from the older sibling. In cases where sisters lose their mother before they become adults, they usually fail to inherit their mother's rank within their kin group, but often succeed in inheriting their mother's rank relative to unrelated females, due to agonistic aids from related females, such as aunts.

As described above, alliance is more likely to occur between closely related individuals than between distantly related individuals; and it is more likely to occur between related individuals than between unrelated individuals. As a result, alliances serve to stabilize the dominance hierarchy among females. This is true not only for Japanese macaques (Watanabe 1979) but also for other monkey species, such as macaques and baboons (Cheney 1977; Bernstein and Ehardt 1985). In fact, 10–40 % of aggressive interactions among members of these species are polyadic (Deag 1977; Walters 1980).

Screams uttered by subordinates and threat calls uttered by dominants play a major role in the recruitment of allies (Cheney 1977; de Waal 1977; Gouzoules and Gouzoules 1989). This means, for example, that a mother may act agonistically toward her older daughter in response to recruiting behaviors, such as screams, displayed by her younger daughter. Unlike macaques and baboons, ring-tailed lemurs (*Lemur catta*), the prosimian primates that first diverged from the common ancestor of primates (see Fig. 4.1), have stable, non-linear dominance hierarchies among adult females. Daughters do not rank immediately below their mothers, and an older sister is not necessarily subordinate to her younger sisters. This dominance structure is probably related to the rarity of alliance. The lack of alliance may be due to the fact that the submissive call by a ring-tailed lemur is too short for supporters to recognize what has happened to the vocalizer and determine its location (Nakamichi and Koyama 1997). In other words, it is difficult for ring-tailed lemurs to form alliance because they do not emit calls or screams to effectively recruit support from allies against opponents, even though allies might provide support anyway. As a result, ring-tailed lemurs do not have linear dominance hierarchies even in a group of less than 10 adult females.

On the other hand, Japanese macaques often emit submissive calls (or threat calls) repeatedly. Calls are not only directed toward their opponents but also used to recruit potential allies (usually closely related individuals) against opponents. The abilities to recruit support from allies and to attract appropriate attention to agonistic interactions must be indispensable for maintaining dominance relationships and making complex social groups stable.

#### 4.2.1.3 Dominance Relationships and Group Size

Kawamura (1958) reported that his rules were based on the observations of a small group including 11 adult females. It has been confirmed that dominance relationships conform to Kawamura's rules in groups of up to 50 adult females (Koyama 1967, 1970). In a group of 62 adult females, Kawamura's rules still apply, but deviations from the rules are relatively common (Takahata 1991). In some groups of approximately 70 adult females, the dominance rank order is almost linear, and the dominance relationships of nearly all adult female dyads are consistent. However, the dominance relationships of a few kin dyads do not adhere to Kawamura's rules, and some females rank separately from other members of their kin groups (Fig. 4.5). Kawamura's rules apply less often in middle- or low-ranking kin groups than in high-ranking groups. No linear dominance rank order has been found in very large groups including more than 100 adult females. However, it may still exist in high-ranking kin groups (Nakamichi et al. 1995a; Nakamichi and Yamada 2010).

The reason that dominance relationships are more likely to deviate from Kawamura's rules as group size increases is closely related to the fact that each member of a group identifies other members by both its own social interactions with them and observation of the interactions among them. As a group becomes larger, it becomes more difficult (or maybe, even impossible) for a member of the group to

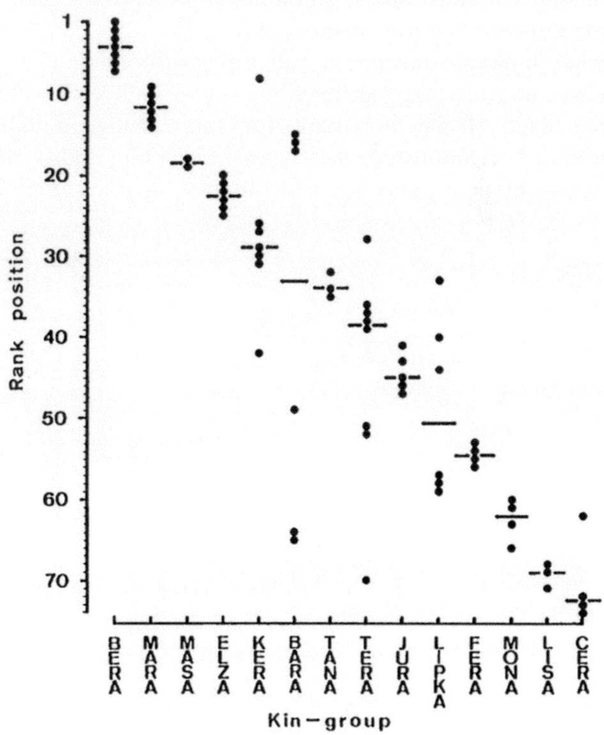


**Fig. 4.5** Rank positions of females aged 6 or more years in the Katsuyama free-ranging, provisioned group of Japanese macaques. Each *dot* represents a female and her rank position among the 74 adult females. Each *horizontal bar* indicates the median rank position of a kin group. The kin groups are arranged in order of dominance rank, with the highest-ranking kin group listed on the *left* (From Nakamichi et al. 1995a, with permission of Wiley-Liss)

directly interact with most of the other members and to observe the interactions among them. Probably, by interacting with others and watching the interactions among them, an individual learns to identify others, understands its relationships with them, and understands the relationships among them. In a large group with more than 100 females, it is impossible for each group member to interact with all of the other group members and to learn the social relationships between them. The difficulty of interaction results in unstable dominance relationships in a large group (Nakamichi and Yamada 2010).

## 4.2.2 Social Grooming Among Adult Females

### 4.2.2.1 Functions of Grooming

In wild monkey parks, you can see monkeys grooming each other (i.e., social grooming, Fig. 4.6, left) and themselves (i.e., self-grooming, Fig. 4.6, right). Grooming is a common and frequently observed behavior of Japanese macaques and most other primate species.

Grooming behavior consists of manual or oral manipulation of the skin and fur (e.g., parting and stroking fur) and picking off things from the skin by hands or mouth. The functions of grooming in primates are presumed to be fur cleaning and the removal of ectoparasites. In fact, lice eggs comprise the majority (98.9 %) of what Japanese macaques eat during grooming (Tanaka and Takefushi 1993). Grooming has a social function besides its hygiene functions. Social grooming is presumed to be important for establishing, maintaining, and strengthening affiliative relationships with others. Grooming is also used to restore social relationships with others after agonistic interactions. Grooming in this context is considered

**Fig. 4.6** Social grooming: a 13-year-old female grooms her 26-year-old mother (*left*). Self-grooming: a 26-year-old female grooms her left hind-leg (*right*)

to be useful for reducing the tension that has increased during agonistic interactions (Schino et al. 1988). In addition, grooming is considered to be a tradable commodity, and exchanged for reciprocal grooming or various kinds of social tolerance (Barrett et al. 1999). For example, in virtue of grooming a higher-ranking individual, a lower-ranking individual is allowed to remain near the higher-ranking individual even after grooming is completed. An individual may also be permitted access to a newborn infant by grooming its mother, and receive support from previously groomed individuals during agonistic interactions.

#### 4.2.2.2 Distribution of Social Grooming Among Females

Grooming has various social functions. Generally, most females have grooming interactions with a relatively small subset of available females. Even in a large group with 50 or more adult females, the number of grooming partners is around 10 on average, and most females tend to devote more than 50 % of their grooming effort to a particular female grooming partner (Nakamichi and Shizawa 2003). Therefore, investigating the frequency with which individuals groom one another, i.e., the distribution of social grooming among group members, is useful for understanding the social organization of a group and the social intelligence of macaques.

As in other macaque species and baboons, social grooming among adult female Japanese macaques is kin biased and mainly directed up the hierarchy. Although related female dyads account for less than 10 % of all possible dyads in a group of Japanese macaques, about 50–70 % of grooming bouts among adult females occur among relatives; these results hold no matter what group they are in and what year grooming data is collected in (Ando 1982; Koyama 1991; Nakamichi and Shizawa 2003). This means that grooming tends to be kin biased at the group level but not at the individual level. It is not the case that all adult females groom related females more often than they do unrelated ones.

Analysis of grooming at the individual level reveals patterns that may be hidden in analysis at the group level. Most kin groups include at least one female who grooms unrelated females more often than she does related females. A similar pattern can be observed for the relationship between grooming and dominance. At the group level, approximately 70 % of grooming between unrelated females is directed up the hierarchy (Ando 1982; Koyama 1991). However, when grooming is analyzed at the dyad level, grooming is either directed down the hierarchy or well-balanced in 75 % of unrelated female dyads (Nakamichi and Shizawa 2003). These results indicate that while some individuals in a group show despotic behavioral tendencies, such as kin-biased grooming and grooming directed up the hierarchy, some other individuals show egalitarian behavioral tendencies, such as preferentially grooming unrelated or subordinate females. Egalitarian adult females may be important for increasing the integrity and cohesion of a large group because they may maintain or extend social networks with members of different kin groups.

#### 4.2.2.3 Grooming Distribution and Cohesiveness Within Kin Groups

As described in Sect. 4.2.1, dominance relationships among adult females in a high-ranking kin group can be collectively ranked in accordance with Kawamura's rules. However, middle- and low-ranking kin groups include related dyads in which dominance relationships do not conform to Kawamura's rules and females tend to rank independently of other members of their kin groups (Nakamichi et al. 1995a). It is very important and interesting to determine why there are such differences in dominance relationship between high-ranking and lower-ranking kin groups. Here, I consider the relations between social grooming and dominance relationships within a kin group.

Figure 4.7a shows sociograms of grooming in the top ranking and middle-ranking kin groups in the Katsuyama group. Both of these kin groups include 8 adult females. In the top-ranking kin group, grooming interactions were observed in 21 of 28 possible dyads, while in the middle-ranking kin group, grooming interactions were observed in only 11 of 28 possible pairs. That is, the females of the top-ranking kin group distributed their grooming interactions among almost all available related females, while the females of the middle-ranking kin group tended to restrict their grooming interactions to a very small number of their relatives. The females of the top-ranking kin group maintained much stronger kin-group cohesiveness than those in the middle-ranking kin group do, by interacting with most of their relatives via social grooming. This strong cohesiveness among related females may account for why their collective ranking accords with the dominance relationships given by Kawamura's rules.

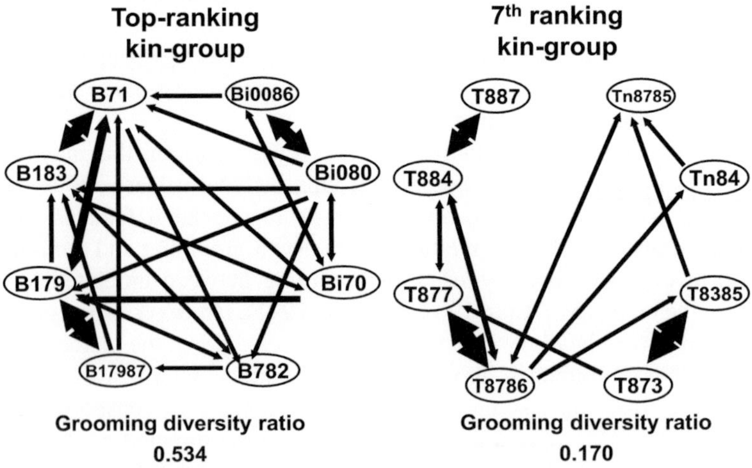


**Fig. 4.7a** Sociograms indicating the social grooming patterns among adult female relatives in the top-ranking kin group (*left*) and the middle-ranking kin group (*right*) of the Katsuyama group of Japanese macaques. The width of each *bar* corresponds to the number of observed grooming bouts. See Fig. 4.7b for grooming diversity ratio

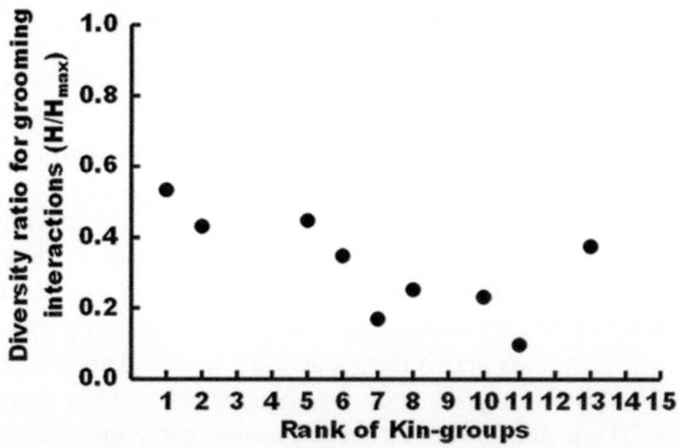


**Fig. 4.7b** The mean diversity ratio for grooming interactions among related females in kin groups consisting of 5 or more adult females. If a female had grooming interactions with all available related females equally, the ratio would be 1. If a female limited her grooming to a small subset of available related females, the ratio would approach 0 (Dunbar 1984). In comparison to the grooming interactions in lower-ranking kin groups, the grooming interactions of females in higher-ranking kin groups were more evenly spread among related females (From Nakamichi and Shizawa 2003, with permission of Springer)

#### 4.2.2.4 Long-Term Grooming Relationships

The duration of grooming partnership is important for understanding macaque relationships, because females usually stay in their natal groups throughout their life and maintain affiliative relationships with some particular females, such as their relatives. Even though wild Japanese macaques have been observed for more than 50 years, data available for determining the duration of grooming partnership is unfortunately very limited. Nakamichi and Yamada (2007) compare the grooming interactions of 18 adult females in a free-ranging group of Japanese macaques with the data of the same females recorded 10 years earlier. These females engage in grooming interactions with some of the unrelated partners with whom they engaged in grooming interactions 10 years ago. This indicates that female Japanese macaques maintain grooming partnerships with the same females for at least 10 years. The 18 females also tend to engage in grooming interactions with some females related to their old grooming partners. The findings about kin-biased grooming tendencies, grooming directed up the hierarchy, and long-term grooming partnerships, indicate that female Japanese macaques are conservative with regard to grooming partner choice. On the other hand, they sometimes form grooming relationships with new partners, indicating that their social relationships can also be progressive and flexible.

### 4.2.3 Special Food-Processing Behavior Performed by Some High- and Middle-Ranking Females

Japanese macaques consume a wide variety of foods, such as leaves, buds, flowers, grass, and invertebrates, and do not have trouble processing most of these food items. Japanese macaques can easily consume most edible foods without using any difficult food-processing methods. They may process foods to make them more palatable (i.e., washing foods to remove the dirt). In the Katsuyama group, adult monkeys of both sexes, but not infants and juveniles, dig up grass roots in the winter and early spring when plants above the ground are scarce. Although most monkeys simply scrape the dirt from grass roots by their hands and eat them, some adult females pull 10–20 cm grass roots from the ground, carry them to the edge of the river, and wash them in the water before eating (Fig. 4.8, Nakamichi et al. 1998). The washing of sweet potatoes and wheat is often observed among Japanese macaques of a free-ranging, provisioned group on Koshima Island, Miyazaki Prefecture (Kawai 1965). Nakamichi et al. (1998) is the first report of Japanese macaque's washing the dirt from natural diets.

Females, when they wash grass roots, usually pull one root at a time and make a pile of several roots. They hold the pile of roots in one hand and carry it to a river, where they wash the roots by immersing them one at a time into the water. When macaques eat seeds from the ground, they normally pick up one seed at a time with one hand and eat it. It has rarely been observed that they pick up seeds with one hand, place them in the other hand, and subsequently consume several seeds at once. Some females carry several roots to the water at a time. This behavior may be considered to be very economical, and may reflect a relatively advanced cognitive ability that has not yet been examined in Japanese macaques (Nakamichi et al. 1998).

The grass root washing behavior is greatly related to dominance relationships among adult females. Less than 10 % of all adult females show root-washing

**Fig. 4.8** Carrying and washing grass roots by adult females of the Katsuyama group of Japanese macaques. An adult female carrying several roots in one hand (*left*) and another adult female washing a root by rolling it along the surface of a flat rock at the edge of a river (from Nakamichi et al. 1998, with permission of Karger)

behavior. All the individuals that wash roots are high- or middle-ranking females; no adult males have been observed washing grass roots. It is very common for macaques to get food and eat it immediately. It is disadvantageous for low-ranking individuals to carry desirable foods because such behavior may increase the likelihood of an antagonistic response by high-ranking individuals (Nakamichi et al. 1998). Since digging up 10–20 cm grass roots from the ground can be a difficult task for immature Japanese macaques, they are unlikely to have the opportunity to wash grass roots. However, juveniles have sometimes been observed to watch their mothers washing and eating grass roots, and sometimes to pick up a half-eaten root to eat. It is probable that young adult females who have watched their mothers washing grass roots show similar behavior after they become mature enough to dig up grass roots without difficulty (Nakamichi and Yamada 2010). In fact, most females observed carrying and washing grass roots have at least one closely related female, e.g., a mother, who also carries and washes grass roots (Nakamichi et al. 1998). It is plausible that mothers transmit the ways of carrying and washing grass roots to their daughters but not to sons, in high- or middle-ranking kin groups. This tendency may be related to the fact that, as compared with juvenile males of the same age, older juvenile females spend more time with their mothers (Nakamichi 1989; Nakamichi and Yamada 2010).

## 4.3 Dominance Relationships Among Adult Males

### *4.3.1 Central and Peripheral Males*

The number of adult males in a group of Japanese macaques is much smaller than the number of females, because males leave the group upon maturity. Thus, the ratio of adult males to adult females is less than 1. In a wild monkey park, there are usually a few adult males and much more adult females, with many juveniles and infants of both sexes. This is because only a limited number of adult males remain with adult females; whereas other adult males, most young adult, and older juvenile males tend to spend their time alone or with a few males. It has been recognized since the beginning of the research into wild Japanese macaques that their groups schematically have a concentric circle structure consisting of central and peripheral areas (Itani 1954). The central area includes a small number of adult males, many adult females, and their immature offspring. The peripheral area includes adult, young adult, and older juvenile males. Peripheral males, therefore, may have infrequent interaction with individuals in the central area.

Infant males remain in the central area with their mothers, but as they age, juvenile males tend to move from the central area to the peripheral area and spend much more time with similar-aged and/or young adult males (Nakamichi 1989). Most peripheral males eventually leave their natal group either by themselves or with other males. The move from the central to the peripheral area and eventual emigration of males contrast with the common tendency of females to stay in the central area throughout their lives. Males leave their natal group usually at 3–6 years of age and sometimes after reaching adulthood. Males of high-ranking kin groups tend to leave the groups when they are older and have a wider range of ages than those of less dominant kin groups (Itoigawa 1975; Kato 2001). Some adult males move farther than 20–30 km to enter a non-natal group (Hazama 1965). However, information regarding how males spend their time after leaving their natal group is still very limited. In free-ranging, provisioned groups, some males may remain in their natal group throughout their life.

### *4.3.2 Dominance Relationships Among Central Males*

Adult males are easily identified in a group of Japanese macaques because they are a little larger than most adult females. It is easy to identify dominance relationships among males as well, by observing supplanting interactions between males: the supplanting male is dominant to the supplanted one. Moreover, unlike females, dominant males are likely to raise their tail while approaching subordinate males. This is a clear dominant behavior for male Japanese macaques (Fig. 4.9). Therefore, having the tail down may signal submissiveness. If a male that has been walking with his tail up lowers his tail just as he passes a sitting male, the walking male is subordinate to the sitting male.

In general, dominance relationships among central adult males are stable, and severe antagonism resulting in rank reversal has rarely been reported. A change in dominance rank among males usually occurs when a higher-ranking male suddenly disappears from the group, and, as a result, males who are subordinate to him rise in rank. This is also true for the change of the top-ranking (alpha) male. After the disappearance of an alpha male, it is very common for the former second-ranking (beta) male to become the new alpha male. On the other hand, when a male newly enters a group, it is common for him to be ranked at the bottom of the resident adult males (Itoigawa 1975, 1993). A newcomer male may acquire the top-ranking

**Fig. 4.9** The second-ranking Japanese macaque male raises his tail (tail-up behavior) when approaching the third-ranking male, who keeps his face down

position if he enters a small group in which there are only a few males (Suzuki et al. 1998).

### 4.3.3 Change of Alpha Males

It is sometimes assumed that fully mature high-ranking males are physically superior to middle- or low-ranking males. This is sometimes but not always true. The following case illustrates how important psychological bonding with certain adult females is for maintaining the alpha male's dominance (Nakamichi et al. 1995b).

*Rikinio* was the 28-year-old alpha male of the Katsuyama free-ranging, provisioned group of Japanese macaques, showing the negative effects of aging, including a bent back, thin body, tooth loss, and relatively poor locomotor activity (Fig. 4.10). He remained the alpha male for 17 years, unusually long time, since 11 years of age. Two and a half months before his death, he suddenly began to drag his hind legs for an unknown reason. One month later, it became difficult for him to move with the group in the forest, and he stayed near the feeding site, not only during daytime but also at night. He interacted with group members only when the group was at the feeding site. However, he sometimes supplanted the beta male even after his locomotor ability decreased further, indicating that he was still dominant to the beta male (Fig. 4.10).

Nine days before his death, *Rikinio* was attacked (e.g., bitten) by the beta male four times. All the attacks occurred when *Rikinio* approached the beta male. *Rikinio* might have approached the beta male to confirm his dominance over the beta male (i.e., *Rikinio* tried to supplant the beta male). Instead of moving away from *Rikinio*, however, the beta male attacked him. In all the four attacks, the beta male made

**Fig. 4.10** The first-ranking (alpha) male of the Katsuyama group of Japanese macaques, *Rikinio*, at 27 years of age. He lost three of four canine teeth, as well as some other teeth (*upper left*), and became thin and bent (*upper right*). When *Rikinio* approached the second-ranking (beta) male who was receiving grooming from an adult female (*lower left*), the second-ranking male quickly left, and then *Rikinio* received grooming from the same female (*lower right*)

aggressive sounds and bit *Rikinio*, and *Rikinio* screamed. The alpha female, *Pet* (her formal name *Be71*), rescued *Rikinio* at the fourth attack. She rushed in and acted aggressively toward the beta male. The beta male then stopped the attack and moved away. After this event, *Rikinio* sometimes approached the beta male and screamed at him when *Pet* was near *Rikinio*. *Pet* usually ran toward them and uttered aggressive sounds. Then *Rikinio*'s scream changed to aggressive sounds, and the beta male moved away from *Rikinio* and *Pet*. On the other hand, when he was not near *Pet*, he grimaced at the beta male with bared teeth (this is a typical submissive facial expression). These episodes indicate that *Rikinio* asked *Pet* to give him agonistic support, and he was able to remain dominant to the beta male only when he received her support.

Four days before his death, maggots had hatched in the bites on *Rikinio*'s back and his physical condition became much worse. When the beta male approached *Rikinio* who was sitting near *Pet*, *Rikinio* did not scream but grimaced at him, and *Pet* also grimaced him. Immediately after this event, the beta male mounted *Pet* three times. This change indicates that *Rikinio* was decisively outranked by the beta male.

Affiliation or psychological bonding with adult females and in particular with the alpha female allowed *Rikinio* to remain dominant over males who might otherwise outrank him. Adult males seem to recognize the significance of the alpha female. New alpha males of other groups, not just the one in the example above, have been observed to mount or attack the alpha female at the center of the group (Nakamichi 1999). It is rarely reported that other females than the alpha female are mounted or attacked by the new alpha male. Mounting or attacking the alpha female may be a ritualized display that functions to make the alpha female and other group members recognize the appearance of the new alpha male (Nakamichi et al. 1995b).

### 4.3.4 Changes in Dominance Rank Among Adult Males

Because the dominance rank change is rarely observed, it is interesting and important to describe an example of it. The following event was observed in a captive group of Japanese macaques. There were approximately 250 animals, including 16 adult males living in a 20 ha enclosure (Nakamichi, unpublished data). The third-ranking (gamma) male (18 years old) suddenly received vigorous challenges (e.g., biting and chasing repeatedly) from three middle-ranking (5th, 6th, and 7th ranking) adult males, and his rank went under them. A few days later, he was attacked by low-ranking males who joined the middle-ranking males in chasing him, and he was ranked at bottom. He rose in dominance rank again to a middle-position by receiving support from a middle-ranking male, one of those middle- and high-ranking males to which he attempted to maintain proximity.

This 18-year-old gamma male maintained relatively close proximity to the 17-year-old alpha male, and both of them had proximity to many adult females.

The affiliation between the two might have been useful for keeping their rank positions. The middle-ranking males' sudden attack on the gamma male happened 11 days before the death of the alpha male. They also attacked or threatened some of the adult females that surrounded the alpha and gamma males when resting or walking to forage. The middle-ranking males might have attacked these individuals to test the power of the alpha male. This is just one of the probable explanations.

It is very difficult to determine the factors that cause complicated social changes, such as dominance rank changes among adult males. This is because social changes tend to occur in a relatively short period, and they are very difficult to reproduce experimentally. The only way to collect detailed observational data on social changes is by observing behaviors of macaques in natural settings, however infrequent they are. Such data may lead to understanding of individual's social intelligence, which might otherwise be hidden.

## 4.4 Mating Behavior

While many primate species inhabiting the tropical regions give birth year-round, Japanese macaques mate and give birth in a certain time of year. The mating season includes 3–4 months in the fall and winter, and the birth season lasts for 3–4 months in the spring and summer. The gestation period is approximately 170–180 days. Females first come into heat and mate with males at 3.5 years of age, but rarely become pregnant at that age. Females usually give birth to their first offspring at 5–7 years of age.

Takahata and his colleagues (Takahata 1982a, b; Takahata et al. 2002) observed the Arashiyama free-ranging, provisioned group for 7 years and reported interesting findings about mating behaviors (Fig. 4.11). Their most salient finding is that individuals rarely mate with each other when they are closely related through maternal lines. Copulation between mother and son has not been observed, and that between siblings of different sexes has been observed only in one dyad. Copulations between uncle and niece and between aunt and nephew have been recorded. The number of the observed copulations, however, is much lower than the expected value. Copulations occur in more distantly related pairs as frequently as in unrelated pairs. This indicates that Japanese macaques avoid copulation in a close-kin dyad, i.e., within the third or less degree of relationship. They mate with more distant relatives as frequently as they do with unrelated individuals. This suggests that Japanese macaques have some mechanism by which they avoid mating with closely related individuals (inbreeding avoidance).

The mechanism by which Japanese macaques avoid inbreeding is probably based on their cognitive ability to memorize their relationships with others. Male offspring maintain relatively frequent interactions (including nurturing-nurtured relations) with their mothers during the first few years after birth. The interactions of male offspring with their mothers usually decrease with increasing age, and they are less frequent than the interactions of female offspring with their mothers are.

**Fig. 4.11** A 12-year-old male is mounting an 18-year-old female in the Katsuyama group of Japanese macaques. The female was his mother's maternal cousin

Probably, the fact that siblings of different sexes suckle from the same teat in different years and have relatively frequent interactions with each other in immature days affects interactions among them in mature days. Frequent affiliative interactions at early stage of life may be responsible for male's psychological bonding with his mother or sisters. This psychological bonding or affiliative relationship may prevent a male from mating with his mother and sister. The reverse is also true. A mother and her daughters may avoid mating with her sons and their brothers, respectively, since frequent interactions occurred between her and her sons and between her offspring of different sexes when they were immature.

Similarly, when an adult male and an adult female maintain relatively frequent proximity to one another (peculiar-proximate relations) in the non-mating season, they tend to avoid mating (Takahata 1982a). Instead, they tend to mate with those with which they are not in frequent proximity. As Takahata (1982a) points out, Japanese macaques tend to select unfamiliar or unknown individuals as mating partners. In other words, psychological bonding or affinity greatly influences mating partner selection in Japanese macaques.

It might appear that higher-ranking males mate with females more frequently and sire more offspring than lower-ranking ones do. In a captive group of Japanese macaques, male dominance rank was positively correlated with the number of copulations with ejaculation, but was not related to the number of offspring (paternity can be determined by DNA analysis). This indicates that having a high rank among males may lead to a large number of opportunities for copulation, but not necessarily to a larger number of offspring (Inoue et al. 1991). Similar results have been reported for a free-ranging, provisioned group: peripheral males sire more infants than central males do (Inoue and Takenaka 2008).

## 4.5 Group Division

A group divides usually when food resources in the home range of the group are insufficient to meet the demands of all individuals of the group. Monkeys travel to a food patch and forage, and then move to another food patch. If many food patches in a group's home range are too small for all members to forage at the same time, it is more difficult for all group members to exploit these food patches, and agonistic interactions among some of them increase in order to make use of limited resources. This may cause some individuals of the group to travel independently of the other group members, leading to the loosening of group cohesion and social fragmentation. This is the beginning of group division. When a subgroup constantly travels and its membership is stable, group division is completed: the original group has divided into two groups. The home ranges of two groups are usually different, and one group may avoid the other group consistently whenever the two groups come into contact. As with the dominance relationship between individuals, a group is likely to be consistently dominant to a neighboring group.

The excessive increase in group size makes it difficult for individuals to interact with one another, not just to forage for themselves. Interactions among group members in general being less frequent may make their dominance relationships ambiguous. While this may increase agonistic interactions, they spend more time independently of one another, and then group division occurs. In other words, the increase in group size seems to be a primary cause of group division.

In addition, social conflicts among group members may cause group division (Koyama 1970). An example of this type of group division is described below (Itoigawa 1993, 1997).

The Katsuyama free-ranging, provisioned group of Japanese macaques consisting of 23 kin groups split into two groups: the main group and the branch group. The main group included one adult female of the 2nd-ranking kin group, most adult females of the 5th-ranking kin group, and all adult females of the 7th- through 23rd-ranking (bottom) kin groups. The branch group included all adult females of the 1st-, 3rd-, 4th-, and 6th-ranking kin groups, all but one adult females of the 2nd-ranking kin group, and one adult female of the 5th-ranking kin group. All infants but one joined the same group as their mothers did. An exceptional 1-year old female remained in the main group even after her mother switched to the branch group (Fig. 4.12).

Three years before the group division, two young adult males (8 and 5 years old) of the 1st-ranking kin group became the alpha and beta males, respectively. They acquired their dominance rank chiefly because of agonistic support from closely

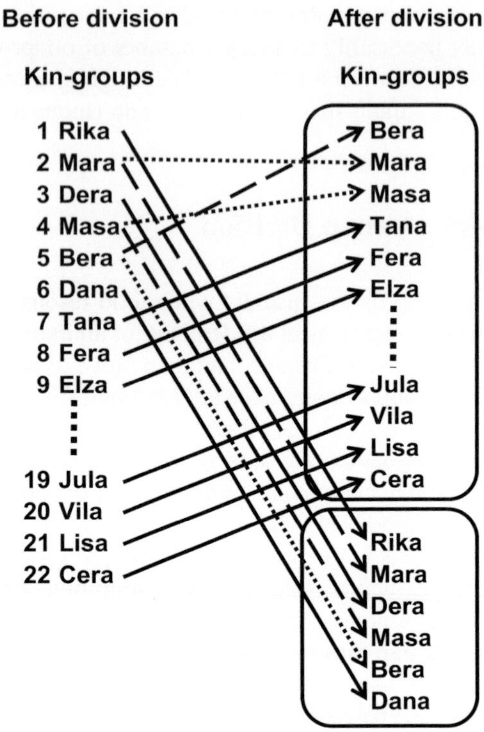


**Fig. 4.12** Kin group composition of the free-ranging, provisioned Katsuyama group of Japanese macaques before and after group division in 1973 (based on Itoigawa 1993, 1997). *Solid lines* indicate that all individuals of a kin group became members of the same group at the time of division; *dotted lines* indicate that one or more members of a kin group became members of one group; and *dashed lines* indicate that the remaining large majority of the kin group become members of the other group

related females, including their mother and grandmother (the latter was the alpha female). A few adult females of the 5th-ranking kin group, by mating with the new alpha male, formed and maintained affiliative relationships with him after the mating season. They then started to behave aggressively toward adult females of the 1st-ranking kin group, especially the alpha male's mother and grandmother. This antagonistic behavior escalated and produced severe social conflicts between adult females of the high-ranking kin groups and those of the middle-ranking kin groups. The latter succeeded in expelling the former after repeated agonistic acts. The 5th-ranking kin group of the original group thus became the 1st-ranking kin group in the newly formed main group. Group members who had occupied high-ranking positions in the original group formed the branch group and began to travel through the new home range.

Group division is always caused by social changes in the original group, such as loosening of cohesion, social fragmentation, and instability of dominance relationships among group members. Such social changes are triggered by an increase in group size and social conflicts between subgroups of adult females. Indeed, adult females play the leading role in group division. Importantly, a group does not divide in a disorderly manner but along kin lines. The importance of relationships through maternal lines for the social lives of female Japanese macaques may be most saliently manifested in group division.

## 4.6 Mother-Immature Offspring Relationships

### *4.6.1 Cognition of Mother by Infant and Cognition of Infant by Mother*

#### 4.6.1.1 Precocial and Altricial Animals

Precocial birds, such as ducks, geese, and chicks, can walk soon after hatching, and hatchlings will follow the first moving object they see after hatching. This phenomenon is called 'imprinting' (Lorenz 1935). Since the first moving objects that hatchlings see are usually their mothers, imprinting is an indispensable mechanism by which precocial hatchlings identify their mothers and thus survive in the wild.

Goats are precocial, and stand and walk soon after birth. They show appropriate maternal behaviors, such as licking and permitting their kids to suckle, even when the kids are separated from them 5 min after birth and returned to them 2 h later. However, mothers do not accept kids that are separated from them immediately after birth (i.e., before they start to lick their kids) and returned to them 2 h later (Klopfer 1971). This shows that mother goats identify their kids in the first 5 min after birth in which they lick their kids.

In precocial birds and mammals, either mothers can identify their infants or infants can their mothers, soon after birth. By contrast, altricial birds and mammals are born too premature to walk, and need not identify mothers or infants soon after birth. Rather, they are assumed to be able to identify mothers or infants by the time infants become able to move independently.

Even though nonhuman primates cannot walk soon after birth, they are usually classified as precocial animals, because their infants are born with their eyes open and their body shapes are similar to adults'. In the majority of primate species, infants are born with grasping reflex, i.e., their fingers on four legs can flex in response to stimuli, allowing them to cling to their mothers' belly or back immediately after birth, without any help from their mothers. Nonhuman primate infants tend to spend most of their time in contact with their mothers during the early stages of development, and as with altricial animal infants, they are presumed to be able to identify their mothers by the time they spend much time away from the mothers.

Infant Japanese macaques start to move by themselves by the end of the first week, albeit very clumsily. In the second month, the movements of infants become increasingly steady. At this time, infants increase the amount of time spent out of contact with their mothers and sometimes interact with other group members in the vicinity of their mothers, such as other infants and young juveniles. Two-month-old infants rarely move very far from their mothers. In the third month, as their locomotor skills develop, infants spend much less time in contact with their mothers and much more time independently. At this stage, infants begin to extend their social interactions with other group members (Itoigawa 1973; Nakamichi 1999, 2001). These developmental processes in infant Japanese macaques are similar to those in closely related species, such as rhesus macaques (Berman 1980).

#### 4.6.1.2 Visual Recognition Between Mothers and Infants

The question here is when mothers and infants become able to discriminate between them and other group members. A pig-tailed macaque (*Macaca nemestrina*) mother is reported to discriminate her own infant from other infants by the end of the second week after giving birth (Jensen 1965). On the basis of this report, it may be deduced (but not experimentally determined) that Japanese macaque mothers can discriminate their own infants from other infants within a few weeks postpartum. In finding their infants in a group of two or more infants, macaque mothers may rely both on visual cues and on other cues, such as olfactory cues. Maestripieri (2001), after reviewing the substantial literature about mother-infant relationships in Old World and New World monkeys, concludes that mothers can discriminate between their offspring and other infants of the same sex and age within 3–4 weeks postpartum. Moreover, in primate species, lactating adult females often adopt an infant within 2–3 postpartum weeks. Experimental cross-fostering of infant rhesus monkeys has a high rate of success within the first 2 postpartum weeks. Maestripieri (2001) therefore remarks that in the early postpartum weeks, monkey mothers are in a maternal sensitive state in which they are highly motivated to take care of infants.

In order to assess the age at which infants discriminate their mothers from other adult females, Nakamichi and Yoshida (1986) conducted experiments with mother-infant pairs of captive Japanese macaques; infants were 12 weeks old or younger. In these experiments, 2 h after each infant was separated from its mother, it was allowed to approach four adult females: its mother and three unknown adult females. Each of them was placed in a small cage at a distance of 150 cm from the infant. Infants were forced to spend 2 h away from their mother before the beginning of an experiment, and not permitted to touch their mother directly during the experiment. Infants at 8–12 weeks of age, but not younger ones, approached and stayed in front of the cage of their mother much longer than they remained near the cages of the other adult females. This clearly demonstrates that the infant Japanese macaques at 8–12 weeks of age have acquired the ability to discriminate their mothers from other adult females. Infants tend to extend their social interactions with group members in the third month of age. The results of the experiments imply that visual recognition of mother precedes infant socialization.

#### 4.6.1.3 Recognition of Infant Calls by Mothers

When observing monkeys at a wild monkey park, you may hear screams from a nearby thicket. You then see a female rush into the thicket and come out with her infant. After observing this scene, you would naturally think that the mother came to help her infant as soon as she heard the infant scream and that she was able to recognize the screams of her infant.

Playback experiments are useful for confirming the mother's ability to recognize infant calls. Shizawa et al. (2005) conducted playback experiments in a Japanese macaque group living in an open enclosure. They first collected the calls infants uttered when they lost sight of their mothers, and then played back the calls made by these infants and by other infants to the mothers. The results demonstrated that each mother was able to distinguish her own infant's calls from those of other infants when the infants were 4–6 months old. Japanese macaque infants after 3 months of age tend to leave their mother's sight much more often than they do earlier. Moreover, they tend to interact with group members much more often than before, thereby increasing the frequency of social encounters in which the infants may need agonistic support from their mothers. It is essential that their mothers correctly recognize the infant calls uttered to ask the mothers for help. Since Shizawa et al.'s (2005) experiments do not involve infants younger than 4 months old, it is possible that Japanese macaque mothers may recognize the calls of their own infants earlier. The age at which immature offspring are able to identify their mother's calls has not been identified.

## 4.6.2 *Maternal Role in Infant Development of Locomotion and Foraging*

### 4.6.2.1 Scaffolding

Japanese macaque newborns are able to cling with all four limbs at birth (see 6.1.1), but cannot walk at all for at least a few days. They are only able to stand by keeping the forearms straight and the hind legs bent at the knee. After infants are able to stand on all four legs with both elbows and knees straight, they begin to walk in a very clumsy manner. In the second and third weeks after birth, the following scene is sometimes observed: a mother puts her infant on the ground, leaves it a few meter away and exhibits lip-smacking, i.e., a friendly behavior of slightly opening and closing the lips several times at a relatively high speed. The infant then approaches the mother, keeping its eyes on her. When the infant reaches its mother, the mother holds it (Fig. 4.13). This suggests that the mother encourages her infant to walk by lip-smacking.

Scaffolding is a process of assistance that enables a human child to achieve a goal that would otherwise be beyond his or her abilities at a given stage of development (Wood et al. 1976). Parents' scaffolding is important for the development of competences in human children. Just as human mothers do, Japanese macaque mothers provide scaffolding for their infants, even though the infant's locomotor competence can develop with increasing age without maternal scaffolding. Scaffolding promotes mutual visual cognition between mother and infant, and strengthens the psychological bonding between them.

If a human mother demonstrates how to draw a circle with a crayon for her child, the child may mimic the mother's behavior and draw a circle, albeit clumsily. This is an instance of *active teaching*. Nonhuman primates rarely engage in active teaching. Even though monkeys of various species and great apes, such as gorillas

**Fig. 4.13** Scaffolding by a Japanese macaque mother. (**a**) The mother encourages her 1–2-week-old male infant to walk independently by lip-smacking. (**b**) The mother lifts up her infant as he approaches her

and chimpanzees, have been studied for years, only one episode of active teaching has been reported: a chimpanzee mother in the wild demonstrated how to effectively crack a nut on an anvil stone with a stone hammer (Boesh 1991).

#### 4.6.2.2 Acquisition of Food Repertoire by Infants

Without the opportunities of receiving active teaching, however, nonhuman infants learn important survival skills by observing their mothers and other group members. The acquisition of food repertoire is an example of this type of learning. Infant Japanese macaques start to lick or chew solid objects, such as grass, twigs, and even small stones, at around 2 weeks of age, and to eat foods in the second month after birth (Hiraiwa 1981). By observing infants and 1-year-old juveniles in a provisioned group of Japanese macaques, Ueno (2005) found that they were more likely to feed when other group members were feeding within 1 m of them than when these neighbors were not feeding. They also tended to eat the same items as their neighbors did. This tendency became more apparent with increasing age during the first 6 months and continued in the 2nd year. Because infants spend much more time with their mothers than with any other group members, mothers have the greatest influence on infant's food repertoire. In fact, mothers and their young offspring often co-feed in the wild (Fig. 4.14). Peers are influential as well, because same-aged young macaques spend a large amount of time playing, travelling, and resting together (Nakamichi 1989).

In addition, Tarnaud and Yamagiwa (2008) found that in a wild group of Japanese macaques, infants in the second half of the first year intensively observe elder group members engaging in plant and invertebrate foraging and learn what are edible for them. Juveniles in the second half of the second year do the same but less often. Intensive observation is defined as the act of turning the head toward the foraged food or the location of a foraging individual. The main target individual of intensive observation is the mother. These findings indicate that infants and juveniles build a similar food repertoire to neighbors' (mother and peers') by observing

**Fig. 4.14** A 9-year-old Japanese macaque mother and her 10-month-old son are eating grass of the same species

their feeding behavior. Food repertoire is very often transmitted from mother to infant and less often from peer to peer, through the observation of feeding behaviors.

## *4.6.3 Flexibility of Maternal Behaviors*

### 4.6.3.1 Maternal Responses to Infants with Congenitally Malformed Limbs

The clinging of infants to their mother has been hypothesized to be important for eliciting maternal care (Hansen 1966). However, experimental studies have cast doubt upon this hypothesis. For example, a macaque monkey mother carries and cradles her anesthetized infant (Rosenblum and Youngstein 1974); and a squirrel monkey mother carries her infant, even when its limbs are immobilized with masking tape (Rumbaugh 1965). These infants are only temporarily unable to cling. It has little implication for the question of how mothers behave toward infants that are unable to cling from birth. Maternal care might cease if infants were persistently unable to cling.

Since the 1950s, Japanese macaques with congenitally malformed limbs have been observed in free-ranging, provisioned groups in Japan. The number of such macaques has recently decreased near to zero in most free-ranging groups (Fig. 4.15). Unfortunately, the causes of limb malformations are largely unknown. The observations of severely malformed monkey infants (e.g., ones with no hands or feet) and their mothers provide an insight into the problem of whether monkey mothers continue maternal care for infants that cannot perform normal clinging behavior. Similarly, the observations are useful for determining how differently mothers of malformed infants behave than those of healthy infants do.

In a free-ranging, provisioned group of Japanese macaques on Awajishima Island, Hyogo Prefecture, mothers of several infants with congenital limb

**Fig. 4.15** Japanese macaques with congenitally malformed limbs of the Awajishima group. A 1-year-old male with no hands is able to walk bipedally (*left*). A 9-month-old male infant with distorted feet and no hands is crawling on elbows and knees (*right*)

malformations cared for their infants so as to complement the infants' disabilities (Nakamichi 1986; Nakamichi et al. 1983a, b; Turner et al. 2005, 2012). The mothers carried their malformed infants in one hand and walked on the other three limbs. They cradled the infants in their forearm(s) and enabled the infants to suckle, because the infants, due to their limbs being malformed, were not able to keep their mother's nipple in their mouth on their own. This indicates that Japanese macaque mothers care for newborns, whether or not they have clinging ability. Infants with no cling ability requires more care than healthy infants do. Nevertheless, healthy and malformed infants have contact with their mothers and suckle for a similar length of time, at least in the first year after birth.

Malformed infants usually develop their own locomotor patterns that are appropriate for their limb malformations. Some infants with no hands walk on their hind limbs, and infants with four severely distorted limbs mainly crawl. As malformed infants age and develop their distinctive means of moving, they spend less time with their mothers and more time with other group members. They tend to spend time with adults or older female juveniles through such behaviors as passive body contact and receiving grooming. By contrast, healthy infants tend to interact with same-aged infants or juveniles that are older by 1 or 2 years, and travel and engage in social play with them. The restricted locomotor abilities of malformed infants often prohibit them from following healthy infants when they move quickly and widely, especially in the trees. Malformed infants usually approach, and if permitted, stay near adults that are less active than infants and juveniles. Adult group members generally behave tolerantly toward malformed infants. Not only do macaque mothers provide appropriate care for their disabled infants, but also other group members provide social support for them. Such maternal care and social support from other group members result in a 1-year survival in the wild of as many as 72 % of infants with congenital limb malformations; the corresponding value for healthy infants is 90 % (Nakamichi et al. 1997).

Mothers with limb malformations can provide appropriate care for their infants in the same ways as healthy mothers do. It is observed in at least one case that a mother with limb malformations is able to care for offspring that are disable to cling just as she is. A female named *Yuki* had normal feet but no hands, and acquired the skill to walk stably on her hind legs by the end of the 2nd year after birth. She gave birth to her first child at 6 years of age. Her first four children had congenital limb malformations, and therefore they had extremely limited or no clinging ability. However, *Yuki* was able to carry these infants in a special manner while walking bipedally: she held the infants to her belly or thigh using her stick-like forearm (Fig. 4.16). Two of her four offspring survived in the group for at least 2 years and the other two survived for 3 years and 6 months. Her 5th offspring was an healthy infant and able to cling with all four limbs. However, *Yuki* did not permit the infant to cling to her belly. Whenever the infant clung to her belly, she put her stick-like forearm between the infant and the belly and removed the infant. On the 4th day of the infant's birth, she abandoned it (Nobuhara, personal observation; Nakamichi 2002).

*Yuki*'s 6th and 7th infants were both congenitally malformed. When the 7th infant clung to her with its normal hand, *Yuki* often attempted to refuse it. She then

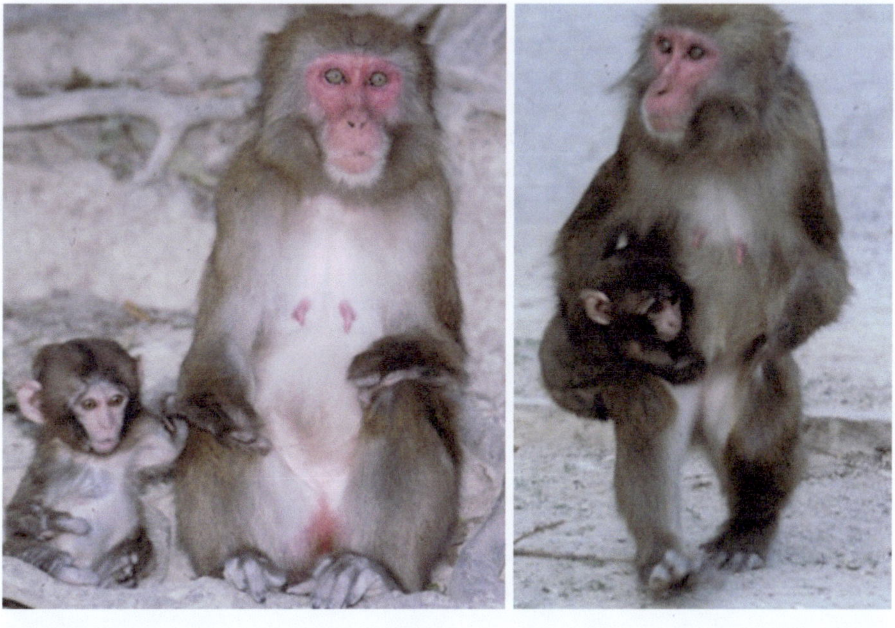

**Fig. 4.16** *Yuki*, an adult female with normal feet but no hands, and her severely malformed infant with no hands and distorted feet (*left*). *Yuki* walks bipedally while carrying the infant in a special manner (*right*) (Courtesy of T. Nobuhara)

gradually started to permit it to cling to her. After this, she permitted other healthy offspring to cling to her body.

The clinging ability has been considered to be essential for infant's survival and maternal care. Healthy mothers are usually able to take care of congenitally malformed infants that are not able to cling by their hands, as well as healthy infants with clinging abilities. Although *Yuki* lacked hands, she was able to develop her own infant-carrying pattern that was appropriate for both her disability and the disabilities of her offspring. However, she was not able to accept her first healthy child that was able to cling tightly to her body by its hands and feet. Disabled mothers who develop their own carrying patterns may not be able to quickly alter their patterns in order to care for infants with normal clinging abilities.

#### 4.6.3.2 Maternal Responses to Dead Infants

It is important to determine whether monkey mothers abandon infants soon after death or maintain maternal behaviors toward dead infants, because the description of maternal behavior toward dead infants may reveal the aspects of the nature of nonhuman primate mothers that might otherwise be concealed.

It has been reported that a number of simian species, such as Japanese macaques, gorillas, and chimpanzees, carry infant corpses (see Fig. 4.1 for simian *vs.* prosimian, and Nakamichi et al. 1996 for a review). On the other hand, in prosimian species, no mothers have been reported to carry their dead infants, either in captivity or in the wild. Although wild ring-tailed lemur mothers do not carry their dead infants (Fig. 4.17), they have been observed to move back and forth between the dead infant that was left on the ground and the group on the day of the infant's death (Nakamichi et al. 1996). This behavior can be interpreted as follows: lemur mothers seek to stay not only with their dead infants but also with their group. However, it is difficult for lemur mothers to carry dead infants by hand. They then

**Fig. 4.17** A ring-tailed lemur mother sitting and holding her dead infant in a clumsy manner (*left*). A Japanese macaque mother carrying her dead infant (*right*)

have a psychological conflict. They want to but cannot both stay near dead infants and remain with the group. It is assumed that the repeated movement between dead infants and the group is a means to resolve the psychological conflict. Although simian and prosimian mothers differ as to whether they carry their dead infants, they may not differ greatly in terms of maternal affection.

Simian mothers can avoid the psychological conflict between the desire to stay near their infants and the desire to stay near their group, because they are able to carry dead infants in their hands. There have been limited studies to address the questions: do all mothers carry their dead infants? How many days do mothers carry their dead infants? How do mothers behave toward their dead infants? How old must dead infants have been before mothers will refrain from carrying their bodies?

Sugiyama et al. (2009) analyzed the quantitative features of dead-infant-carrying behavior of Japanese macaque mothers in free-ranging, provisioned groups at Takasakiyama, Ohita Prefecture, based on data recorded over a 24-year period. They found that the dead-infant-carrying rate (dead-infant carrying cases to deaths within a year of birth) was as small as 10 %. In 26 % of all dead-infant-carrying cases, infants died on the day of birth, and in 78 % of the cases, infants died within 30 days of birth. In only 3 % of all dead-infant-carrying cases, infants died after 100 days of age. The oldest infant that was carried after death was 253 days old. Moreover, 28 %, 15 %, and 24 % of mothers abandoned carried infants within 1, 2, and 3 days of death, respectively. That is, 67 % of mothers abandoned their dead infants within 3 days of death. About 9 % of mothers carried their dead infants for more than a week after death. The longest carrying period was 17 days.

These findings indicate that dead-infant-carrying behavior is most likely to occur during the maternal sensitive period in the early postpartum weeks, during which mothers are highly motivated to take care of their infants (Maestripieri 2001). However, Sugiyama et al. (2009) reported that some mothers carried infants who died at 1 month of age or older, and Izawa (1987) observed that a wild Japanese macaque mother carried her 1-year old dead infant. Taking all these data together, it can be concluded that, whether or not they are in the maternal sensitive period, Japanese macaque mothers sometimes have strong attachment to infants, and carry them for a while even after death.

Even though Japanese macaque mothers may bond strongly with their infants, most mothers abandon their dead infants within a few days. According to Sugiyama et al. (2009), most corpses decompose rapidly due to hot and humid weather conditions. The corpses then smell bad and attract many flies, and this may lead mothers to abandon the corpses.

Rarely do group members other than mothers approach, touch, or sniff infant corpses. In contrast to the behavior of other group members, mothers carry and groom their dead infants. It has even been observed that they shoo flies that are circling or swarming the body of a dead infant even after the body has lost its original form (Nakamichi, personal observation). Moreover, it is difficult to find mothers behave toward dead infants in the way that can be interpreted as aversion (I never did). Therefore, decomposed corpses do not necessarily cause mothers to abandon their dead infants.

Mothers are likely to place dead infants on the ground and move away from them, when they feed on the feeding ground or forage for natural food in the vicinity of the feeding site. Mothers can very easily find dead infants left on the feeding ground, but not in the bushes. In fact, mothers have been observed to search for their dead infants in the bushes (Nakamichi, personal observations). When a mother cannot find her dead infant, her dead-infant-carrying behavior has to end unexpectedly. In other words, an explanation of mothers ceasing to carry dead infants is that they have lost sight of dead infants. This is just one of the probable explanations. Unfortunately, there is not enough data available to determine the factors responsible for mother's abandonment of dead infants.

Some mothers carry their stillborn, premature babies. A mother was observed to carry a hairless premature baby that was estimated to be approximately 90–100 gestation days old (the gestation period of a Japanese macaque is 170–180 days) (Nobuhara, personal communication). A parous mother was observed to lick and carry a hairless baby she gave birth to in a cage (Nakamichi, unpublished data). The hairless baby was different in appearance from a full-term newborn baby. These observations do not support the idea that mothers only carry infants who die after birth.

## 4.7 Social Developments of Immature and Adolescent Individuals

### *4.7.1 Relationships of Immature and Adolescent Males*

In parallel with decreased time they spend with their mothers, infant Japanese macaques tend to increase their interactions with other group members, especially with age-mates (i.e., same-aged individuals). Japanese macaques have a birth season of 3–4 months in the spring and summer. As a result, infants have age-mates, and their locomotor abilities are very similar, even between sexes. They can interact with each other through social play and other behaviors. At the beginning in the 2nd year of life, however, clear differences between immature males and females appear and become important for partner selection (Nakamichi 1989).

Generally, younger juvenile males tend to spend time in the central area of the group. When they get older, they spend most of the time at the periphery of the group. That process is called 'peripheralization.' When juvenile males are 1 year old (i.e., in the 2nd year after birth), they live in the central area, and are very likely to maintain proximity to similar-aged males. When juvenile males are 2–4 years old and live at the periphery of the group, most of their social partners are similar-aged males. This is because older juveniles and young adult males occupy the peripheral area. The close relationships with similar-aged peers continue until males leave their natal group. There are many females of various age classes, many infant and younger juvenile males, and several adult males, in the central area. The preference of juvenile males for associating with individuals of the same sex and similar age appears in the 2nd year of life and precedes peripheralization (Nakamichi 1989). Juvenile males are not likely to change their social partners drastically in the process of peripheralization.

Nakamichi (1996) report that most immature males in the first 4 years of life tend to show a consistent preference for proximity (within 2 m) to certain same-sex individuals in their cohort that have dominance ranks adjacent to their own. Mothers with infants are likely to stay near each other, and sometimes interact in social grooming and other behaviors. Usually, such mothers are closely ranked individuals. The proximity relationships between mothers may provide the opportunity for their infant offspring to spend a large amount of time in proximity to each other. The proximity relationships thus formed may continue after infants grow and spend less time near their mothers. This suggests that the prolonged proximity relationships between immature males are largely a reflection of those between their mothers.

Most males leave their natal group upon maturity (i.e., 3–6 years of age; Itoigawa 1975). The frequency with which males are found at the periphery of the group or in the home range of the group gradually decreases until they leave the group. If males are not observed with their group for several months, they are considered to have left the group. However, the following questions have yet to be answered: what makes males leave the group, whether they leave with similar-aged males with whom they are closely associated throughout immaturity, and where and how they live after leaving the group.

### 4.7.2 Social Relationships of Immature and Adolescent Females

In contrast to immature males, immature females do not go through the process of peripheralization, and remain in the central area. They tend to maintain proximity to females of various ages and infants of both sexes that inhabit the central area of the group. Females maintain relationships with such individuals through various social behaviors from the 2nd year of age. For example, juvenile females tend to interact with infants by touching, grooming, and holding, and may also actively groom adult females (Nakamichi 1989).

As immature males do with closely ranked males, immature females tend to maintain prolonged proximity relationships with closely ranked females throughout the first 4 years, and such relationships are largely a reflection of the relationships between their mothers (Nakamichi 1996).

Most immature females have affiliative relationships with infants and females of various age classes in their group. However, the most frequent grooming occurs between mothers and their daughters (Koyama 1991; Nakamichi and Shizawa

2003), and agonistic alliances are most often formed by closely related females, and typically by mothers and their daughters (Watanabe 1979). These observations indicate that a female's most important social partner is her mother from immaturity to adulthood.

The loss of mother may have a great influence on young adult female's group social life. Yamada et al. (2005) observed the grooming relationships of 5–7-year-old adolescent orphan females and compared them with those of adolescent females with surviving mothers. Orphan females engaged in grooming interactions with other group members as much as females with mothers do. Sisters who lost their mother tended to develop grooming relationships with each other. Orphan females without sisters tended to develop grooming relationships with same-aged females and unrelated adult females. This study suggests that adolescent females have sufficient social abilities to overcome the loss of their mothers.

Most Japanese macaque females have the first pregnancy at 5 or 6 years of age, and their relationships with other group members may change after they produce their first offspring. A hypothesis, although there is not enough evidence to fully support it, is that newborns are attractive to some female group members, and this allows mothers with newborns to develop new social networks. The question of what social changes first childbirth give rise to may be investigated by observing young adult females who remain nulliparous and older than the average age of females at first childbirth. Katsu et al. (2013) observed young nulliparous adult females at 6–9 years of age and found that most of them maintained close relationships with their mothers. More specifically, most of them were observed to receive grooming from their mothers more frequently than the mothers do from them. That is, young adult female Japanese macaques may maintain close relationships with their mothers at least until first childbirth. Little data are available to examine whether or not young adult females change relationships with their mothers after first childbirth.

## 4.8 Aging

### *4.8.1 Physical Aging*

Most Japanese macaques over 20 years of age develop physical declines: tooth loss, incomplete change from winter fur to summer fur, and a bent back. In addition to these changes, the vision of monkeys may become dim with age. Old monkeys may suffer from presbyopia and farsightedness. In grooming, young animals put their eyes close to the body part they are grooming and pick up small items, such as lice eggs. By contrast, some old animals tend to put their eyes away from the body part they are grooming. This is probably because old monkeys require a greater distance to bring an item into focus than young monkeys do (Fig. 4.18) (though no ophthalmological diagnosis of aged monkeys has been conducted). Moreover, locomotor

**Fig. 4.18** Presbyopia in the Japanese macaque. At 21 years of age, this female (formal name, *Lipkira72'83*) grooms others, putting her eyes relatively close to the grooming fingers (*left*). At 25 years of age, the distance between her eyes and fingers is much greater than before, indicating a decrease in eyesight with age (*right*)

activities of monkeys tend to decline with increasing age: as they age, they are likely to spend more and more time sitting or lying motionlessly (Nakamichi 1984).

Reproductive abilities also decline with age. While the birth rate among 10–19-year-old Japanese macaques is as high as 50–65 %, the value decreases drastically with age after 20 years old, and approaches zero at 25–26 years of age (Itoigawa et al. 1992). The oldest recorded age at which a free-ranging, provisioned Japanese macaque female has given birth is 26 years (Itoigawa et al. 1992; Koyama et al. 1992). The average age of death for females in the Katsuyama group of Japanese macaques is 21.2 years (Itoigawa 1982). Based on these findings, female Japanese macaques older than 20 years of age can be regarded as aged.

### *4.8.2 Social Aging*

Old age also affects social behavior of monkeys. Nakamichi (1984, 1991) observed a cross-section of Japanese macaque females ranging in age from 11 to 29 years in a free-ranging, provisioned Arashiyama group in Kyoto Prefecture. He also observed females of the same age in a group living in a 20-hactare open enclosure. The captive group was originally a branch group of the Arashiyama group that was transplanted to Texas, USA (Nakamichi 1984, 1991). Similar behavioral changes with increasing age were found in the two groups, even though their living conditions were different. As compared with younger females, older females, especially those over 25 years of age, tended to spend much less time in contact with or grooming other group members, and much more time alone. Moreover, old females tended to concentrate their social interactions on their youngest daughters. Some aged females were outranked by their adult daughters, but most showed no marked decline in rank. Other studies have shown that old age is more likely to affect the social interactions of low-ranking females than it does those of high-ranking females. Pavelka et al. (1991) found that high-ranking females in a captive group of Japanese macaques did not decrease social contact even when they were aged, while low-ranking ones did. Kato (1999) reported that old high-ranking females in the Katsuyama group of Japanese macaques maintained proximity to a larger number of unrelated adult females and immature individuals than old low-ranking females did. Nakamichi (2003) observed grooming interactions in the same group and found that the frequency of grooming bouts toward unrelated females decreased for both high- and low-ranking females with age, while the frequency of receiving grooming from unrelated adult females decreased with age for low-ranking females but not for high-ranking females. These results indicate the following: as they age, both high-and low-ranking females become more and more prone to select related individuals as recipients of grooming; and old high-ranking females, as opposed to old low-ranking females, remain as socially attractive as young high-ranking females and continue to receive grooming from others.

In extreme cases of social aging, old female Japanese macaques may leave the group. Some old females have been observed to leave a free-ranging, provisioned group, and to live alone for a few days or more (Nakamichi, unpublished data). It is unclear why they left the group. No information is available about how long they survived and how they spent their time alone.

The tendency toward social withdrawal and social dependency on related individuals becomes apparent with increasing age, especially for females over 20 years of age. Old high- and low-ranking individuals exhibit different social behaviors. Indeed, all females are different in terms of social factors, such as the number of adult daughters, their ages, and the number of grand offspring. Differences among old animals in both social environment and physical condition may cause different social behaviors. Therefore, the process of social aging should be described for each female, and special attention should be paid to idiosyncrasies.

### *4.8.3 Grandmother Hypothesis*

One of the primary functions of social groups is to enhance reproductive efficiency. Therefore, almost all group members have reproductive abilities, or they are able to acquire them as they mature. Post-reproductive members do not usually stay in the group. Humans, unlike other species, usually remain in the group even after losing their reproductive abilities. The grandmother hypothesis is proposed to explain the evolutionary meaning behind the survival of post-reproductive women. According to the hypothesis, post-reproductive females can no longer produce their own offspring, but can contribute to the survival of their grandchildren. The survival of their grandchildren, in turn, increases the reproductive success of their offspring and thus their own inclusive fitness. Some studies provide empirical evidence that the grandmother hypothesis is true for humans (Hawkes et al. 1997; Mace 2000; see Fedigan and Pavelka 2007 for a review).

Fairbanks and McGuire (1986) report that young adult females with mothers, as compared with those without mothers, in captive groups of vervet monkeys (*Cercopithecus aethiops*), produce significantly more surviving offspring and have a lower level of infant mortality. This may be related to the tendency of grandmothers to stay near and take care of their infant grandchildren. The increased survival of infants living with grandmothers may also be related to the tendency of infants with grandmothers to be more independent of their mothers, as compared with infants without grandmothers (Fairbanks 1988a, b). However, it is not specified whether the grandmothers described in these studies are post-reproductive.

According to Takahata et al. (1995), Japanese macaque females at 20 years of age in a provisioned group have a post-reproductive lifespan of 4.5 years. This post-reproductive lifespan is defined as a time span in the following way: it is the sum of the time length between final parturition and disappearance from the group, (regarded as death) minus 1.5 years (the time length for an infant to become independent of its mother). However, there is no qualitative studies of the interactions of post-reproductive Japanese macaque grandmothers with their infant grandchildren. There are a few reported cases of social groups of nonhuman primates that support the grandmother hypothesis: for example, an old grandmother without dependent children (i.e., children who depend on her for care) continuously provides essential care for the survival of her dependent grandchild.

Nakamichi et al. (2010) is the first reported case to demonstrate the contribution of a grandmother to the survival of her grandchild. A 24-year-old grandmother provided essential care for the survival of her 2-month-old granddaughter; she held and carried the granddaughter for at least 6 days while the mother was temporarily absent from the group. The grandmother kept the granddaughter warm at night by holding her, but could not supply sufficient nutrition. She was too young to acquire food on her own, and would not have survived if her mother had been absent for a little more days. Another 23-year-old grandmother permitted the 14-month-old granddaughter to suck her nipples sometime within 6 weeks after her mother gave birth to a younger child. The grandmother started to produce milk within 1 month after the granddaughter's sucking acts were first observed. She also held and sometimes carried the granddaughter. Although 1-year-old Japanese macaque infants usually survive without maternal care, such as nursing and carrying, they suffer psychologically from the absence of maternal affection (Schino and Torisi 2001; Nakamichi et al. 2004). The grandmother satisfied the psychological need of the granddaughter by cuddling and carrying her and permitting her sucking behavior. The 1-year-old granddaughter could not have overcome her psychological distress without her grandmother's appropriate care.

These two cases show that old (but healthy) Japanese macaque females without dependent offspring can directly contribute to the survival of their physically and psychologically dependent granddaughters in the wild. This result accords with the grandmother hypothesis.

**Acknowledgements** I sincerely thank the staff working for the provisioned groups of Japanese macaques at Arashiyama, Awajishima, and Katsuyama for their help and encouragement throughout various studies conducted by my colleagues and myself. I would not have been able to complete this manuscript without observing Japanese macaques in these three provisioned groups for years. I am grateful to Ishiguro, H., Asada, M., Osaka, M. Fujikado, T. for providing the opportunity to write the present paper. I also thank the members of Department of Ethology and Ethological Studies Laboratory, Graduate School of Human Sciences, Osaka University, for their valuable comments and encouragement on my studies.

## Exercise

List the probable benefits and costs to individuals living in social groups.

## References

Ando, A.: Analysis of grooming relationships in a free-ranging group of Japanese macaques (*Macaca fuscata*). Annu. Anim. Psychol. **32**, 59–71 (1982) (in Japanese with English summary)

Barrett, L., Henzi, S.P., Welngrill, T., Lycett, J.E., Hill, R.A.: Market forces predict grooming reciprocity in female baboons. Proc. R. Soc. London, Ser. B **266**, 656–670 (1999)

Berman, B.M.: Mother-infant relationships among free-ranging rhesus monkeys on Cayo Santiago: a comparison with captive pairs. Anim. Behav. **28**, 860–873 (1980)

Bernstein, I.S., Ehardt, C.L.: Agonistic aiding: kinship, rank, age, and sex influences. Am. J. Primatol. **8**, 37–52 (1985)

Boesch, C.: Teaching among wild chimpanzees. Anim. Behav. **41**, 530–532 (1991)

Boyd, R., Silk, J.B.: How Human Evolved, 2nd edn. W. W. Norton & Company, New York (2000)

Cheney, D.L.: The acquisition of rank and the development of reciprocal allianaces among free-ranging immature baboons. Behav. Ecol. Sociobiol. **2**, 303–318 (1977)

de Waal, F.B.M.: The organization of agonistic relations within two captive groups of Java-monkeys (*Macaca fascicularis*). Z. Tierpsychol. **44**, 225–282 (1977)

Deag, J.M.: Aggression and submission in monkey societies. Anim. Behav. **25**, 465–474 (1977)

Dunbar, R.I.M.: Reproductive Decisions: An Economic Analysis of Gelada Baboon Social Strategies. Princeton University Press, Princeton (1984)

Dunbar, R.I.M., Barrett, L.: Cousins; Our Primate Relatives. Dorling Kindersley, London (2000)

Fairbanks, L.A.: Vervet monkey grandmothers: interactions with infant grandoffspring. Int. J. Primatol. **9**, 425–441 (1988a)

Fairbanks, L.A.: Vervet monkey grandmothers: effects on mother-infant relationships. Behaviour **104**, 176–188 (1988b)

Fairbanks, L.A., McGuire, M.T.: Age, reproductive value and dominance-related behavior in vervet monkey females: cross-generational influences on social relationships and reproduction. Anim. Behav. **34**, 1710–1721 (1986)

Fedigan, L.M., Pavelka, M.S.M.: Reproductive cessation in female primates: comparison of Japanese macaques and humans. In: Campbell, C.J., Fuentes, A., MacKinnon, K.C., Panger, M., Bearder, S.K. (eds.) Primates in Perspective, pp. 437–447. Oxford University Press, New York (2007)

Gouzoules, H., Gouzoules, S.: Design features and developmental modification pigtail macaque, *Macaca nemestrina*, agonistic screams. Anim. Behav. **32**, 182–193 (1989)

Hansen, E.W.: The development of maternal and infant behaviour in the rhesus monkey. Behaviour **27**, 107–149 (1966)
Hanya, G.: Ecological adaptations of temperate primates: population density of Japanese macaques. In: Nakagawa, N., Nakamichi, M., Sugiura, H. (eds.) The Japanese Macaques, pp. 79–97. Springer, Tokyo (2010)
Hawkes, K., O'Connell, J.F., Blurton Jones, N.G.: Haza women's time allocation, offspring provisioning, and the evolution of long postmenopausal life spans. Curr. Anthropol. **38**, 551–577 (1997)
Hazama, N.: Wild Japanese monkeys of Mt. Hiei. In: Kawamura, S., Itani, J. (eds.) Monkeys and Apes Sociological Studies, pp. 375–401. Chuokorin-Sha, Tokyo (1965) (in Japanese)
Hill, D., Okayasu, N.: Absence of 'youngest ascendancy' in the dominance relations of sisters in wild Japanese macaques (*Macaca fuscata yakui*). Behaviour **132**, 367–379 (1995)
Hiraiwa, H.: Maternal and alloparental care in a troop of free-ranging Japanese monkeys. Primates **22**, 309–329 (1981)
Inoue, E., Takenaka, O.: The effect of male tenure and female mate choice on parternity in free-ranging Japanese macaques. Am. J. Primatol. **70**, 62–68 (2008)
Inoue, M., Mizunaga, M., Ohsawa, H., Takenaka, A., Sugiyama, Y., Gaspard, S.A., Takenaka, O.: Male mating behavior and paternity discrimination by DNA fingerprinting in a Japanese macaque group. Folia Primatol. **56**, 202–210 (1991)
Itani, J.: The Japanese Monkeys of Takasakiyama. Kobunsha, Tokyo (1954) (in Japanese)
Itoigawa, N.: Group organization of a natural troop of Japanese monkeys and mother-infant interaction. In: Carpenter, C.R. (ed.) Behavioral Regulators of Behavior in Primates, pp. 229–250. Bucknell University Press, Lewisburg (1973)
Itoigawa, N.: Variables in male leaving a group of Japanese macaques. In: Kondo, S., Kawai, M., Ehara, A., Kawamura, S. (eds.) Proceedings from the Symposia of the Fifth Congress of the International Primatological Society, pp. 233–245. Japan Science Press, Tokyo (1975)
Itoigawa, N.: Aging and behavior in Japanese monkeys. Jpn. J. Ethnology **46**, 376–389 (1982) (in Japanese with English summary)
Itoigawa, N.: Social conflict in adult male relationships in a free-ranging group of Japanese monkeys. In: Mason, W.A., Mendoza, S.P. (eds.) Primate Social Conflict, pp. 145–169. State University of New York Press, Albany (1993)
Itoigawa, N.: The History of the Katsuyama Japanese Monkey Group. Dobutsusha, Tokyo (1997) (in Japanese)
Itoigawa, N., Tanaka, T., Ukai, N., Fujii, H., Kurokawa, T., Koyama, T., Ando, A., Watanabe, Y., Imakawa, S.: Demography and reproductive parameters of a free-ranging group of Japanese macaques (*Macaca fuscata*) at Katsuyama. Primates **33**, 49–68 (1992)
Izawa, K.: The ecological study of wild Japanese monkeys living in Kinkazan, Miyagi prefecture: some discussions on the infant returned to the wild. Bull. Miyagi Univ. Educ. **22**, 27–49 (1987) (in Japanese with English summary)
Jensen, G.D.: Mother-infant relationship in the monkey *Macaca nemestrina*: development of specificity of maternal response to own infant. J. Comp. Physiol. Psychol. **59**, 305–308 (1965)
Kato, E.: Effects of age, dominance, and seasonal changes on proximity relationships in female Japanese macaques (*Macaca fuscata*) in a free-ranging group at Katsuyama. Primates **40**, 291–300 (1999)
Kato, E.: Is observability a good indicator of male Japanese monkeys leaving their natal group? Primate Res. **17**, 39–50 (2001) (in Japanese with English summary)
Katsu, N., Yamada, K., Nakamichi, M.: Social relationships of nulliparous young adult females beyond the ordinary age of the first birth in a free-ranging troop of Japanese macaques (*Macaca fuscata*). Primates **54**, 7–11 (2013)
Kawai, M.: On the rank system in a natural group of Japanese monkey (I) the basic and dependent rank. Primates **1**, 111–130 (1958) (in Japanese with English summary)
Kawai, M.: Newly-acquired pre-cultural behavior of the natural troop of Japanese monkeys on Koshima Islet. Primates **6**, 1–30 (1965)

Kawamura, S.: The matriarchal social order in the Minoo-B group. Primates **1**, 149–156 (1958) (in Japanese with English summary)
Klopfer, P.H.: Mother love: what turns it on? Am. Sci. **59**, 404–407 (1971)
Koyama, N.: On dominance and kinship of a wild Japanese monkey troop in Arashiyama. Primates **8**, 189–215 (1967)
Koyama, N.: Changes in dominance rank and division of a wild Japanese monkey troop in Arashiyama. Primates **11**, 335–390 (1970)
Koyama, N.: Grooming relationships in the Arashiyama group of Japanese monkeys. In: Fedigan, L.M., Asquith, P.J. (eds.) The Monkeys of Arashiyama: 35 Years of Research in Japan and the West, pp. 211–226. State University of New York Press, Albany (1991)
Koyama, N., Takahata, Y., Huffman, M.A., Norikoshi, K., Suzuki, H.: Reproductive parameters of female Japanese macaques: thirty years data from the Arashiyama troops, Japan. Primates **33**, 33–47 (1992)
Kutsukake, N.: Matrilineal rank inheritance varies with absolute rank in Japanese macaques. Primates **41**, 321–335 (2000)
Lorenz, K.: Der Kumpan in der Umwelt des Vogels. Journal für Ornithogie **83**, 137–413 (1935)
Mace, R.: Evolutionary ecology of human life history. Anim. Behav. **59**, 1–10 (2000)
Maestripieri, D.: Is there mother-infant bonding in primates? Dev. Rev. **21**, 93–120 (2001)
Martin, R.D.: Primate Origin and Evolution: A Phylogenetic Reconstruction. Chapman & Hall, London (1990)
Nakamichi, M.: Behavioral characteristics of old female Japanese monkeys in a free-ranging group. Primates **25**, 192–203 (1984)
Nakamichi, M.: Behavior of infant Japanese monkeys (*Macaca fuscata*) with congenital limb malformations during their first three months. Dev. Psychobiol. **19**, 335–341 (1986)
Nakamichi, M.: Sex differences in social development during the first 4 years in a free-ranging group of Japanese monkeys, *Macaca fuscata*. Anim. Behav. **38**, 737–748 (1989)
Nakamichi, M.: Behavior of old females: comparison of Japanese monkeys in the Arashiyama East and West groups. In: Fedigan, L.M., Asquith, P.J. (eds.) The Monkeys of Arashiyama: 35 Years of Research in Japan and the West, pp. 175–193. State University of New York Press, Albany (1991)
Nakamichi, M.: Proximity relationships within a birth cohort of immature Japanese monkeys (*Macaca fuscata*) in a free-ranging group during the first four years of life. Am. J. Primatol. **40**, 315–325 (1996)
Nakamichi, M.: Japanese Monkey Mothers and Offspring. Fukumura-shuppann, Tokyo (1999) (in Japanese)
Nakamichi, M.: Mother-offspring relationships in macaques. In: Matsuzawa, T. (ed.) Primate Origins of Human Cognition and Behavior, pp. 418–440. Springer, Tokyo (2001)
Nakamichi, M.: Maternal behaviors in Japanese monkeys. Shusanki Igaku (Perinatology) **32** (Suppl), 355–359 (2002) (in Japanese)
Nakamichi, M.: Age-related differences in social grooming among adult female Japanese monkeys (*Macaca fuscata*). Primates **44**, 239–246 (2003)
Nakamichi, M., Koyama, N.: Social relationships among ring-tailed lemurs (*Lemur catta*) in two free-ranging troops at Berenty Reserve, Madagascar. Int. J. Primatol. **18**, 73–93 (1997)
Nakamichi, M., Shizawa, Y.: Distribution of grooming among adult females in a large, free-ranging group of Japanese macaques. Int. J. Primatol **24**, 607–625 (2003)
Nakamichi, M., Yamada, K.: Long-term grooming partnerships between unrelated adult females in a free-ranging group of Japanese monkeys (*Macaca fuscata*). Am. J. Primatol. **69**, 652–663 (2007)
Nakamichi, M., Yamada, K.: Lifetime social development in female Japanese macaques. In: Nakagawa, N., Nakamichi, M., Sugiura, H. (eds.) The Japanese Macaques, pp. 241–270. Springer, Tokyo (2010)
Nakamichi, M., Yoshida, A.: Discrimination of mother by infant among Japanese macaques (*Macaca fuscata*). Int. J. Primatol. **7**, 481–489 (1986)

Nakamichi, M., Fuji, H., Koyama, T.: Behavioral development of a malformed infant in a free-ranging group of Japanese monkeys. Primates **24**, 52–66 (1983a)
Nakamichi, M., Fuji, H., Koyama, T.: Development of a congenitally malformed Japanese monkey in a free-ranging group during the first four years of life. Am. J. Primatol. **5**, 205–210 (1983b)
Nakamichi, M., Itoigawa, N., Imakawa, S., Machida, S.: Dominance relations among adult females in a free-ranging group of Japanese monkeys at Katsuyama. Am. J. Primatol. **37**, 241–251 (1995a)
Nakamichi, M., Kojima, Y., Itoigawa, N., Imakawa, S., Machida, S.: Interactions among adult males and females before and after the death of the alpha male in a free-ranging troop of Japanese macaques. Primates **36**, 385–396 (1995b)
Nakamichi, M., Koyama, N., Jolly, A.: Maternal responses to dead and dying infants in wild troops of ring-tailed lemurs at the Berenty Reserve, Madagascar. Int. J. Primatol. **17**, 505–523 (1996)
Nakamichi, M., Nobuhara, H., Nobuhara, T., Nakahashi, M., Nigi, H.: Birth rate and mortality rate of infants with congenital malformations of the limbs in the Awajishima free-ranging group of Japanese monkeys (*Macaca fuscata*). Am. J. Primatol. **42**, 225–234 (1997)
Nakamichi, M., Kato, E., Kojima, Y., Itoigawa, N.: Carrying and washing of grass roots by free-ranging Japanese macaques at Katsuyama. Folia Primatol. **69**, 35–40 (1998)
Nakamichi, M., Yamada, K., Ohtsuka, N., Imakawa, S., Yasuda, J., Shizawa, Y.: Case reports on parturition and maternal behavior in a free-ranging group of Japanese monkeys at Katsuyama. Primate Res. **20**, 31–43 (2004) (in Japanese with English summary)
Nakamichi, M., Onishi, K., Yamada, K.: Old grandmothers provide essential care to their young granddaughters in a free-ranging group of Japanese monkeys (*Macaca fuscata*). Primates **51**, 171–174 (2010)
Pavelka, M.S.M., Gillespie, M.W., Griffin, L.: The interacting effect of age and rank on the sociability of adult female Japanese monkeys. In: Fedigan, L.M., Asquith, P.J. (eds.) The Monkeys of Arashiyama: 35 Years of Research in Japan and the West, pp. 194–207. State University of New York Press, Albany (1991)
Rosenblum, L.A., Youngstein, K.P.: Developmental changes in compensatory dyadic responses in mother and infant monkeys. In: Lewis, M., Rosenblum, L.A. (eds.) The Effects of the Infant on Its Caregiver, pp. 141–161. Wiley, New York (1974)
Rowe, N.: The Pictorial Guide to the Living Primates. Pognias Press, New York (1996)
Rumbaugh, D.M.: Maternal care in relation to infant behavior in the squirrel monkey. Psychol. Rep. **16**, 171–176 (1965)
Schino, G., Troisi, A.: Relationship with the mother modulates the response to yearling Japanese macaques (*Macaca fuscata*) to the birth of a sibling. J. Comp. Psychol. **115**, 392–396 (2001)
Schino, G., Scucchi, S., Maestripieri, D., Turillazzi, P.G.: Allogrooming as a tension-reduction mechanism: a behavioral approach. Am. J. Primatol. **16**, 43–50 (1988)
Shizawa, Y., Nakamichi, M., Hinobayashi, T., Minami, T.: Playback experiment to test maternal responses of Japanese macaques (*Macaca fuscata*) to their own infant's call when the infants were four to six months old. Behav. Processes **68**, 41–46 (2005)
Sugiyama, Y., Kurita, H., Matsui, T., Kimoto, S., Shimomura, T.: Carrying of dead infants by Japanese macaque (*Macaca fuscata*) mothers. Anthropol. Sci. **117**, 113–119 (2009)
Suzuki, S., Hill, D.A., Sprague, D.S.: Intertroop transfer and dominance rank structure of nonnatal male Japanese macaques in Yakushima, Japan. Int. J. Primatol. **19**, 703–722 (1998)
Takahata, Y.: The socio-sexual behavior of Japanese monkeys. Z. Tierpsychol. **59**, 89–108 (1982a)
Takahata, Y.: The reproductive biology of free-ranging troop of Japanese monkeys. Primates **21**, 303–329 (1982b)
Takahata, Y.: Diachronic changes in the dominance relations of adult female Japanese monkeys of the Arashiyama B group. In: Fedigan, L.M., Asquith, P.J. (eds.) The Monkeys of Arashiyama: 35 Years of Research in Japan and the West, pp. 123–139. State University of New York Press, Albany (1991)

Takahata, Y., Koyama, N., Suzuki, S.: Do old aged females experience a long post-reproductive life span?: the cases of Japanese macaques and chimpanzees. Primates **36**, 169–180 (1995)

Takahata, Y., Huffman, M., Bardi, M.: Long-term trends in matrilineal inbreeding among the Japanese macaques of Arashiyama B troop. Int. J. Primatol. **23**, 399–410 (2002)

Tanaka, I., Takefushi, H.: Elimination of external parasites (lice) is the primary function of grooming in free-ranging Japanese macaques. Anthropol. Sci. **101**, 187–193 (1993)

Tarnaud, L., Yamagiwa, J.: Age-dependent patterns of intensive observation on elders by free-ranging juvenile Japanese macaques (*Macaca fuscata yakui*) with foraging context on Yakushima. Am. J. Primatol. **70**, 1103–1113 (2008)

Turner, S.E., Gould, L., Duffus, D.A.: Maternal behavior and infant congenital limb malformation in a free-ranging group of *Macaca fuscata* on Awaji Island, Japan. Int. J. Primatol. **26**, 1435–1457 (2005)

Turner, S.E., Fedigan, L.M., Nobuhara, H., Matthews, H.D., Nakamichi, M.: Disability, compensatory behavior, and innovation in free-ranging adult female Japanese macaques (*Macaca fuscata*). Am. J. Primatol. **74**, 788–803 (2012)

Ueno, A.: Development of co-feeding behavior in young wild Japanese macaques (*Macaca fuscata*). Infant. Behav. Dev. **28**, 481–491 (2005)

Walters, J.R.: Intervention and the development of dominance relationships in female baboons. Folia Primatol. **34**, 61–89 (1980)

Watanabe, K.: Alliance formation in a free-ranging troop of Japanese macaques. Primates **20**, 459–474 (1979)

Wood, D., Bruner, J.S., Ross, G.: The role of tutoring in problem-solving. J. Child. Psychol. Psychiatry **17**, 89–100 (1976)

Yamada, K., Nakamichi, M., Shizawa, Y., Yasuda, J., Imakawa, S., Hinobayashi, T., Minami, T.: Grooming relationships of adolescent orphans in a free-ranging group of Japanese macaques (*Macaca fuscata*) at Katsuyama: a comparison among orphans with sisters, orphans without sisters, and females with a surviving mother. Primates **46**, 145–150 (2005)

Yamagiwa, J.: Research history of Japanese macaques in Japan. In: Nakagawa, N., Nakamichi, M., Sugiura, H. (eds.) The Japanese Macaques, pp. 3–25. Springer, Tokyo (2010)

# Chapter 5
# Adaptation and Psychological Disorders

**Osamu Imura**

**Abstract** This chapter reviews the issues of adaptation and psychological disorders. First, the definitions of adaptation and psychological disorders are stated, and the criteria for psychological disorders are given. Depression is introduced as a common and typical example of psychological disorder. You can check your depressive tendency by the self-rating scale presented here. Second, after discussing an example of a person with depression, the biological (neurotransmitter) and psychological mechanisms of depression are explained. Theory of learned helplessness and attribution theory reveal the mechanisms of depression. You can check your attribution type and cognitive style and recognize your susceptibility to depression by following the instruction given here. Third, antidepressant medication and cognitive behavioral therapy (CBT) are each considered as an effective means for the treatment of depression. Fourth, historically famous persons who suffered from depression are introduced, and the risk of suicide in depressive patients is discussed.

**Keywords** Adaptation • Psychiatric and psychological disorders • Depression • Self-rating depression scale • Antidepressants • Learned helplessness • Theory of attribution • Cognitive behavioral therapy • Five column method • Suicide

## 5.1 Introduction

The term 'adaptation' has two meanings: it means either biological adaptation or psychological adaptation. Our ancestors (Homo sapiens) appeared five million years ago. We developed languages and various kinds of tools. We invented letters and symbols, and managed to control fire. Today, we can fly like a bird using an air-plane. Useful tools provided us with comfortable life, and contributed to our adaptation. For instance, we could not have survived in Siberia without wearing clothes and producing heat. We have developed culture to adapt to and modify the external environment. Human beings live almost all over the world, except for the Antarctic. Culture made it possible for us to live in severe natural environments,

O. Imura (✉)
Graduate School of Human Sciences, Osaka University, Suita, Osaka, Japan
e-mail: osamui@hus.osaka-u.ac.jp



M. Kasaki et al. (eds.), *Cognitive Neuroscience Robotics B*,
DOI 10.1007/978-4-431-54598-9_5

such as cold areas and deserts. The question of whether culture is a biological or psychological adaptation is very difficult to answer, because we could not have developed our culture without skillful hands and verbal communication. The handicraft and communicative abilities are unique to humans, and they are mainly based on human's highly developed brain. For this reason, it is difficult to discriminate between biological and psychological adaptation; the two types of adaptation are related to each other. However, I wish to separate them in order to clarify my discussion about adaptation.

### 5.1.1 Biological Adaptation

Biological adaptation is an evolutionary process, and usually called 'phylogenesis.' Animals and plants, bacteria, even viruses, rely on DNA (deoxyribonucleic acid) for reproduction. DNA is a code which consists of four bases: A (adenine), T (thiamine), G (guanine), and C (cytosine). The combinations of four bases (A, T, G, and C) determine the complex components of proteins (Sadava et al. 2008). Mutations in DNA code are responsible for the variation of the spices. Nearly all mutations produce abnormal proteins and cause illness, but a few contribute to adaptive changes for survival. If an animal is born with a longer neck than the necks of other animals, it has a greater chance of survival. For the animal can eat leaves on high branches. The giraffe's neck is a typical example of biological adaptation. Adapted animals can survive. The history of evolution is filled with adaptive changes of forms and functions in animals and plants. Homo sapiens are no exception. We have acquired skillful hands and a large brain by mutation of genes in the process of natural selection.

### 5.1.2 Psychological Adaptation

Psychological adaptation is an ontogenetic process, and involves biological and psychological developments. The neural and physical development of fetus in uterus is part of biological adaptation. New born babies go through the stages of psychological developments. We are born with immature physical abilities. We cannot walk and eat on our own immediately after birth. We need maternal care and education over 10–20 years before we can live independently. There are many things to learn, e.g., standing, walking, speaking, use of the toilet, and arithmetic. Learning ability is the basic element of psychological adaptation. It may be very difficult to achieve psychological adaptation without learning ability. A simple and broad definition of psychological adaptation is that it is a synonym for learning ability. We learn how to control our body and mind in order to survive, and our learning continues from the cradle to the grave. In a narrow definition, psychological adaptation means mental health, anger and anxiety control, and social skills.

### 5.1.3 What Is Psychological Disorder?

Psychological disorder is simply defined as a state in which we fail to adapt psychologically. This simple definition is not enough to cover the entire variety of psychological disorders. Psychological disorders can be caused by brain dysfunction, psychological and environmental stressors, and even one's cognitive style. Moreover, culture has an influence on what counts as a psychological disorder. For example, homosexuality was a psychological disorder half a century ago. Today, however, no homosexual person is recommended to see a counsellor unless he/she wants. The definition of psychological disorder changes with culture and age. Several criteria for psychological disorder will be presented in what follows.

1. Deviation from mean value
   Intelligence quotient (IQ) is a measure of intellectual ability, and it is presumed to show normal distribution. If a person's IQ score is far below the normal range, he/she may be regarded as abnormal. Generally, a low IQ (under 70) is defined as mental retardation (DSM-IV-TR). IQ is measured by an intelligence test, and hence based on an operational definition. One of the limitations of this criterion for psychological disorder is that low IQ persons do not always have psychological disorder. Many persons with mental retardation have a job, enjoy spare time, and contribute to the society.
2. Violation of social rules
   When one violates social rules and standards, one makes others disturbed and anxious. Persons with mania and antisocial personality disorder tend to bring about one or another trouble with other people. A person with mania tends to ignore traffic signals and spends excessive money. A person with antisocial personality disorder is likely to commit a crime. Are persons who violate social rules all abnormal and have psychological disorder? Almost everyone violates speed limit when they drive. It is illegal in many countries to drink alcohol in a park, but it is not, e.g., in Japan; many people drink under the cherry blossoms. Social rules vary from culture to culture.
3. Subjective distress
   Patients with pain disorder experience serious distress about their bodies. Some patients have a headache, and others have pain in joints or stomach. Pain is subjective experience. It is very difficult to evaluate pain objectively. If subjective distress disturbs our daily life, it may provide a criterion for abnormality and psychological disorder. But cases of malingering imply that one's subjective distress may not be a reliable criterion. If one can get some benefits, e.g., social insurance, by complaining about pain, one's subjective distress may continue while one has no physical cause of pain.
4. Malfunction
   If a person cannot perform desired behavior, he/she may suffer from malfunction. Examples of malfunction are plane phobia, social phobia, and obsessional disorder. A person with plane phobia wants to use an airplane to go abroad,

but he/she cannot enter it as his/her anxiety increases. Patients with obsessional disorder check the safety of gas and electric appliances repeatedly, when they are about to leave home; they have to spend long before they leave home. In the most serious case, a person with obsessional disorder cannot even leave home. Such a malfunction is usually accompanied by distress.

These criteria are defined operationally and behaviorally. This means that they are not as absolute standards as measurable quantities are. Moreover, criteria for psychological disorder may change, depending on culture and age. Either way, these criteria are useful to give a diagnosis of patients with psychiatric and psychological disorders.

## 5.2 Psychiatric and Psychological Disorders

There are a number of psychiatric and psychological disorders. DSM-IV-TR (American Psychiatric Association 2000) has 17 large categories of psychiatric disorders: schizophrenia and other psychiatric disorders, mood disorders, anxiety disorders, and so on. Each large category has many diagnostic subcategories. For example, anxiety disorders have 11 diagnostic subcategories, including panic attack, agoraphobia, social phobia, and so on. More than five hundreds of psychiatric disorders are described in DSM-IV-TR. However, school non-attendance is not included in DSM-IV-TR, because, only in Japan, school non-attendance is a serious social problem and associated with a psychiatric disorder. School non-attendance is defined as a refusal to go to school for more than 30 days a year, for a psychological reason, as opposed to a physical or financial reason. The number of cases of school non-attendance is estimated about 130,000 in elementally and junior high schools in Japan. Interestingly, there were few cases of school non-attendance in the 1960s, and the number of cases blew up in the 1970s and 1980s. The education system and child-parents relationship are taken to have an influence on the increase in school non-attendance in Japan. If culture-based psychological disorders like school non-attendance are added to DSM-IV-TR, there are innumerable psychiatric and psychological disorders in the world.

### *5.2.1 Depression*

Depression is a mental illness categorized as a mood disorder. Mood disorders have 15 diagnostic subcategories, such as major depressive disorders (single episode), dysthymic disorder, bipolar disorders, and so on. The prevalence rate of major depressive disorder (MDD) is 12–17 % (Angust 1997). This rate is higher than that of schizophrenia (0.7 %). Depression has a crucial impact on our daily life: it deteriorates quality of life (QOL), makes us feel sad and hopeless, and limits our activity; and it jeopardizes our employment, academic activity, and family relationship. Depression is caused by biological, psychological, and environmental risk factors. The prevalence rate of bipolar disorder is 69.3 % in monozygotic twins and 20 % in dizygotic twins (Sevy et al. 1995). The difference in prevalence rate between the two groups means that genetic factors are very important in depression, and more generally in mood disorders. In addition, psychological and environmental stressors influence the onset of depression. For example, a person is likely to suffer from depression after the death of the person's spouse, since the death of a spouse is one of the most stressful events in life. Even a job promotion may be a stressor for depression because it increases one's responsibility. Many cases of depression including this one are reviewed in the subsequent sections.

The diagnostic criteria for major depressive disorder are as follows (DSM-IV-TR).

A. At least five of the following symptoms have been present during the same 2-week period and represent a change from previous functioning: at least one of the symptoms is either (1) Depressed mood or (2) Loss of interest or pleasure.

   1. Depressed mood most of the day, nearly every day, as indicated either by subjective report (e.g., feels sad or empty) or observation made by others (e.g., appears tearful)
   2. Markedly diminished interest or pleasure in all, or almost all, activities most of the day, nearly every day (as indicated either by subjective account or observation made by others)
   3. Significant weight loss when not dieting or weight gain (e.g., a change of more than 5 % of body weight in a month), or decrease or increase in appetite nearly every day
   4. Insomnia or hypersomnia nearly every day
   5. Psychomotor agitation or retardation nearly every day (observable by others, not merely subjective feelings of restlessness or being slowed down)
   6. Fatigue or loss of energy nearly every day
   7. Feelings of worthlessness or excessive or inappropriate guilt (which may be delusional) nearly every day (not merely self-reproach or guilt about being sick)
   8. Diminished ability to think or concentrate, or indecisiveness, nearly every day (either by subjective account or as observed by others)
   9. Recurrent thoughts of death (not just fear of dying), recurrent suicidal ideation without a specific plan, or a suicide attempt or a specific plan for committing suicide

B. The symptoms do not meet criteria for a mixed episode.

C. The symptoms cause clinically significant distress or impairment in social, occupational, or other important areas of functioning.

D. The symptoms are not due to the direct physiological effects of a substance (e.g. a drug of abuse, a medication) or a general medical condition (e.g., hypothyroidism).

E. The symptoms are not better accounted for by bereavement, i.e., after the loss of a loved one, the symptoms persist for longer than 2 months or are characterized by marked functional impairment, morbid preoccupation with worthlessness, suicidal ideation, psychotic symptoms, or psychomotor retardation.

### 5.2.2 *Check Yourself by SDS (Self-Rating Depression Scale)*

The Zung Self-Rating Depression Scale (SDS) was designed by W.W. Zung (1965) as a short self-administered survey to quantify the depressed status of a person. There are 20 items on the scale that rate four common characteristics of depression: the pervasive effect, the physiological equivalents, other disturbances, and psychomotor activities. There are 10 positively worded and 10 negatively worded questions. Each question is scored on a scale of 1–4 (a little of the time, some of the time, good part of the time, and most of the time) (Table 5.1).

Questions 2, 5, 6, 11, 12, 14, 16, 17, 18 and 20 are reversed questions. Scores on these questions are converted as follows: $1 \Rightarrow 4$, $2 \Rightarrow 3$, $3 \Rightarrow 2$, and $4 \Rightarrow 1$. For example, 1 point on question 2 is converted to 4 points. The sum of original scores (Q1, 3, 4, 7, 8, 9, 10, 19) and converted scores (Q2, 5, 6, 11, 12, 14, 16, 17, 18, 20) is assessed as follows:

The scores range from 25 to 100.
25–49 Normal Range
50–59 Mildly Depressed
60–69 Moderately Depressed
70 and above Severely Depressed

SDS is useful not only for the evaluation of depression severity but also for the screening of depressive disorder. If one gets a high score on SDS, 70 and above, one is recommended to see a mental health professional. Since SDS is a self-rating scale, rater bias is unavoidable. To make a correct diagnosis of depression, it is necessary to be interviewed by a psychiatrist.

## 5.3 Case Example of Major Depressive Disorder

In order to understand the illness course and treatment details of depression, it is helpful to look at a case example. Mr. A was in his 40s, and he worked for a company as a specialist. He had been in a depressive state when he was a college student. He recovered from depression without seeing a doctor. He got a job and his job performance was good, although he sometimes had difficulties with his supervisors. He was promoted to a manager before his 40th birthday. He was required to move his office. But then, he found himself not be able to move; he could not help standing in front of cardboard boxes filled with papers. A colleague recommended

**Table 5.1** Self-rating depression scale (Zung 1965)

| Place evaluate yourself in correct column on a scale of 1–4. | A little of the time (1) | Some of the time (2) | Good part of the time (3) | Most of the time (4) |
|---|---|---|---|---|
| 1. I feel down-hearted and blue | | | | |
| 2. Morning is when I feel the best | | | | |
| 3. I have crying spells or feel like it | | | | |
| 4. I have trouble sleeping at night | | | | |
| 5. I eat as much as I used to | | | | |
| 6. I still enjoy sex | | | | |
| 7. I notice that I am losing weight | | | | |
| 8. I have trouble with constipation | | | | |
| 9. My heart beats faster than usual | | | | |
| 10. I get tired for no reason | | | | |
| 11. My mind is as clear as it used to be | | | | |
| 12. I find it easy to do the things I used to | | | | |
| 13. I am restless and can't keep still | | | | |
| 14. I feel hopeful about the future | | | | |
| 15. I am more irritable than usual | | | | |
| 16. I find it easy to make decisions | | | | |
| 17. I feel that I am useful and needed | | | | |
| 18. My life is pretty full | | | | |
| 19. I feel that others would be better off if I were dead | | | | |
| 20. I still enjoy the things I used to do | | | | |

him to see an industrial psychologist. She referred him to a psychiatrist, who diagnosed Mr. A with depression.

Mr. A was allowed to take a medical leave for his illness. He did not accept that he was suffering from depression, and refused to take antidepressants. As a result, his symptoms were not improved; He had difficulty in sleeping, and was not able to enjoy watching TV anymore. He stayed home all day, doing nothing. He was in a depressive mood, although he did not plan to commit suicide. Mr. A and his wife were impatient with this situation. The company extended the medical leave for another 3 months. He felt uneasy and lonely during his leave. He thought that it would be impossible to go back to work. At the end of the extended leave, he decided to go to his office and work, ignoring his psychiatrist's advice that he needed more time for treatment. He was not able to work as well as before. Being flooded with complaints from customers, he broke down again in 2 weeks. Then, his psychiatrist advised him to change his lifestyle.

He now accepted that he was a patient with depression. He went to a mental clinic regularly and took antidepressants. He became to be able to walk around home in the evening. His sleeping problem became less serious, and his depressive mood and other symptoms were improved gradually. Then, his psychiatrist and office manager recommended him to return to work. He started to work again, but not too hard; he worked at office only in the morning, and went home in the afternoon. He became more and more confident in his work abilities, and finally decided to work full time. He is now an active specialist in a responsible position. He enjoys everyday life with his wife.

### *5.3.1 Mr. A's Comments About His Experience*

- Although he felt happy about the promotion, he was afraid of increased responsibility.
- He made a small mistake before the promotion.
- He lost his self-confidence in his ability as a specialist.
- He did not recognize the presence of depression. He thought that he was just tired.
- When he was advised to take antidepressants, he was afraid that it might change his personality.
- If he took the drug, he thought, he would have to accept that he was the patient.
- The failure to return to work helped him to admit his illness.
- He was encouraged by the warm support of his wife and the manager.

Mr. A was good at his job. He was diligent and earnest, but nervous about evaluation of his work. He made a small mistake before his promotion. He felt responsible for too much and thought that it would be very difficult to work in the promoted position as the manager expected. He became depressive and suffered from a sleeping problem. He was reluctant to accept that he needed psychiatric treatment. These kinds of rejection often occur in psychiatric situations, because psychiatric patients are sometimes socially stigmatized. Many people do not have an enough understanding about antipsychotic medication, and they are usually worried about the side effects of antipsychotic medications (for this reason, psychoeducation is very important). Mr. A changed his negative attitude toward psychiatric treatment after he failed to go back to work. He accepted the medical approach and agreed to pursue the goal offered by his psychiatrist. This pattern of attitude change is typically common among patients with major depressive disorder.

### 5.3.2 *Diagnosis of Mr. A*

The symptoms of Mr. A meet seven criteria for major depressive disorder. He was depressive for more than 2 weeks (criterion 1), and showed markedly diminished interest or pleasure (criterion 2) (for example, he did not want to watch TV). He suffered from insomnia (criterion 3) and psychomotor retardation (criterion 5). He felt fatigued and inactive almost every day (criterion 6). He felt guilty for his small mistake in work, and found himself worthless when he was absent from work (criterion 7). He was not able to make appropriate decisions and lost concentration when he worked (criterion 8). Nevertheless, he did not have recurrent thoughts of death. His appetite was normal, and his weight was stable.

## 5.4 Biology and Psychology of Depressive Disorders

The etiology of depressive disorders has been developed since the last century. Although a large body of knowledge of causal risk factors and mechanisms of depression has accumulated, more research is needed for answering every question about depression. In this section, the biological and psychological mechanisms of depression are introduced and discussed.

Figure 5.1 shows what kinds of risk factors affect depressive disorders. Biological factors are larger than psychological factors on the left-hand side, and the latter are larger than the former on the right-hand side. This means that biological factors are more likely to cause depression, as compared to psychological factors. On the other hand, psychological factors are more likely to cause depressive responses and temporal depressive emotion when negative events, e.g., a failure to meet a person on time, happen. Persons with depressive neurosis have a negative cognitive style; they negatively interpret situations and relationships with others. It is assumed that their negative cognitive style is developed through their parents-child relationship and environment of growing up. Persons with adjustment disorder have a normal

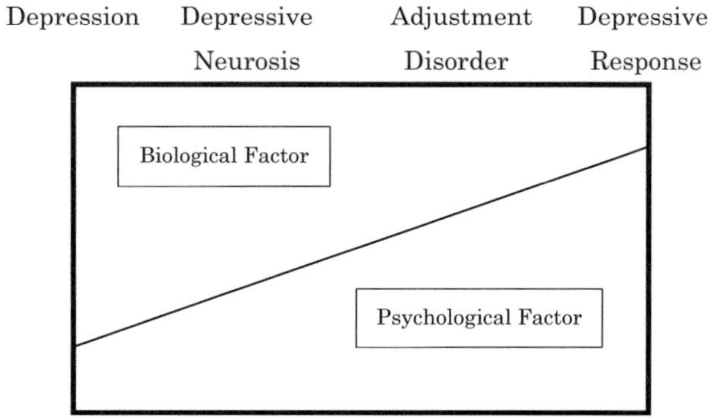


**Fig. 5.1** Risk factors of depressive disorders

life before the onset of disorder. An extremely stressful life event, e.g., a spouse being killed by a traffic accident, may trigger adjustment disorder. Biological risk factors, i.e., abnormalities in genes and the brain, are responsible for certain disorders. It is beyond the scope of this chapter to discuss abnormalities in genes in detail. If the reader wants to know more about the genes and brains of patients with depression, I recommend to read Clinical Psychology 4-Abnormal Psychology II (Kameya 2002, pp. 127–146). Abnormalities of neurotransmitters in the brain of patients with depression are introduced in the next section.

### 5.4.1 Biological Mechanism of Depression

The decrease in serotonin in the synaptic cleft is presumed to be one of the main biological causes of depression and mood disorders. The role of serotonin as a neurotransmitter is depicted in Fig. 5.2. SSRIs (selective serotonin reuptake inhibitors), such as Fluvoxamine and Paroxetine, are used for the treatment of depressive mood, because they block the reuptake of serotonin by presynaptic neurons. SSRIs, then, increase the amount of serotonin in the synaptic cleft, and thereby promote synaptic communication in the nervous system, resulting in the improvement of depressive mood.

Freudenrich (2007) explains the process in a neuron that uses neurotransmitter serotonin, as follows:

1. The presynaptic cell (sending cell) makes serotonin (5-hydroxytryptamine, 5HT) from the amino acid tryptophan and packages it in vesicles in its end terminals.
2. An electrochemical nerve signal passes down the presynaptic cell into its end terminals.
3. The nerve signal stimulates the vesicles containing serotonin to fuse with the cell membrane and dump serotonin into the synaptic cleft.
4. Serotonin passes across the synaptic cleft, binds with special proteins called receptors on the membrane of the postsynaptic cell (receiving cell) and sets up a new electrochemical signal in that cell (the signal can stimulate or inhibit the postsynaptic cell). Serotonin fits with its receptor like a lock and key.
5. The remaining serotonin molecules in the cleft and those released by the receptors after use get destroyed by enzymes in the cleft (monoamine oxidase (MAO) and catechol-o-methyl transferase (COMT)). Some get taken up by specific transporters on the presynaptic cell (reuptake). In the presynaptic cell, the absorbed serotonin molecules get destroyed by MAO and COMT. This enables the nerve signal to be turned "off."

A similar process occurs for norepinephrine, which is also implicated in mood, emotions and MDD. Serotonin, norepinephrine and dopamine are chemically similar and belong to a class of neurotransmitters called 'monoamine

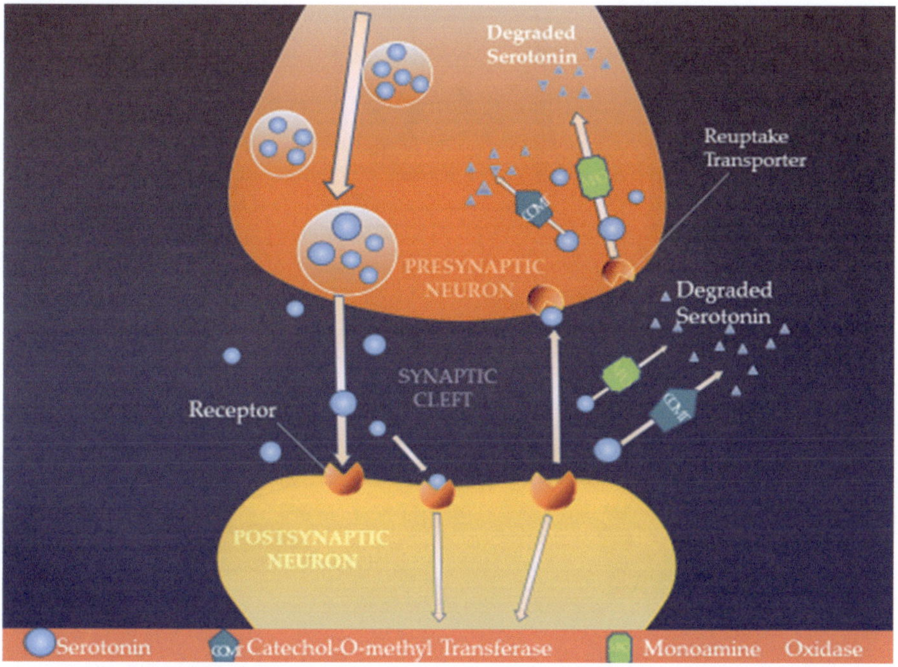


**Fig. 5.2** Serotonin in the synaptic cleft (Illustrated by Sugao Shoko)

neurotransmitters.' Because these chemicals are structurally similar, they are all recognized by MAO and COMT.

## *5.4.2 Psychological Mechanism of Depression*

### 5.4.2.1 The Theory of Learned Helplessness as an Animal Model of Depression

Overmier and Seligman (1967) and Seligman and Maier (1967) found that dogs, when given unavoidable electric shocks repeatedly, did not respond appropriately even in an escapable situation. They called the behavior that dogs showed 'learned helplessness.' It is an animal model of depression. The experimental procedure for this finding is described as follows (Fig. 5.3).

There were three dog groups: contingent, non-contingent (yoked), and control groups. Each group consisted of eight dogs. The dogs in the contingent group were in a position to escape from electric shocks by switched it off. The dogs in the non-contingent group were in no position to escape from electric shocks. They might touch the switch, but it was not wired. Both groups received electric shocks of the same frequency and intensity. The dogs in the control group did not receive

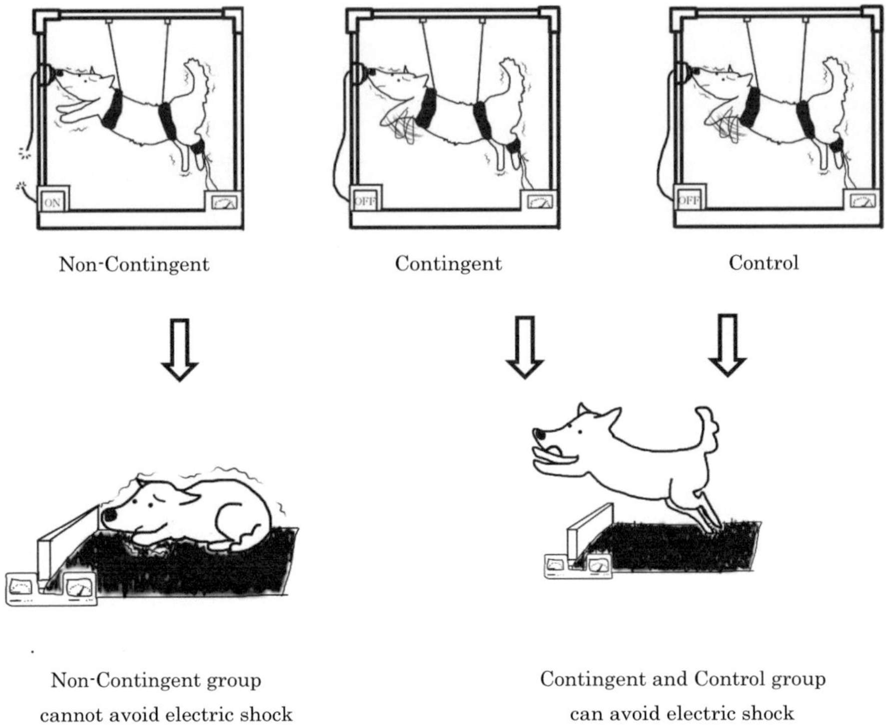


**Fig. 5.3** Experiment of learned helplessness (Illustrated Sugao Shoko)

electric shocks. After these conditioning procedures are completed, each of the three groups was moved to a two-way shuttle box, which had two compartments separated by a barrier. Electric shocks were administered through the grid floor of the shuttle box. The barrier was not so tall; dogs could easily jump over it and escape from electric shocks. The dogs in the contingent and control groups jumped over the barrier, whereas the dogs in the non-contingent group did not; they kept sitting, and showed no avoidance behavior.

#### 5.4.2.2 Depression and Theory of Attribution

When we face with a trouble or a negative life event, we usually seek for a reason. Why did I make such a mistake? What is the cause of the accident? Why did I fail the exam? To seek for a reason why a negative event has happened seems to be one of the means to keep our mind stable. Once we find a reason, we feel at ease, even though that reason may not be appropriate or rational. In other words, we are strongly motivated to find a reason why we fail or suffer. Moreover, if the reason we find is appropriate, it is possible for us to avoid similar negative events in the future.

#### 5.4.2.2.1 Check Your Cognitive Style at a Negative Event

If you are a university student, remember your high school days. Imagine the situation in which you failed a math exam. Your score is much lower than you expected. Why did you think you got such a low score? Write down your answer in the column below.

Your reason

Analyze your answer by following the instruction given here (cf. Table 5.2). There are three dimensions of attribution. The first dimension is internal–external. If you think that you failed due to a lack of effort, your attribution is internal. If you think that you failed due to other or environmental factors, your attribution is external. The second dimension is general–specific. General attribution appeals to factors that cause the same event in different situations. Attribution in terms of ability, personality, or effort is general. Specific attribution appeals to factors that are specific to the situation at hand. Attribution in terms of mood, atmosphere, or incompetence in math is specific. The third dimension is stable–unstable. If you appeal to factors that do not change easily, e.g., your ability, your attribution is stable. If you appeal to factors that you can change, e.g., a lack of effort, your attribution is unstable. Fix your responses along the three dimensions. If you attribute your failure to an internal, general and stable reason, you have a tendency to have depression. Internal attributions lower your self-esteem; stable attributions prolong your depressive mood; and general attributions make you feel hopeless in a wide range of situations.

### 5.4.2.3 Depressive Cognitive Style

It had been assumed that emotional problems were primarily, and cognitive and motivational problems were secondary, in depression. Beck (1976) proposed a new theory of depression against this assumption. According to his theory, cognitive distortion is primarily, and it is the cause of emotional problems. His theory was very unique, because most clinicians and researchers believed that depressive mood caused pessimistic thoughts and lowered motivations. On the basis of Beck's theory,

**Table 5.2** Three dimensions of attribution

| | | Internal | External |
|---|---|---|---|
| General | Stable | I am not smart | The exam was difficult |
| | Unstable | Lack of effort | It was Friday 13th |
| Specific | Stable | I am poor at math | The teacher is harsh |
| | Unstable | Bad physical condition | The room was too hot |

he and other clinicians developed cognitive therapy and cognitive behavioral therapy (CBT) for patients with depressive disorders (Shimoyama and Tannno 2002).

Depressive persons have a specific cognitive style, the features of which are summarized as follows:

- Over-generalization: "If I fail A, I will fail B."
- Selective abstraction: "The teacher's advice means that I am incompetent."
- All or nothing: "If I fail the exam, my life will be over.
- Negative thinking: "Her proposal implies that she is teasing me."
- Arbitrary inference: "The absence of Mr. C means that he hates me."
- Over-connection to the self: "I made Miss D ill"
- Over-interpretation: "My mistake reduced the profit of the company."

The depressive cognitive style makes good things smaller and bad things larger. It is very important for a person with depression to change her negative cognitive style to positive one. The five column method is a useful intervention technique to change cognitive styles. The detail of the method is described in the next section.

#### 5.4.2.4 Psychological Treatment for Persons with Depression

Table 5.3 shows the thought recording approach (five column method) in CBT. This approach provides useful procedures for identifying and changing automatic thoughts. In the first column, a patient (or a client) is required to state a stressful event she has experienced. In the second column, the automatic thoughts that precede the negative emotions toward the event are recorded, and the degrees of the patient's beliefs in the automatic thoughts are rated. In the third column, the emotions that are experienced after the occurrences of the automatic thoughts are specified, and the degrees of the emotions are rated. In the fourth column, the patient is asked to identify her cognitive errors, such as listed in the previous section; and then to write rational responses to the automatic thoughts, and to rate the degrees of beliefs in them. In the fifth column, re-evaluation of the emotions toward the event in question is performed. Patients with depressive disorders are required to record their dysfunctional thoughts in their daily life and to modify their cognition and behavior. Please challenge to improve your own depressive mood using the five column method.

#### 5.4.2.5 Effectiveness of Antidepressants and CBT for Patients with Depression

CBT is an effective treatment that enables patients to correct false self-beliefs. The fundamental assumption of CBT is that thought precedes emotion; therefore, to learn healthy ways of thinking will improve a person's mood. There is a good deal of evidence that CBT reduces relapse rates in patients with depression. It is comparable in effectiveness to antidepressants. The combination of CBT and

**Table 5.3** Five column method to change cognitive styles

| **Event** | **Automatic thoughts** (Rate beliefs in automatic thought, 0–100) | **Emotions** (Rate emotions, 0–100) | **Rational thoughts** (Rate beliefs in rational responses, 0–100) | **Outcomes** (Specify and rate subsequent emotions, 0–100) |
|---|---|---|---|---|
| Failure of exam | I am not smart (90) | Regrettable (90) Hopeless (75) | I want to study math more (95) | Regrettable (40) Hopeless (20) |
| Breakup | I am not attractive (85) | Angry (85) Depressive (80) | She has other priorities (80) | Angry (25) Depressive (30) |
| **Write your event** | **Write your automatic thoughts** | **Write your emotions** | **Write your rational thoughts** | **Write your outcomes** |
| | | | | |

antidepressants has been shown to effectively manage severe or chronic depressions.

As Rupke et al. (2006, pp. 83–86) reported:

> Many studies and meta-analyses show that cognitive therapy or CBT effectively treats patients with unipolar major depression. Several studies have pointed out that cognitive therapy is superior to no treatment or to placebo. Two comprehensive meta-analyses showed that cognitive therapy is as effective as interpersonal or brief psychodynamic therapy in managing depression. They also showed that cognitive therapy is as effective as and possibly more effective than pharmacotherapy in managing mild to moderate unipolar depression.

CBT is an established therapy for depression today. How effective the combination of CBT and antidepressants is in comparison with antidepressants or CBT alone is shown in Table 5.4.

**Table 5.4** Summary of the efficacy study by Keller et al. (2000)

| Age | Diagnosis | Treatment | Num. of patients | Period (week) | Drop out | HRSD[a] | Efficacy (%) | |
|---|---|---|---|---|---|---|---|---|
| | | | | | | | Remission | Effective |
| 18–75 | Chronic major depression (DSM-IV) & HRSD≧20 | Antidepressant (nefazodone) 200 mg/day Max: 600 mg/day | 226 | 12 | 6 | 26.8 → 14.7 | 29 | 48 |
| | | CBT | 228 | 12 | 12 | 26.4 → 15.1 | 33 | 48 |
| | | Combination (AD + CBT) | 227 | 12 | 1 | 27.4 → 9.7 | 48 | 73 |

[a]*HRSD* Hamilton rating scale for depression

#### 5.4.2.5.1 Epilogue: Notable Persons with Depression

According to the biographies and historical records (Wikipedia 2014), there were many notable persons who suffered from depression and bipolar disorder. Bipolar disorder is a type of mood disorder in which depressive and manic episodes occur alternatively. Abraham Lincoln, Fyodor Dostoyevsky, Marilyn Monroe, John Lennon, and others had hard time due to this disorder. In the last part of this chapter, let us review the life of Ernest Miller Hemingway (1899– July 2, 1961), who was an American author and journalist. He won the Nobel Prize in Literature in 1954. (While I was a high school student, I read his masterpiece, *The Old Man and the Sea*. It describes the struggle between a big fish and an old fisherman. The story was boring for me at that time, because I was too young to understand the mental state of an old man.) Hemingway committed suicide and died in 1961. (Fig. 5.4 is a photo of E. M. Hemingway. He looks tough and stable.)

> He was a very active man and loved fishing and hunting.
> Yet at the same time, he sometimes became gloomy and depressive.
> He had typical symptoms of bipolar disorder.

He was the second child and first son, born to Clarence and Grace Hemingway. His father, Clarence, was a physician, and his mother, Grace, was a musician. After high school he reported for a local newspaper for a few months, and serve on the Italian front in World War I. In 1918, he was seriously wounded and sent back home. In 1921, he married Hadley Richardson, the first of his four wives. The couple lived in Paris, where he was influenced by the modernist writers and artists of the "Lost Generation" expatriate community. Hemingway's first novel was published in 1926. After working as a journalist during the Spanish Civil War, he wrote *For Whom the Bell Tolls*. At the end of World War II, he experienced the Normandy Landings and the liberation of Paris. Shortly after the publication of *The Old Man and the Sea* in 1952, Hemingway went to Africa, where he was almost killed in two successive plane crashes; he was left in pain or ill health for much of the rest of his life. In the late 1950s, Hemingway suffered from depression and other

**Fig. 5.4** Ernest Hemingway (1899–1961) (Photoshot/AFLO)

conditions, such as high blood pressure and liver disease. He committed suicide at his house in Ketchum, Idaho, in the summer of 1961.

Mariel Hemingway is a granddaughter of Ernest Hemingway. She is a famous actress, and worked on the documentary film *Running from Crazy*, directed by Barbara Kopple and shown at the Sundance Film Festival in 2013. The film chronicles problems of Hemingway's family: suicide, substance abuse, and mental illness. Seven members of Hemingway's family have killed themselves, including Ernest and Mariel Hemingway's older sister Margaux, who was an American fashion model and actress. Margaux seemed to suffer from bipolar disorder as Ernest did. She overdosed antianxiety drugs and died on the same day as her grandfather shot himself by a gun 35 year ago. As mentioned in Sect. 5.2.2, bipolar disorder has higher genetic risk factors than other mood disorders. It may be conjectured that the Hemingway family is genetically susceptive to bipolar disorder. Mariel Hemingway said that they had many family conflicts, despite their family fame. The reader who wants to know more about the Hemingway family is recommended to watch the documentary film *Running from Crazy*.

The risk of committing suicide is very high in patients with mood disorders. The Ministry of Health, Labour and Welfare of Japan (2010) reported that the number of suicides in Japan was 32,845 in 2009, and it included 6949 persons (21 %) with depression, a very high number as compared with the number of persons with other psychiatric disorders. As you have seen in the Hemingway family, suicide is a serious problem for patients and their family. Psychoeducation for patients and family members is important for preventing suicide. Adequate information, proper intervention including medication and psychological treatments, and social support will save their life. The government needs to take actions to reduce the number of suicides (some actions have already been implemented). However, Japan has a serious shortage of psychiatric and psychological professionals and especially CBT experts. An accredited training and license system for clinical psychologists is necessary for teaching and maintaining proper knowledge and skills in their practical fields.

## Exercises

1. Choose the correct answer.

   (a) Eating problem is a symptom of depression.
   (b) Visual hallucination is a symptom of depression.
   (c) Sleeping problem is not a symptom of depression.
   (d) Thought disorder is a symptom of depression.
   (e) Lack of interest is not a symptom of depression.

2. Read the following description on serotonin and choose the correct order in which serotonin works as a transmitter.

   1. An electrochemical nerve signal passes down the presynaptic cell into its end terminals.
   2. Serotonin passes across the synaptic cleft, binds with special proteins called receptors on the membrane of the postsynaptic cell (receiving cell) and sets up a new electrochemical signal in that cell (the signal can stimulate or inhibit the postsynaptic cell). Serotonin fits with its receptor like a lock and key.
   3. The remaining serotonin molecules in the cleft and those released by the receptors after use get destroyed by enzymes in the cleft (monoamine oxidase (MAO) and catechol-o-methyl transferase (COMT)). Some get taken up by specific transporters on the presynaptic cell (reuptake).
   4. The presynaptic cell (sending cell) makes serotonin (5-hydroxytryptamine, 5HT) from the amino acid tryptophan and packages it in vesicles in its end terminals.
   5. The nerve signal stimulates the vesicles containing serotonin to fuse with the cell membrane and dump serotonin into the synaptic cleft.

3. Choose the correct answer.

   (a) Contingent aversive stimuli (electric shock) strengthen learned helplessness.
   (b) Non-contingent aversive stimuli weaken learned helplessness.
   (c) Non-contingent aversive stimuli strengthen learned helplessness.
   (d) Contingent aversive stimuli weaken learned helplessness.
   (e) Aversive stimuli always strengthen learned helplessness.

4. Choose the correct answer.

   (a) Internal attributions strengthen depression.
   (b) External attributions strengthen depression.
   (c) Internal attributions weaken depression.
   (d) Stable attributions weaken depression.
   (e) Unstable attributions strengthen depression.

5. Choose the cognitive style that depressive persons usually do not tend to have.

   (a) Over-generalization: "If I fail A, I will fail B."

(b) Selective abstraction: "The teacher's advice means that I am incompetent."
(c) All or nothing: "If I fail the exam, my life will be over.
(d) Negative thinking: "Her proposal means that she is teasing me."
(e) Rationalization: "I do not like the movie, because I cannot get a ticket."

## References

American Psychiatric Association: Quick Reference to the Diagnostic Criteria from DSM-IV-TR. American Psychiatric Association, Washington, DC (2000)

Angust, J.: Epidemiology of depression. In: honing, A., Van Prag, H.M. (eds.) Depression: Neurobiological, Psychopharmacological and Therapeutic Evidence, pp. 17–29. Wiley (1997)

Beck, A.T.: Cognitive Therapy and the Emotional Disorders. International University Press, New York (1976)

Freudenrich, C.: How Nerves Work. 19 October 2007. HowStuffWorks.com. http://health.howstuffworks.com/human-body/systems/nervous-system/nerve.htm. 24 March 2015 (2007)

Kameya, H.: Depression: biological aspects. In: Shimoyama, H., Tanno, Y. (eds.) Clinical Psychology 4-Abnormal Psychology II, pp. 127–146. University of Tokyo Press, Tokyo (2002)

Keller, B.M., McCullough, P.J., Klein, N.D., Arnow, B., Dunner, L.D., Gelenberg, J.A., Markowitz, C.J., Nemeroff, B.C., Russell, M.J., Thase, E.M., Trivedi, H.M., Blalock, A.J., Borian, E. F., Jody, N.D., DeBattista, D.M.H.C., Koran, M.L., Schatzberg, F.A., Jan Fawcett, J., Hirschfeld, M.A.R., Keitner, G., Miller, I., Kocsis, H.J., Kornstein, G.S., Manber, R., Ninan, T.P., Rothbaum, B., John Rush, A., Vivian, D., Zajecka, J.: A comparison of nefazodone, the cognitive behavioral-analysis system of psychotherapy, and their combination for the treatment of chronic depression. N. Engl. J. Med. **342**(20), 1462–1470 (2000)

Ministry of Health, Labour and Welfare of Japan.: http://www.mhlw.go.jp/seisaku/2010/07/03.html (2010)

Overmier, J.B., Seligman, M.E.P.: Effects of inescapable shock upon subsequent escape and avoidance responding. J. Comp. Physiol. Psychol. **63**, 28–33 (1967)

Rupke, J.S., Blecke, M.D., Renfrow, M.: Cognitive therapy for depression. Am. Fam. Physician **73**, 83–86 (2006)

Sadava, H.D., Heller, C., Orians, H.G., Purves, K.W., Hills, M.D.: Life, 8th edn. Sinauer Associates, Sunderland (2008)

Seligman, M.E.P., Maier, S.F.: Failure to escape traumatic shock. J. Exp. Psychol. **74**, 1–9 (1967)

Sevy, S., Mendlewicz, J., Mendelbaum, K.: Genetic research in bipolar illness. In: Bechham, E.E., Leber, W.R. (eds.) Handbook of Depression, 2nd edn, pp. 203–212. Guilford Press (1995)

Shimoyama, H., Tanno, Y.: Clinical Psychology 4-Abnormal Psychology II. University of Tokyo Press, Tokyo (2002)

Wikipedia.: Ernest Hemingway. http://en.wikipedia.org/wiki/Ernest_Hemingway (2014)

Zung, W.W.: A self-rating depression scale. Arch. Gen. Psychiatry **12**, 63–70 (1965)

# Chapter 6
# Mechanisms of Pain

**Aya Nakae**

**Abstract** No one can deny that pain serves as an alarm. Sometimes pain informs us about severe abnormalities inside our body. However, pain (in particular, chronic pain) does not always have the role of an alarm. Pain itself can be harmful, as in the case of patients with chronic pain. Thus, we should reveal the mechanisms of pain and control it even though this is a challenging task. Pain is always subjective to individuals, and we have no definitive and objective evaluation methods for measuring pain. Indeed, pain is very complex and difficult to understand. In this chapter, in order to better understand the complex phenomenon of pain, I first discuss the definitions and classifications of pain. Second, I explain methods for evaluating pain, including the methods of potential objective evaluation. Lastly, I describe the relationship between pain and the central nervous system, especially with respect to cognitive function, emotion, and psychiatric illness. I expect that there will be a paradigm shift in pain treatment in the future; studies of the affective components of pain will greatly progress, and drug discovery research will specifically aim at reducing pain.

**Keywords** Pain mechanisms • Emotion • Psychiatric disease • Pain behavior • Pain evaluation

## 6.1 Definition and Classification of Pain

The International Association for the Study of Pain endorses the following definition of pain: "an unpleasant sensory and emotional experience that is associated with actual or potential tissue damage or that is described in terms of such damage."

An important point about this definition is that pain does not always come with injuries. Even when patients do not have any tissue damage, if they say that they feel pain, then pain exists for them. Pain is a complex phenomenon involving multiple factors (Fig. 6.1). It is difficult to scientifically approach to pain and even to central pain mechanisms. Nevertheless, it is possible to define and classify different kinds of pain in terms of time course, pathology and underlying

A. Nakae (✉)
Immunology Frontier Research Center, Osaka University, Suita, Osaka, Japan
e-mail: anakae@anes.med.osaka-u.ac.jp



M. Kasaki et al. (eds.), *Cognitive Neuroscience Robotics B*,
DOI 10.1007/978-4-431-54598-9_6

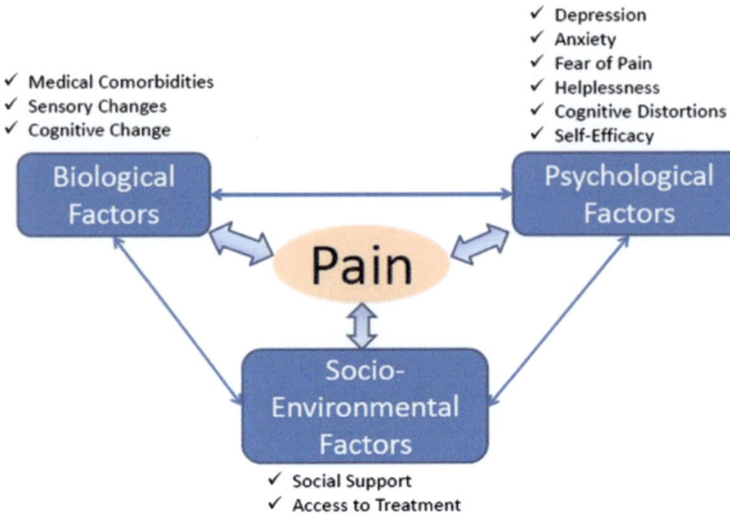


**Fig. 6.1** Causes of pain

mechanism. The definitions and classifications provide the first and necessary step toward understanding and treating pain.

### 6.1.1 *Classification Based on Time Course*

Time course is an important factor in understanding the origin of the conditions in which subjects suffer from pain. Acute pain has an alarm function, which is essential to survive. However, chronic pain does not always have biological significance. The causes of chronic pain are mainly in the brain, and thus pain itself is not so important for survival, except in cases of severe depression in which it could be fatal.

Acute pain is generally regarded as lasting for no more than a month, whereas chronic pain is as lasting longer than 6 months. Bonica (1977) has suggested that chronic pain is defined as persisting for longer than a month beyond the normal healing period or as being associated with a pathological process that causes continuous or recurrent pain over months to years.

Pain that lies somewhere between acute and chronic pain is called 'subacute.' As shown in Fig. 6.2, it is not always easy to distinguish subacute pain from the other two types of pain. Subacute pain may require almost the same treatment as chronic pain.

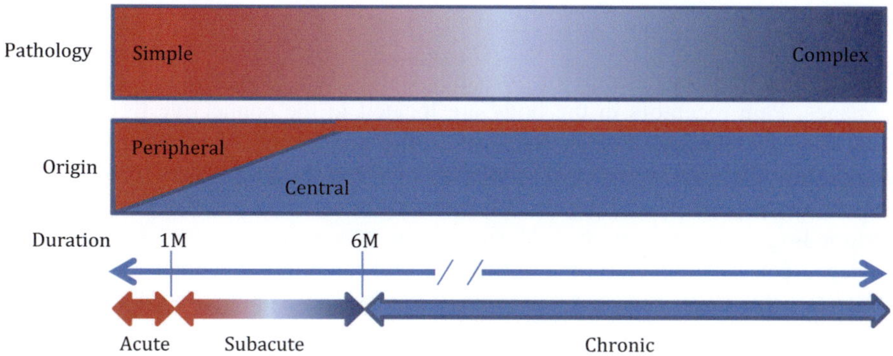


**Fig. 6.2** Classification of pain in typical cases. *M* month

### *6.1.2 Classification Based on Pathology (Table 6.1, Fig. 6.3)*

**Table 6.1** Mechanistic classification of pain

| Types of pain | Origin | Example |
|---|---|---|
| Nociceptive | Somatic | Fracture, postoperative pain |
| | Visceral | Colitis, cystitis |
| | Inflammatory | Skin region, arthritis |
| Neuropathic | Peripheral | Neuralgia, sciatic pain |
| | Spinal cord | Spinal cord injury, spinal canal stenosis |
| | Stroke | Thalamic pain, phantom pain |
| Central-induced | Deafferentation pain | |
| | Psychiatric | Depression, fibromyalgia |

**Fig. 6.3** Classification of pain

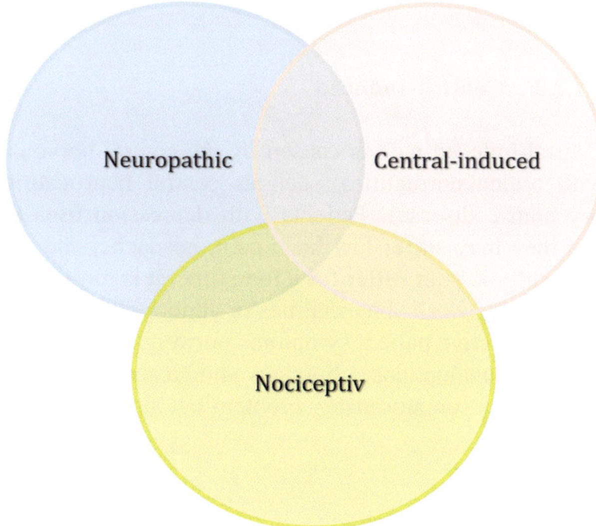

#### 6.1.2.1 Nociceptive

Nociceptive pain is visceral or somatic, and results from normal diseases or damages. Pain receptors in the body are located in the connective tissues of the skin, the subcutaneous tissues, the fascias, the periosteum, the inner bone membranes, and the joint capsules. Stimulation of these receptors leads to a localized dull or sharp pain. Visceral pain receptors are located in most organs and their surrounding connective tissues. Visceral pain, when induced by a hollow organ damage, feels severely cramping or tingling, and sometimes becomes referred pain. Visceral pain, when caused by a damage to the deep connective tissue or other visceral capsule may feel as a sharp localized pain.

#### 6.1.2.2 Neuropathic

Neuropathic pain is caused by injury or dysfunction in the central nervous system or peripheral nervous system rather than by stimulation of pain receptors. Its diagnosis is made by evaluating the disproportionate pain relative to tissue damage, signs of nerve damage (e.g., burning or tingling pain), and abnormal sensations, in neurological examinations. Peripheral nerve damages or dysfunction can cause neuropathic pain. More specifically, the typical causes of neuropathic pain are nerve compression (e.g., neuroma or tumors due to disc herniation) and various metabolic neuropathies. Mechanisms underlying neuropathic pain are probably varying. Central neuropathic pain syndrome is involved in the process of the reorganization of central somatosensory processing. Deafferentation pain is due to partial or complete interruption of the afferent nerve activity of the peripheral or central nervous systems. The afferent nerve activity conveys information from the distal to the proximal.

#### 6.1.2.3 Central-Induced

Central-induced pain is caused by the central nervous system, mainly in patients with brainabnormalities, such as central neuropathic pain and some kinds of psychiatric diseases. Patients with depression usually have complaints of pain, and they may suffer low back pain, headache, and neck stiffness. When patients with chronic pain suffer for a long time, it is mostly central-induced and its causes are brain network abnormalities or abnormalities in some brain regions.

If subjective patient symptoms outweigh what is consistent with the findings of objective evaluations, clinicians should suspect that the pain might not just be caused by a somatosensory problem but also be central-induced.

## 6.2 Classification Based on Region and Each Mechanism

### *6.2.1 Peripheral Mechanisms of Nociception (Fig. 6.4)*

Mechanical, chemical, or thermal nociceptive stimulation activates nociceptors: they conduct pain signals through the primary neurons to the dorsal horns of the spinal cord. Once information conducted by neurotransmitters, neuropeptides, or else arrives in the spinal cord, they make synaptic contact with the secondary neurons. The secondary neurons immediately cross the spinal cord by passing under the central canal to form the spinothalamic tract. The spinothalamic tract is located in a ventrolateral position relative to the spinal cord. It conducts information to different regions of the ventrobasal and the centromedian complex in the somatosensory thalamus. The secondary neurons also make synaptic contacts in different regions of the brain stem, e.g., with the periaqueductal grey matter (PAG) and the nucleus raphe magnus.

The initial nociceptive stimulus is not the only factor that contributes to perceived pain. There are four different steps that occur through a series of chemical and electrical reactions: transduction, transmission, modulation, and perception. In other words, even when a noxious signal is transmitted to the brain, if the signal is modulated as a non-painful state, people do not feel pain. Conversely, as is the case of chronic pain patients, if there are some brain abnormalities, when a non-noxious signal is received in the brain, people may feel pain.

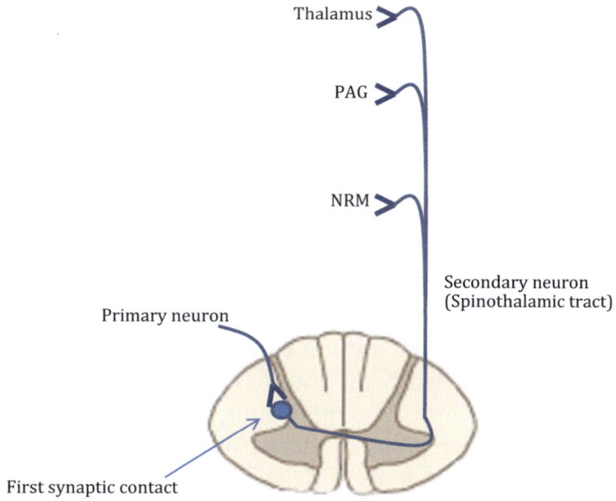


**Fig. 6.4** Pain pathways

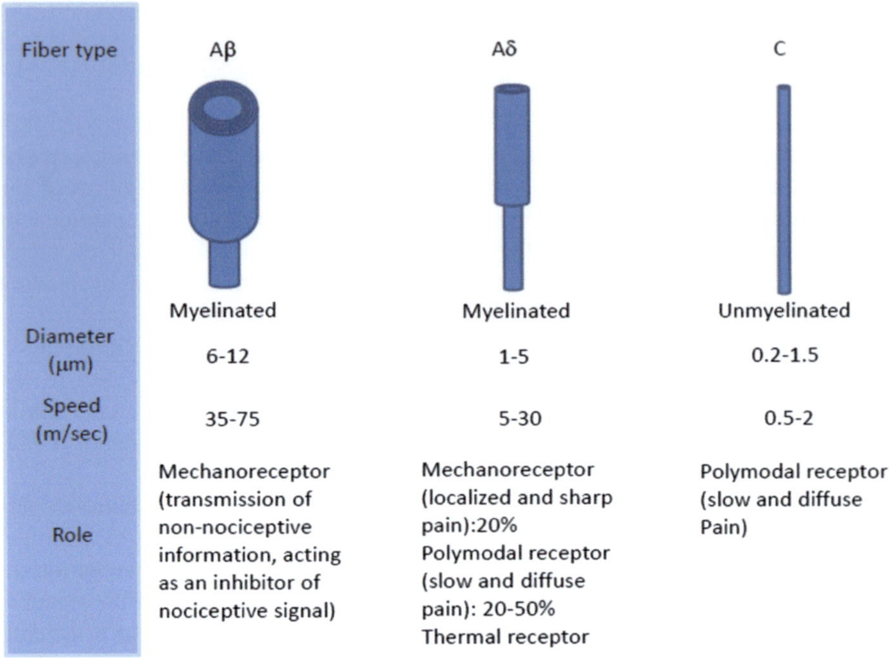


**Fig. 6.5** Sensory nerve fibers

#### 6.2.1.1 Pain Fibers

There are many types of receptors in the skin. The free nerve endings are mainly responsible for pain perception. When signals are received by the nerve endings, they are transferred through the nerve fibers. The nerve fibers are divided into three classes: A-beta, A-delta, and C fibers (Fig. 6.5).

A-beta fibers are mechanoreceptors with a very low response threshold, and they respond to slight touch. Under persistent pain conditions, central nervous system sensitization will occur and result in abnormal connectivity between A-beta fibers and other fibers. The afferent impulses of A-beta fibers can be recognized as pain (allodynia). A-beta fibers have characteristic features, e.g., large diameter (from 6 to 12 μm) and significant myelination. These two features contribute to the rapid conduction capability of A-beta fibers; they convey impulses at speeds of around 35–75 m/s. A-beta fibers also have a significant inhibitory role in transferring nociceptive information. Selective stimulation of these fibers will lead to the recruitment of inhibitory interneurons in the substantia gelatinosa of the dorsal horn of the spinal cord, and these inhibitory interneurons block the transfer of nociceptive information originating from the same segment of the spinal cord.

A-delta fibers transfer sharp pain. They are myelinated, and most of them are nociptive in that most A-delta fibers have the essential characteristics that are necessary to be considered nociceptive. Their diameters range from 1 to 5 μm, and their conduction speeds are between 5 and 30 m/s. They are responsible for first pain, i.e., brief and extremely localized pain felt at the onset of stimulation. A-delta fibers are not spontaneously active. They are divided into two groups: mechanical receptors and mechanothermal receptors (Treede et al. 1998). Mechanonociceptors make up 20 % of all cutaneous A-delta fibers (Besson and Chaouch 1987). They are receptive to certain kinds of stimulation, such as stings and pinching. Under normal circumstances, they do not respond to chemical or thermal stimulation that is less than 53 °C. Mechanothermal nociceptors respond to both mechanical and thermal stimuli, and sometimes to chemical stimuli as well. Twenty percent to fifty percent of A-delta nociceptors are regarded to be mechanothermal, and some of them respond to cold. The latency is lower for mechanical nociceptors than for mechanothermal nociceptors (1 s vs. 0.2 s). This fact suggests that it is mainly mechanothermal nociceptors that are responsible for first pain (Treede et al. 1998).

C fibers are unmyelinated. Their diameters range from 0.2 to 1.5 μm, and their conduction speeds are slow (between 0.5 and 2 m/s). They are responsible for second pain, i.e., later and more diffuse pain than first pain. They play an important role in intensifying of pain (Ringkamp and Meyer 2008). C fibers represent nearly three-quarters of all peripheral nerve fibers, and more than 90 % of them are nociceptors. C fibers are polymodal receptors, and respond to mechanical, chemical, or thermal stimuli. In particular, they are excited by intense stimuli from sharp objects. C fibers can also be involved in non-nociceptive sensations (e.g., pruritus (itch) (Stander et al. 2003)) and in non-nociceptive sensations with a strong emotional component (e.g., caress). A study of patients with complete deafferentation of somesthetic myelinated fibers has shown that the patients do not feel the touch of a hand when gently stroked, while they report a pleasant sensation (Olausson et al. 2002).

## *6.2.2 Central Mechanisms of Pain (Fig. 6.6)*

The brain mechanisms of pain may be investigated with regard to two different aspects: the emotional aspect and the sensory aspect (or the affective component and the sensory component). There are very complex mechanisms involved in human pain, and their interactions are also complex. Human functional brain imaging has shown that pain affects various brain regions as well as the so-called pain matrix (Baranauskas and Nistri 1998).

### 6.2.2.1 Somatosensory Cortex

The somatosensory cortex is responsible for the sensory-discriminative aspect of pain. The entire noxious signal is transmitted from the peripheral nerves to the brain through the thalamus. Sensory information is transmitted to the primary

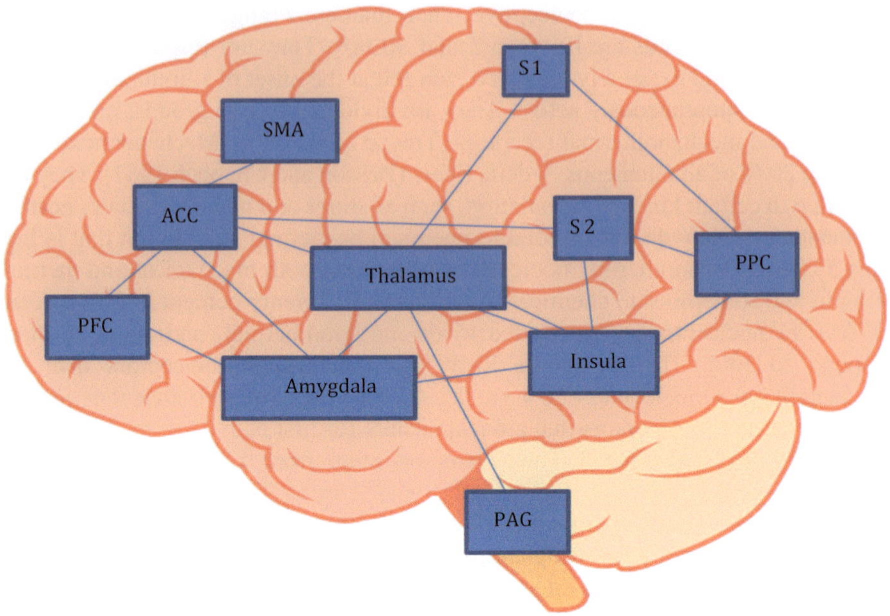


**Fig. 6.6** Brain regions that are activated in response to painful stimulation. This diagram shows the so-called pain matrix

somatosensory cortex (S1) and the secondary somatosensory cortex (S2) through a site called the 'ventral basal complex' in the thalamus.

#### 6.2.2.2 Amygdala

The amygdala is located inside the temporal lobe, and it has the role of integrating emotional and instinctive behavior. On the basis of sensory inputs, the amygdala determines whether its stimulation is comfortable or uncomfortable. It then causes negative emotions, such as fear and anger. The amygdala retrieves memorized information in accordance with nociceptive information. After pain is experienced, the information it carries is evaluated as negative or affective discomfort. Then, when pain is experienced again, negative emotional behaviors, e.g., fear, anger, anxiety, freezing or struggle, or escape responses, are induced.

#### 6.2.2.3 Insula

The insular cortex (or the insula) is located on the outer surface of the brain in the outer groove of the lower halves of the parietal lobe and in the temporal lobe. The insular cortex receives a major input from the thalamus and S2, and it is in contact with the amygdala. Thus, the insula cortex is regarded to be responsible for both the sensory aspect and the emotional aspect of pain. The insular cortex is important and useful in many ways. First, it is regarded to be responsible for integrating and processing higher cognitive and emotional information that is related to the general (healthy) condition. Second, the insular cortex is related to the autonomic nervous system to maintain homeostasis. Third, the insula cortex is the part of the circuit system that is involved in the aversion and avoidance of painful situations. Last, the insula cortex is responsible for the prediction of pain, and it is also involved in placebo analgesia.

#### 6.2.2.4 Anterior Cingulate Cortex (ACC)

The cingulate cortex is the cortical tissue that surrounds the corpus callosum. It is divided into three parts: ACC, midcingulate cortex, and posterior cingulate cortex. The ACC receives signal inputs not only from the amygdala but also from the thalamus, the PAG, and the locus coeruleus. The cingulate cortex is the central region of the noradrenergic descending pain inhibitory system, and anatomically divided into four regions: executive, evaluative, cognitive, and emotional. The emotional region, which is located in the ventral area, receives signal input from the amygdala. The cognitive part, which is located in the dorsal area, projects the signal to the spinal cord, and it is also involved in the avoidance of pain. The ACC plays a role in controlling emotional arousal produced in the limbic system.

#### 6.2.2.5 Prefrontal Cortex (PFC)

The PFC is considered one of the regions in the so-called pain matrix, and it is connected with both the insular cortex and the amygdala. It is involved in the emotional aspect of pain.

#### 6.2.2.6 Orbitofrontal Cortex (OFC)

The OFC is located at the back of the retina, and connects the PFC and the limbic system. It is involved in cognitive processing like decision-making and in the emotion and reward system. It is connected with the amygdala, which, too, is involved in the emotional system.

### *6.2.3 Pain and Cognitive Function*

Pain and cognitive functions correlate with each other because of pain perception in the brain (Fig. 6.7). Suffering from pain for a long time may result in irreversible

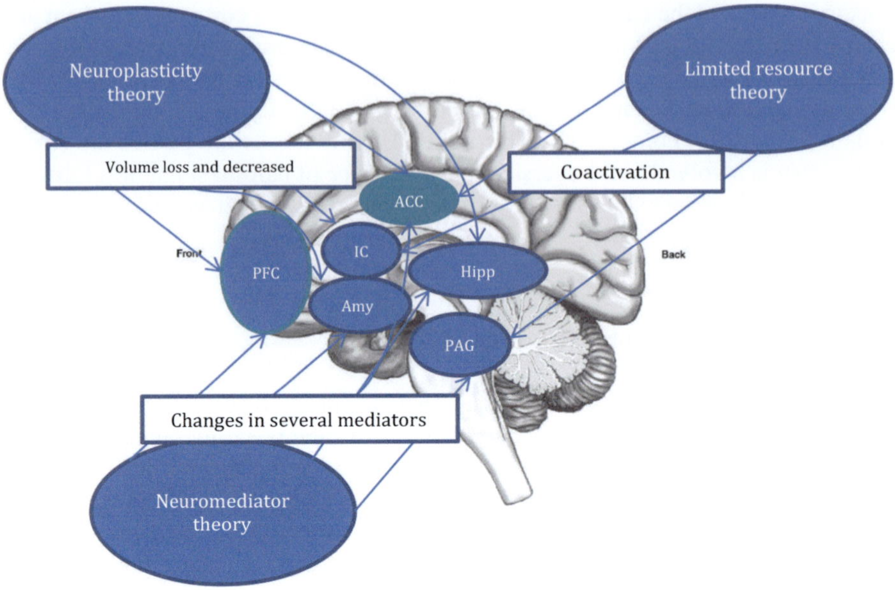


**Fig. 6.7** A theoretical model of pain-related cognitive impairment. *PFC* prefrontal cortex, *IC* insular cortex, *Hipp* hippocampus, *Amy* amygdala, *PAG* periaqueductal grey, *ACC* anterior cingulate cortex

changes in neuroplasticity and/or activation or suppression of several neuromediators. Such results restrict the brain's functional capacities.

## 6.3 Study Tools of Experimental Pain

Substantial evidence has shown that human pain perception can be revealed in experimental pain studies. Experimental pain studies use different patterns of stimulations: tonic pain and phasic pain. The strategy is used not only for examining subjective pain cognition, but also in physiological and imaging studies. Tonic stimulation can also be applied with evoked pain in functional magnetic resonance imaging (MRI) studies. The methods of stimulation include thermal and pressure. Phasic stimulation can also be used with event-related potentials, which are used to analyze the conditions by analyzing the brain. The methods of stimulations are varied and include thermal, electrical, and laser. When relatively higher temperatures of thermal phasic stimulation are used, the A-delta fibers are stimulated, while lower temperature around 42 °C stimulate C fibers. Laser stimulation is known to activate A-delta stimulation. In addition, electrical stimulation can selectively stimulate the three types of A-delta, A-beta, and C fibers separately (Fig. 6.8).

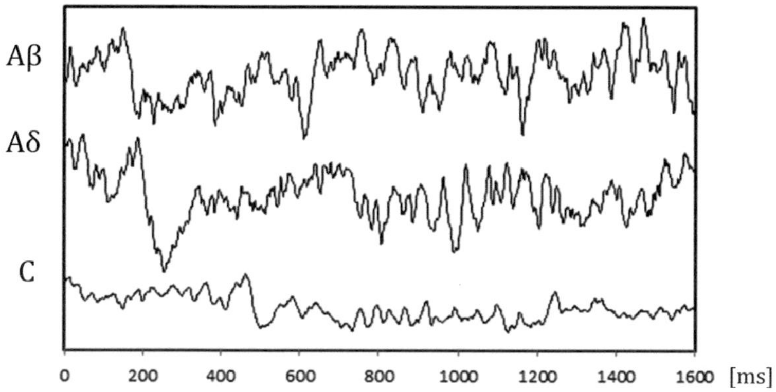


**Fig. 6.8** Electrical stimulation to the three types of fibers

## *6.3.1 Quantitative Sensory Testing (QST)*

Negative and positive sensory phenomena are assessed by neurologic bedside examinations and Quantitative Sensory Testing. Quantitative Sensory Testing (QST) is a psychophysical test of sensory perception during the administration of stimuli with predetermined physical properties and following specific protocols. QST is able to capture and quantify stimulus-evoked negative and positive sensory phenomena and, as such, should become a standard, if not critical, tool in neuropathic pain research and practice.

When we stimulate the injured area, we may find abnormal sensations or allodynia, and we can then diagnose the nerve injury. When we stimulate intact areas, we may find potential abnormalities (in cases of healthy subjects) or secondary or systemic hypersensitivities (in cases of patients with chronic pain).

### 6.3.1.1 Stimulation Detection Threshold

The stimulation detection threshold is the lowest intensity of stimulation that is perceptible to a subject. If the subject has an abnormal or damaged nerve, her pain threshold and pain tolerance may be affected. Thus, before measuring parameters, the detection threshold should be measured. In general, we can measure perception thresholds using mechanical and thermal stimulation.

### 6.3.1.2 Pain Detection Threshold

The pain detection threshold is the lowest level of stimulation that a subject feels as pain. In psychophysics, the pain threshold is measured by identifying the lowest intensity of stimulation that produces pain.

#### 6.3.1.3 Tolerance Threshold

The tolerance threshold is the highest level of stimulation of pain that a subject is able to endure.

### *6.3.2 Event-Related Potential (ERP)*

Event-related potentials (ERPs) are defined as the waveforms recorded on the scalp as a series of positive and negative peaks, which vary in polarity, amplitude, and duration over time. Because ERP data are objective, it is a promising tool for diagnosing neuropathy or other neurogenic disorders.

As we do in pain research, we can use ERPs as measures of brain reactions to noxious stimuli. ERP patterns differ from the ways that are used to stimulate fibers: A-delta, A-beta, and C-fibers. ERP latency varies in these fibers. As shown in Fig. 6.8, we are able to obtain separate ERP data for three different nerve stimulations, but, in fact, some of them are stimulated at the same time.

The problem with ERPs is how to determine the method for analyzing data. If the amplitudes of ERPs are small, we sometimes need to perform repeated stimulations and use computer summation techniques. However, latencies may be different even when the same stimulation is given. If we sum ERP data with the different latencies, the amplitudes of the ERPs are relatively smaller than the actual amplitudes.

## 6.4 Pain Evaluation Tools

### *6.4.1 Questionnaires*

Pain is always subjective. Today, there are many tools that are used to evaluate answers to questionnaires. The most common way is to use a visual analog scale (VAS). A VAS is a 10-cm bar with the left end indicating "no pain" and the right end "the worst pain imaginable." Participants rate the level of their pain on this scale. The results show a normal distribution, and parametric analyses can be performed on them. We commonly use a VAS to evaluate the intensity and unpleasantness of pain (Fig. 6.9).

In order to evaluating patients with chronic pain, the McGill Pain Questionnaire was developed by Melzack and Torgerson (1971). The Short Form McGill Pain questionnaire (SF-MPQ-2) is commonly used in clinical situations (Melzack 1987). Recently, the SF-MPQ-2 is expanded and revised so as to evaluate patients with neuropathic pain symptoms (Dworkin et al. 2009). The SF-MPQ-2 can divide the patients' painful conditions into four categories: intermittent, continuous, affective, and neuropathic.

**Fig. 6.9** Examples of visual analog scales for pain intensity and the unpleasantness of pain

The Pain Catastrophizing Scale is known to be effective for evaluating patients' abnormal thinking (Sullivan et al. 1995) The catastrophizing scale is a 13-item self-report scale created by Sullivan et al. (1995), and used to measure pain catastrophizing. Items are rated on a scale from 0 to 4, and they have three different categories: Rumination, Magnification, and Helplessness. It has been hypothesized (Osman et al. 1997) that pain catastrophizing is related to various levels of pain, physical disabilities, and psychological disabilities in clinical and nonclinical populations.

The Hospital Anxiety and Depression scale (HAD) is effective for evaluating depression and anxiety with physical symptoms (Zigmond and Snaith 1983). This scale consists of seven questions for anxiety and seven questions for depression.

### *6.4.2 The Amount of Pain as a Number by Electrical Stimulation*

When patients evaluate their pain with a VAS, it is sometimes difficult for them to evaluate the effects of treatment. PainVision (Nipro Corporation, Osaka, Japan) is a pain-evaluating tool (Fig. 6.10). The machine applies gradually increasing electrical stimulation to A-beta fibers. When the patients feel the same pain as the electrical stimulation is meant to give, the patients inform us of the amount of the pain by pushing a button. As far as people can use this machine precisely, it can quantify the pain as a number. This is a breakthrough of pain treatment. However, it is not an objective tool, because the machine requires pushing the button intentionally; but if the patients understand how to use this machine, the data would be objective.

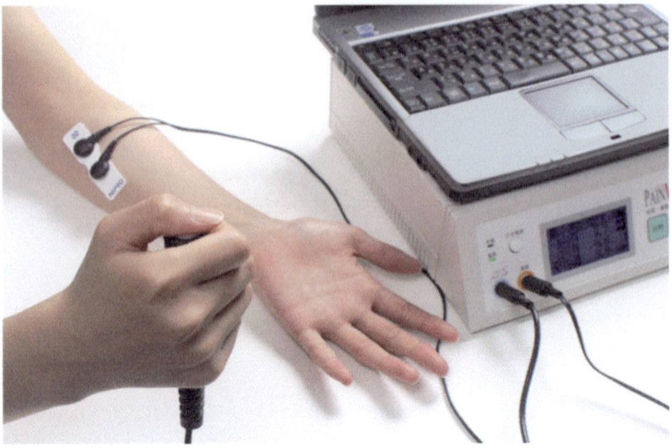


**Fig. 6.10** The PainVision machine (Nipro Corporation, Osaka, Japan)

## 6.5 Pain and Emotion

### *6.5.1 What Is Emotion?*

The definition of emotion is different in different academic fields, whereas emotion may be defined in a simple way. In his book *Psychology*, Schacter et al. (2011) defines emotion as "*a positive or negative experience that is associated with a particular pattern of physiological activity*".

Emotions induce autonomic responses, such as hypertension, tachycardia, endocrinologic changes, and stress hormone responses. As shown in Fig. 6.11, kinds of emotions can be visualized with a hierarchical structure. It is divided into two broad categories of pleasant and unpleasant emotions, dominating smaller sub-categories of emotions.

### *6.5.2 Pain and Emotion*

Felt pain intensity may vary from person to person, even when the amount of stimulation is the same. This is because emotion can modulate the perception of pain. There have been many studies of pain and emotion in various experimental and clinical situations (Table 6.2).

#### 6.5.2.1 Anxiety and Pain

When one has anxiety about pain, one feels more pain. It is difficult to distinguish anxiety about pain from fear of pain. In other words, it is difficult to determine whether fear or anxiety reduces a pain threshold. It has been reported of surgical

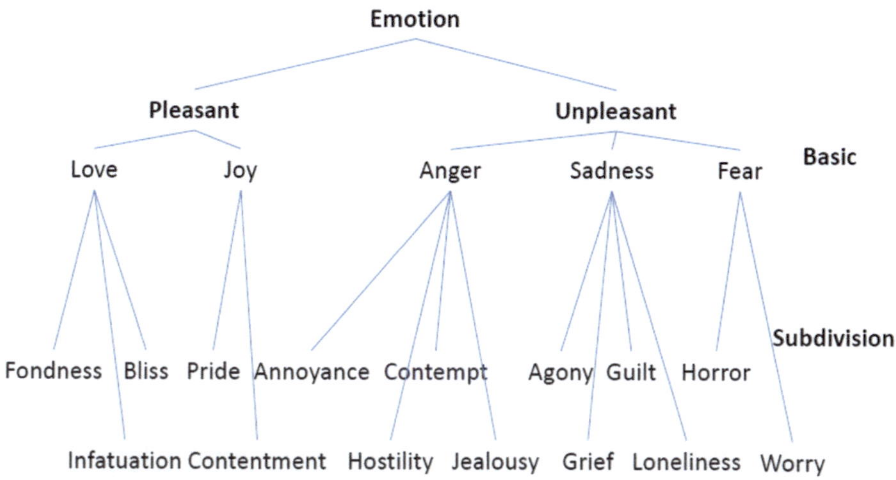


**Fig. 6.11** The structure of emotion

patients that when they are anxious, they experience more severe pain (Lautenbacher et al. 2010). It also has been reported that 7–28.8 % of chronic pain patients have comorbid anxiety disorders (Asmundson and Katz 2009). In an animal model of chronic pain, pain induces anxiety-like behavior (Narita et al. 2006; Suzuki et al. 2007). It also has been reported that it requires fewer amount of drugs to reverse anxiety-like behavior than to reverse pain-related behavior (Munro et al. 2007). These results suggest that anxiety and pain may be based on common mechanisms.

#### 6.5.2.2 Fear and Pain

There are three stages of fear. At the first stage, we interpret stimulation as a threat. At the second stage, we strongly feel that the interpretation is correct, and, at the third stage, we exhibit defensive behaviors (Rachman and Hodgson 1974). Fear and anxiety interact with each other, and both have negative impacts on pain. In particular, for musculoskeletal pain, there is an excellent model called the "Fear-Avoidance Model" (Leeuw et al. 2007). The model explains how a disease or trauma-caused pain fails to become chronic.

In an experimental study of the relation of human pain to anxiety and fear, the pain threshold is increased with fear but decreased with anxiety (Rhudy and Meagher 2000). It is difficult to distinguish anxiety and fear as both may influence pain intensity. In an animal experiment of fear, defensive behaviors appear when an animal only sees a fearful scene; moreover, defensive behaviors are suppressed by experimentally restricting the role of the ACC (Jeon et al. 2010). When observing scary scenes, animals feel anxiety. Animal behaviors are controlled by the common

**Table 6.2** Pain and emotion

| Emotion | | Experimental pain | Clinical pain | References |
|---|---|---|---|---|
| Negative | Fear | ↓ ↑ | ↓ | Fone and Porkess (2008) |
| | | | | Franek et al. (2010) |
| | Anxiety | ↓ | ↓ | Chou and Shekelle (2010) |
| | | | | Defrin et al. (2008) |
| | Sadness | ↓ | ↓ | Abram and Haddox (2000) |
| | Loneliness | ↓ | ↓ | Jeon et al. (2010) |
| | | | | Klossika et al. (2006) |
| | Anger | ↓ | ↓ | Baranauskas and Nistri (1998) |
| | Attention | ↓ | ↓ | Beecher (1946) |
| | | | | Hirofumi and Yuji (2007) |
| Positive | Bliss, etc. | ↑ | ↓ | Kundermann et al. (2009) |
| | | | | Lautenbacher et al. (2010) |
| | | | | Leeuw et al. (2007) |
| | | | | Loeser (1980) |

brain region that regulates pain and fear. Another interesting point is that fear and anxiety interact with each other in animals as well.

#### 6.5.2.3 Attention and Pain

In general, humans tend to forget pain when they concentrate on something else. However, patients with chronic pain feel more pain than it is predicted, because their attention is drawn to pain. In other words, the problem of chronic pain is the attention being directed toward it. When subjects with experimental pain are asked to concentrate on a task, a functional brain imaging study shows, they feel less pain than they do in the no-task condition. When they feel pain, the OFC and ACC (the areas related to emotion) are activated. This suggest that pain is reduced through the emotional aspect of pain (Bantick et al. 2002). In our recent experimental pain study, we assign a cognitive, working-memory task to subjects. The high performance subjects do not reduce their pain intensity when they concentrate on the task. By contrast, the low-performance subjects reduce their pain intensity (Nakae et al. 2013). The effects of attention to pain may be subject to individual differences, at least in normal subjects.

#### 6.5.2.4 Social Isolation and Pain

It has been suggested in animal experiments that social isolation affects brain development and is involved at the onset of mental illness (Fone and Porkess 2008). An animal experiment shows that animal's heat pain threshold decreases after being exposed to the stress of social isolation (Loeser 2000). At least, it seems certain that the stress of social isolation may influence pain in some sense.

#### 6.5.2.5 Anger and Pain

Pain clinicians may see a patient who is angry with someone; in some cases, the patient may be a victim of an unfortunate accident. How does anger influence pain? Substantial evidence has revealed that the expression of anger affects the intensity of pain. One possible mechanism to explain this is the deficiency of the endogenous opioid system with regard to reduction of pain. In a functional brain imaging study, pain and anger both activate the rostral ACC, the OFC, the insular cortex, and the amygdala (Bruehl et al. 2009).

#### 6.5.2.6 Sadness and Pain

Sadness and depression correlate well clinically. An experimental pain study shows that when participants see sad face pictures, they feel more pain, and the increase in pain activates the ACC (Yoshino et al. 2010).

#### 6.5.2.7 Positive Emotion and Pain

How do humans feel experimental pain when they see a pleasant picture? A study has examined this question and obtained the results that when subjects see a pleasant picture, they feel less pain (Rhudy et al. 2010).

In addition, there is a study that analyzes the effects of pleasant emotion by distinguishing them from the effects of attention. The effects of pleasant emotion alleviate pain not just by distracting subjects to focus on comfortable stimuli. Pleasant emotion itself has the effect of pain relief (Villemure and Bushnell 2009). Positive emotion has been shown to influence coping efficacy and social functioning in patients with chronic pain (Park and Sonty 2010).

## 6.6 Mental Disorders as Pathological Models of Pain (Table 6.3)

The prevalence of mental illness has been shown to be significantly higher in patients with chronic pain (Turk et al. 2010). There is evidence that if patients' functions are decreased by pain, they are more likely than otherwise to have comorbid mental illness (Chou and Shekelle 2010). The comorbidity of chronic pain and depression is especially high, and pain symptoms can be a main symptom of depression. It has been revealed in an animal study that when animals feel pain for a long time, they become depressive (Suzuki et al. 2007).

Pain is perceived in the brain. Thus, it is particularly necessary to consider the emotional aspect of pain in relation to the brain. There are some psychiatric illnesses that change pain sensitivity, and this fact suggests that the brain mechanisms of pain perception are involved in the mechanisms of psychiatric illnesses. If this is the case, then by studying alterations in pain sensitivity in psychiatric disorders, clues may appear that elucidate the relationship between pain sensitivity and emotional aspects of pain (Barlow 2001; Klossika et al. 2006).

### *6.6.1 Depression and Pain*

Depression involves the symptoms of depressive mood and loss of appetite, and it remarkably decreases the entire brain function. Depression has a high incidence, and it is caused by genetic and environmental factors. Patients with depression have dysfunctional thoughts, emotions, and motivations. While one can be depressed due to a long exposure to pain, one of the typical symptoms of depression is pain. A positive correlation between depression and chronic post-surgical pain is reported to exist, and the correlation indicates that postoperative pain prolongs depressive states (Hinrichs-Rocker et al. 2009). A factor that is considered responsible for depression is dysfunction in noradrenergic and serotonergic systems. Serotonergic and noradrenergic systems are important systems that contribute to the descending pain inhibitory system in the spinal cord; dysfunction in serotonergic and noradrenergic systems may affect sensitivity to pain (Kundermann et al. 2009).

### *6.6.2 Borderline Personality Disorder and Pain*

Borderline personality disorder is a mental illness that is known to be caused by emotional and cognitive abnormalities and characterized by frequently repeated self-injurious behavior. The main symptoms of borderline personality disorder are instability, impulsive actions, and feelings of depression accompanied by feelings of emptiness. In patients with borderline personality disorder, experimental pain

**Table 6.3** Association between psychiatric disorders with emotional and cognitive disturbances and pain thresholds

| Disease | Clinical symptoms | Sensitivity to pain | | Causes of abnormality | | |
|---|---|---|---|---|---|---|
| | | Experimental | Clinical | Sensory | Emotional | Cognitive |
| Schizophrenia | Cognitive decline, attention deficits, emotional disturbances, abnormal thinking | Reduced or no alterations | Reduced or Enhanced | Unclear | Alteration suggested | Alteration suggested |
| Depression | Decline in motivation, poor concentration, depressive mood | Enhanced | Enhanced | No alteration | Alteration | Unclear |
| Personality disorder | Impaired control of thoughts and emotions | Reduced | Reduced | Unclear | Alteration | Alteration |
| Pervasive developmental disorder | Communication impairments, social disorder, imagination failure | Reduced | Reduced or enhanced | Alteration | Alteration | Alteration |
| Posttraumatic stress disorder | Apathy, hyperarousal | Reduced | Enhanced | Unclear | Alteration | Alteration |
| Eating disorder | Abnormalities of eating behaviors, changes in body image, communication impairments | Reduced | Unclear | Alteration suggested | Unclear | Unclear |

thresholds are higher than normal, and there are no abnormal sensory-discriminative aspects. A functional brain imaging study on experimental heat pain has shown that patients with borderline personality disorder exhibit strong activation in the dorsolateral prefrontal cortex and weak activation in the amygdala and ACC. For this reason, the suppression of emotional aspects of pain is regarded as a cause of the repetition of self-injurious behavior (Schmahl et al. 2004).

### *6.6.3 Schizophrenia and Pain*

Schizophrenia is a syndrome the main symptoms of which are delusions and hallucinations. Schizophrenic patients show decompensation when facing adverse life events. Substantial evidence has shown that schizophrenic patients have less sensitivity to pain. For example, it is reported that patients with schizophrenia do not recognize pain as an alarm signal (Murakami et al. 2010). The mechanisms underlying the decrease in pain sensitivity are unclear, but a phenomenon linked with attention dysfunction and impaired cognitive function in schizophrenia, as some have reported it, is relevant. There are two groups of schizophrenic patients: those with an extremely high pain threshold and those with a normal one. Pain thresholds do not correlate with the amount of antipsychotic medication and disease severity, according to our latest knowledge. A previous study has ruled out the possibility that high pain thresholds in schizophrenia are caused only by antipsychotic drugs (Potvin and Marchand 2008).

Even though some schizophrenic patients have normal sensitivity to experimental pain, it tends to feel less unpleasant to them, due to an abnormality of the emotional aspect of pain. In a recent functional MRI study, presenting experimental heat pain to schizophrenic patients, as compared to healthy volunteers, has resulted in less activation of the insular cortex and the other areas that have roles for the emotional aspect of pain (de la Fuente-Sandoval et al. 2010). In addition, patients have exhibited many other abnormalities in the insula (Wylle and Tregellas 2010). An animal model has been developed in schizophrenia pain research, with an expectation to clarify the mechanisms underlying the abnormal sensitivity to pain.

### *6.6.4 Pervasive Developmental Disorder and Pain*

Developmental disorders are classified in complex ways. Autism, the most popular developmental disorder, exhibits three disorders: social disorders, communication disorders, and imagination disorders; behavioral disorders arise from these disorders. The communication impairments are mild in autism. Pervasive developmental disorders are those with relatively high performance. In epidemiological studies, the frequency of autism is reported to be as high as 1.7 %. Patients with autism spectrum disorders are known to have significant discomfort when they are touched; their sensory systems are generally sensitive, and more specifically, tactual and audiovisual systems are sensitive. Patients with autism have been subject to confusing reports on their sensitivity to pain; they have been reported to show hypersensitivity, to have normal sensation, and to have low sensitivity. It has been suggested that abnormal C-fiber pathways are explanatory of their sensitivity to pain, but the details of the explanation remains unrevealed. Mirror neurons are involved in the ability to sympathize with others, and a dysfunction in minor neurons may be responsible for autism. This explains the fact that autistic patients cannot empathize with the pain of others. Autism is also a syndrome, and its pathogenesis is influenced by a wide variety of biological backgrounds. It is possible that there are various types of these disorders of pain, such as hypersensitivity to pain due to abnormal neurotransmission or lower sensitivity to pain due to abnormal processing of the emotional aspects of pain.

### 6.6.5 Posttraumatic Stress Disorder (PTSD) and Pain

Posttraumatic stress disorder (PTSD) is a type of anxiety disorder. This disorder has been known since the 1970s as a syndrome associated with Vietnam veterans and rape victims. PTSD is caused after a serious, typically traumatic, experience, and it has three clusters of mixed symptoms: (1) symptoms of re-experiencing, e.g., nightmare flashbacks; (2) avoidance of trauma reminders and mental paralysis; and (3) hyperarousal, comorbid depression, alcohol and drug dependence, and anxiety. Generalized anxiety disorder is often comorbid with chronic pain (Asmundson and Katz 2009). Many patients with chronic pain may have comorbid PTSD. In imaging studies of patients with PTSD, the amygdala and the ACC are found to be hyperfunctional. In an experimental pain study of patients with PTSD, they feel more pain, while their sensitivity to experimental pain stimulation has declined. Thus, patients may fail to process the emotional aspects of pain normally (Defrin et al. 2008; Kraus et al. 2009).

### 6.6.6 Eating Disorder and Pain

An eating disorder is a disease that causes abnormal eating behavior. Patients with bulimia nervosa show overeating, while patients with anorexia nervosa show restricted eating. A variety of factors, such as genetic and social backgrounds, stress, and physical and mental changes during adolescence, are considered to be the causes of eating disorders. Children and young women are prone to eating disorders. It is reported that eating disorders change the functions of the ACC. There have been a few reports on the pain sensitivity of patients with eating disorders, and according to one of them, their pain sensitivity is reduced (Florin et al. 1988). The causes of the reduction of pain sensitivity is taken not to be a peripheral problem but to be a cognitive perceptual distortion (Klossika et al. 2006).

## 6.7 Conclusion

The conventional approach to pain treatment has been preoccupied with eliminating the cause of pain, because pain is regarded as a danger signal. However, the mechanisms of pain are more complex than has been imagined, and there is currently no pain relief drug that works for all kinds of pain. The current diagnostic system is insufficient to treat all patients suffering from pain. Currently, we have not established objective evaluation systems that use blood tests or physiological tests for pain. Furthermore, there is no system to evaluate the emotional aspects of pain. I expect that studies of the affective component of pain will greatly progress and advance also with regard to the affective component of pain, and drug discovery research will aim specifically at reducing pain suffering, resulting in a paradigm shift in pain treatment.

## Exercises

1. The figures below show the waveforms of evoked potentials.
   What are the latency and amplitude of each evoked potential?
   What fibers are stimulated?

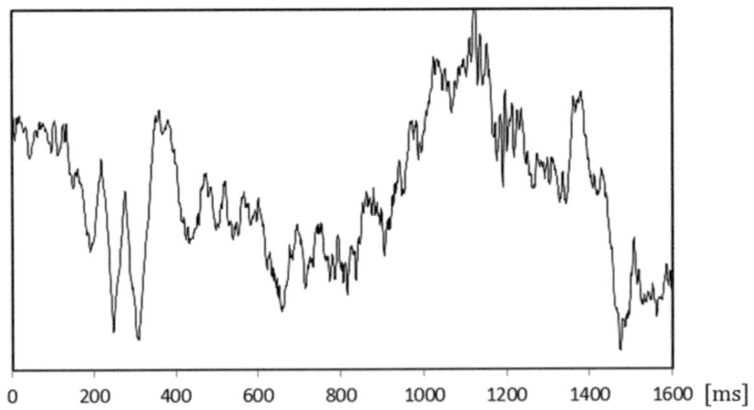

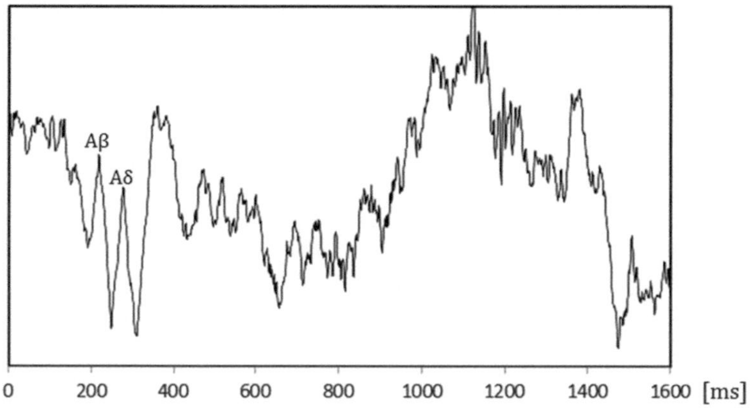


2. State the correlation between pain perception and cognitive impairment.
3. State the correlation between emotion and pain perception.
4. What do you think about the next generation of analgesics?

## References

Abram, S.E., Haddox, J.D.: The Pain Clinic Manual, 2nd edn. Lippincott, Williams, and Wilkins, Philadelphia (2000)

Asmundson, G.J., Katz, J.: Understanding the co-occurrence of anxiety disorders and chronic pain: state-of-the-art. Depress. Anxiety. **26**, 888–901 (2009)

Bantick, S.J., Wise, R.G., Ploghaus, A., Clare, S., Smith, S.M., Tracey, I.: Imaging how attention modulates pain in humans using functional MRI. Brain **125**, 310–319 (2002)

Baranauskas, G., Nistri, A.: Sensitization of pain pathways in the spinal cord: cellular mechanisms. Prog. Neurobiol. **54**, 349–365 (1998)

Barlow, D.H.: Clinical Handbook of Psychological Disorders: A Step-by-Step Treatment Manual, 3rd edn. The Guilford Press, New York (2001)

Beecher, H.K.: Pain in men wounded in battle. Ann. Surg. **123**, 96–105 (1946)

Besson, J.M., Chaouch, A.: Peripheral and spinal mechanisms of nociception. Physiol. Rev. **67**, 67–186 (1987)

Bonica, J.J.: Neurophysiologic and pathologic aspects of acute and chronic pain. Arch. Surg. **112**, 750–761 (1977)

Bruehl, S., Burns, J.W., Chung, O.Y., Chont, M.: Pain-related effects of trait anger expression: neural substrates and the role of endogenous opioid mechanisms. Neurosci. Biobehav. Rev. **33**, 475–491 (2009)

Chou, R., Shekelle, P.: Will this patient develop persistent disabling low back pain? JAMA **303**, 1295–1302 (2010)

de la Fuente-Sandoval, C., Favila, R., Gomez-Martin, D., Pellicer, F., Graff-Guerrero, A.: Functional magnetic resonance imaging response to experimental pain in drug-free patients with schizophrenia. Psychiatry Res. **183**, 99–104 (2010)

Defrin, R., Ginzburg, K., Solomon, Z., et al.: Quantitative testing of pain perception in subjects with PTSD – implications for the mechanism of the coexistence between PTSD and chronic pain. Pain **138**, 450–459 (2008)

Dworkin, R.H., Turk, D.C., Revicki, D.A., Harding, G., Coyne, K.S., Peirce-Sandner, S., Bhagwat, D., Everton, D., Burke, L.B., Cowan, P., Farrar, J.T., Hertz, S., Max, M.B., Rappaport, B.A., Melzack, R.: Development and initial validation of an expanded and revised version of the Short-form McGill Pain Questionnaire (SF-MPQ-2). Pain **144**, 35–42 (2009)

Florin, I., Franzen, U., Meier, M., Schneider, S.: Pressure sensitivity in bulimic women; a contribution to research in body image distortion. J. Psychosom. Res. **32**, 439–444 (1988)

Fone, K.C., Porkess, M.V.: Behavioural and neurochemical effects of post-weaning social isolation in rodents-relevance to developmental neuropsychiatric disorders. Neurosci. Biobehav. Rev. **32**, 1087–1102 (2008)

Franek, M., Vaculin, S., Yamamotova, A., Stastny, F., Bubenikova-Valesova, V., Rokyta, R.: Pain perception in neurodevelopmental animal models of schizophrenia. Physiol. Res. **59**, 811–819 (2010)

Hinrichs-Rocker, A., Schultz, K., Jarvinen, I., Lefering, R., Simanski, C., Neugebauer, E.A.: Psychosocial predictors and correlates for chronic post-surgical pain (CPSP) – a systematic review. Eur. J. Pain **13**, 719–730 (2009)

Hirofumi, M., Yuji, S.: Assessment of cognitive aspect of pain: development, reliability, and validation of Japanese version of pain catastrophizing scale. Jpn. J. Psychosom. Med. **47**, 95–102 (2007)

Jeon, D., Kim, S., Chetana, M., et al.: Observational fear learning involves affective pain system and Cav1.2 $Ca^{2+}$ channels in ACC. Nat. Neurosci. **13**, 482–488 (2010)

Klossika, I., Flor, H., Kamping, S., et al.: Emotional modulation of pain: a clinical perspective. Pain **124**, 264–268 (2006)

Kraus, A., Geuze, E., Schmahl, C., et al.: Differentiation of pain ratings in combat-related posttraumatic stress disorder. Pain **143**, 179–185 (2009)

Kundermann, B., Hemmeter-Spernal, J., Strate, P., et al.: Pain sensitivity in major depression and its relationship to central serotoninergic function as reflected by the neuroendocrine response to clomipramine. J. Psychiatr. Res. **43**, 1253–1261 (2009)

Lautenbacher, S., Huber, C., Schofer, D., et al.: Attentional and emotional mechanisms related to pain as predictors of chronic postoperative pain: a comparison with other psychological and physiological predictors. Pain **151**, 722–731 (2010)

Leeuw, M., Goossens, M.E., Linton, S.J., Crombez, G., Boersma, K., Vlaeyen, J.W.: The fear-avoidance model of musculoskeletal pain: current state of scientific evidence. J. Behav. Med. **30**, 77–94 (2007)

Loeser, J.D.: Perspectives on pain. In: Padgham, C., Hedges, A., Turner, P. (eds.) Clinical Pharmacology and Therapeutics, p. 314. University Press, Baltimore (1980)

Loeser, J.D.: Pain and suffering. Clin. J. Pain **16**, S2–S6 (2000)

Melzack, R.: The short-form McGill pain questionnaire. Pain **30**, 191–197 (1987)

Melzack, R., Torgerson, W.S.: On the language of pain. Anesthesiology **34**, 50–59 (1971)

Munro, G., Erichsen, H.K., Mirza, N.R.: Pharamacological comparison of anticonvulsant drugs in animal models of persistent pain and anxiety. Neuropharmacology **53**, 609–618 (2007)

Murakami, H., Tamasawa, N., Yamashita, M., et al.: Altered pain perception in schizophrenia. Lancet **375**, 864 (2010)

Nakae, A., Endo, K., Adachi, T., Ikeda, T., Hagihira, S., Mashimo, T., et al.: The influence of working memory capacity on experimental heat pain. J. Pain **14**(10), 1088–1096 (2013)

Narita, M., Kaneko, C., Miyoshi, K., et al.: Chronic pain induces anxiety with concomitant changes in opioidergic function in the amygdala. Neuropsychopharamacology **31**, 739–750 (2006)

Olausson, H., Lamarre, Y., Backlund, H., Morin, C., Wallin, B.G., Starck, G., et al.: Unmyelinated tactile afferents signal touch and project to insular cortex. Nat. Neurosci. **5**, 900–904 (2002)

Osman, A., Barrios, F.X., Kopper, B.A., Hauptmann, W., Jones, J., O'Neill, E.: Factor structure, reliability, and validity of the pain catastrophizing scale. J. Behav. Med. **20**, 589–605 (1997)

Park, S.H., Sonty, N.: Positive affect mediates the relationship between pain-related coping efficacy and interference in social functioning. J. Pain **11**, 1267–1273 (2010)

Potvin, S., Marchand, S.: Hypoalgesia in schizophrenia is independent of antipsychotic drugs: a systematic quantitative review of experimental studies. Pain **138**, 70–78 (2008)

Rachman, S., Hodgson, R.: I. Synchrony and desynchrony in fear and avoidance. Behav. Res. Ther. **12**, 311–318 (1974)

Rhudy, J.L., Meagher, M.W.: Fear and anxiety: divergent effects on human pain thresholds. Pain **84**, 65–75 (2000)

Rhudy, J.L., Bartley, E.J., Williams, A.E.: Habituation, sensitization, and emotional valence modulation of pain responses. Pain **148**, 320–327 (2010)

Ringkamp, M., Meyer, R.A.: Physiology of nociceptors. In: Basbaum, A.I., Bushnell, C. (eds.) Science of Pain, pp. 105–107. Amsterdam, Elsevier (2008)

Schacter, D.L., Gilbert, D.T., Wegner, D.M.: Psychology, p. 310. Worth Publishers, New York (2011)

Schmahl, C., Greffrath, W., Baumgartner, U., et al.: Differential nociceptive deficits in patients with borderline personality disorder and self-injurious behavior: laser-evoked potentials, spatial discrimination of noxious stimuli, and pain ratings. Pain **110**, 470–479 (2004)

Stander, S., Steinhoff, M., Schmelz, M., Weisshaar, E., Metze, D., Luger, T.: Neurophysiology of pruritus: cutaneous elicitation of itch. Arch. Dermatol. **139**, 1463–1470 (2003)

Sullivan, M.J.L., Bishop, S.R., Pivik, J.: The pain catastrophizing scale: development and validation. Psychol. Assess. **7**, 524–532 (1995)

Suzuki, T., Amata, M., Sakaue, G., et al.: Experimental neuropathy in mice is associated with delayed behavioral changes related to anxiety and depression. Anesth. Analg. **104**, 1570–1577 (2007)

Treede, R.D., Meyer, R.A., Campbell, J.N.: Myelinated mechanically insensitive afferents from monkey hairy skin: heat-response properties. J. Neurophysiol. **80**, 1082–1093 (1998)

Turk, D.C., Audette, J., Levy, R.M., Mackey, S.C., Stanos, S.: Assessment and treatment of psychosocial comorbidities in patients with neuropathic pain. Mayo Clin. Proc. **85**, S42–S50 (2010)

Villemure, C., Bushnell, M.C.: Mood influences supraspinal pain processing separately from attention. J. Neurosci. **29**, 705–715 (2009)

Wylle, K.P., Tregellas, J.R.: The role of the insula in schizophrenia. Schizophr. Res. **123**, 93–104 (2010)

Yoshino, A., Okamoto, Y., Onoda, K., et al.: Sadness enhances the experience of pain via neural activation in the anterior cingulate cortex and amygdala: an fMRI study. NeuroImage **50**, 1194–1201 (2010)

Zigmond, A.S., Snaith, R.P.: The hospital anxiety and depression scale. Acta Psychiatr. Scand. **67**, 361–370 (1983)

# Chapter 7
# Magnetic Resonance Imaging (MRI) and Magnetic Resonance Spectroscopy (MRS)

**Yuki Mori, Ikuhiro Kida, Haruyuki Fukuchi, Masaki Fukunaga, and Yoshichika Yoshioka**

**Abstract** Magnetic resonance imaging (MRI) has a wide range of applications in medical diagnosis and preclinical research. MRI was invented about 40 years ago, and there are currently estimated to be over 25,000 scanners in the world. In general, contrast agents are not necessary for MRI and the soft tissue contrast of MRI is better than other imaging techniques. The important point is that MRI intensity depends on not only the concentration but also physico-chemical properties of molecules in tissues. In the first part of this chapter, several kinds of MRI techniques are described. Magnetic resonance spectroscopy (MRS) is an application of magnetic resonance. The second part of this chapter is concerned with MRS. This technique provides information in metabolism non-invasively, and obtains spectra from a region of interest two- and three-dimensionally. Some physiological parameters, such as pH and temperature, can be estimated by the spectra. Applications of MRI and MRS are very broad, since many factors affect MRI signals. Functional MRI (fMRI) is an important application used widely in the neurosciences, human sciences, and economics, as well as in medical sciences. The major restriction of MRI is its long scan time. An accelerated technique is described in the last part of the chapter.

**Keywords** Magnetic resonance imaging • Magnetic resonance spectroscopy • fMRI • Cell tracking • Compressed sensing

## 7.1 Introduction of Magnetic Resonance

The magnetic resonance method to measure the nuclear magnetic moment was discovered in the laboratory of Rabi (Rabi et al. 1938). This discovery was made in the context of molecular beam experiments in which individual, isolated atoms or

---

Y. Mori • I. Kida • H. Fukuchi • M. Fukunaga • Y. Yoshioka (✉)
Immunology Frontier Research Center, Osaka University, Suita, Osaka, Japan

Center for Information and Neural Networks, National Institute of Information and Communications Technology, Suita, Osaka, Japan
e-mail: yoshioka@fbs.osaka-u.ac.jp



M. Kasaki et al. (eds.), *Cognitive Neuroscience Robotics B*,
DOI 10.1007/978-4-431-54598-9_7

molecules were the objects of investigation. The nuclear magnetic resonance (NMR) in a bulk sample was first reported independently by two groups (Purcell et al. 1946; Bloch et al. 1946). The chemical shift, which measures resonance frequency variations according to the nuclear chemical environment, was subsequently discovered (Arnold et al. 1951), and the NMR method came to be used for chemical structure analysis and research into physical properties. In 1971, extensions of longitudinal and transverse relaxation times, which are physical properties of the NMR phenomenon, were discovered in pathological tissues and compared with normal tissues (Damadian 1971). This suggested the possibility of applying NMR for pathological analysis. In 1973, the first MR image was reported (Lauterbur 1973). This was the starting point of magnetic resonance imaging (MRI) techniques. During the 1970s and 1980s, the development of MRI for clinical use in medical practice was pushed forward extensively by universities and companies. As a result, image quality improved dramatically and clinical scanners proliferated worldwide. Overall, this rapidly growing field encompasses an array of disciplines, and many new developments continue to emerge. The introduction of MRI and magnetic resonance spectroscopy (MRS) to brain research has dramatically altered our understanding of the brain in recent decades. Today, MRI and MRS techniques have become indispensable tools for both the clinical diagnosis of brain disorders and basic neuroscience research. Recent developments in MRI and MRS (i.e., ultra-high field MRI, multi-channel RF systems, etc.) allow measuring high resolution in vivo human brain images on the order of several hundred microns; however, such high resolution MRI requires unreasonably long scanning times for whole brain acquisition. Over the last decade, efficient data sampling schemes and high precision data reconstruction methods (i.e., sparse sampling or compressed sensing) were introduced in MRI, which brought about a dramatic reduction in acquisition times.

In this chapter, we discuss the basics of MRI and MRS, especially for brain research, and compressed sensing in MRI.

## 7.2 Anatomical Information Obtained by MRI

The magnetic resonance imaging (MRI) method is used to depict the anatomical structure of the brain non-invasively. Various brain structures, such as the corpus callosum, pituitary, medulla, cerebellum, caudate nucleus, putamen, and so on, are easily identified on the magnetic resonance images in Fig. 7.1. The corpus callosum displays low signal intensity, and the cerebrospinal fluid around the medulla and cerebellum displays high intensity.

MRI signal intensity is determined by the concentration and physico-chemical properties of water in the region of interest. For example, the signal intensity of the cerebrospinal fluid is higher than that of brain tissues because the water concentration of the former is higher than that of the latter. The concentration and physico-chemical properties of water depend not only on its structure but also on many other factors, such as inflammation, edema, lipid content, susceptibility, and so on. For this reason, MRI is sensitive to pathological changes in disease. MRI has become a

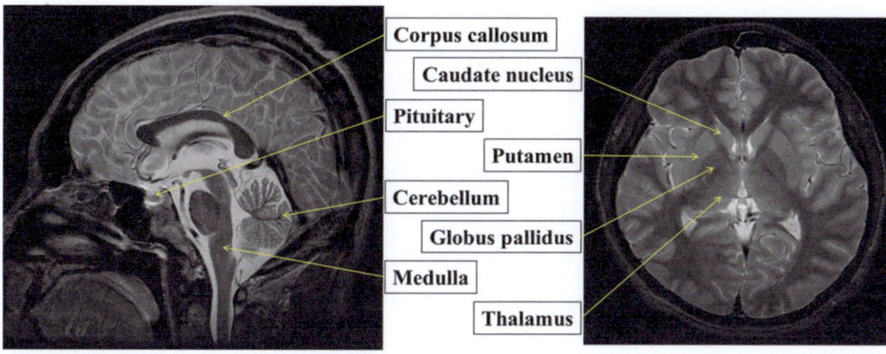


**Fig. 7.1** MRI anatomy of the normal adult human head

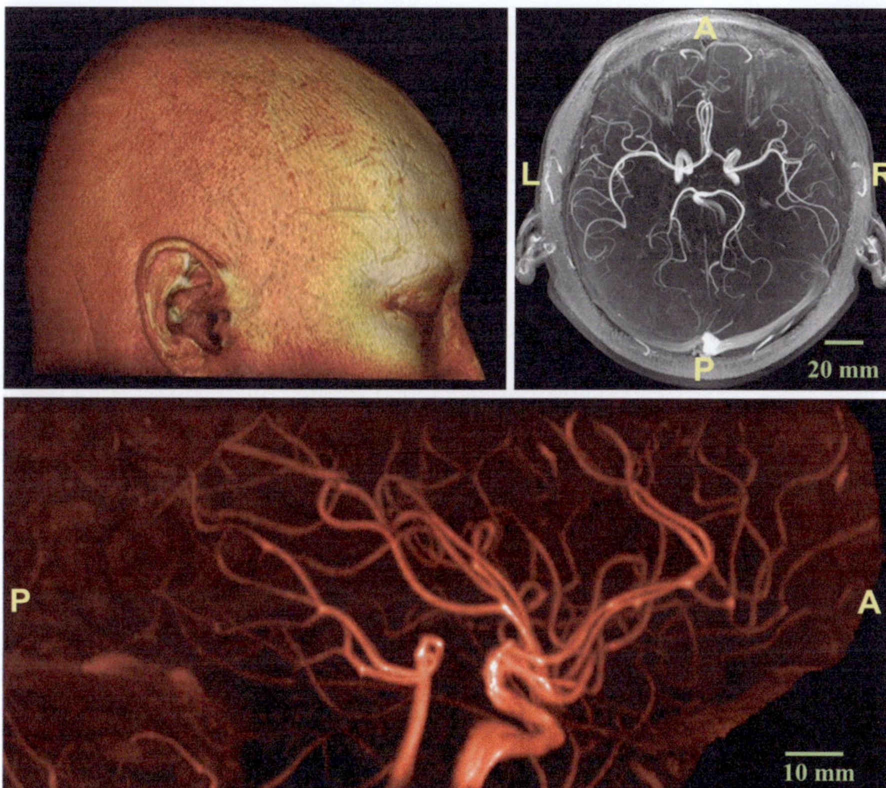


**Fig. 7.2** Blood vessels in the healthy human brain. *Upper right*: top-down view of brain blood vessels. *Lower*: side view of brain blood vessels. Nearly all vessels visualized in this case are arteries. Veins can be visualized by other MRI techniques (MR venography or susceptibility weighted image (SWI))

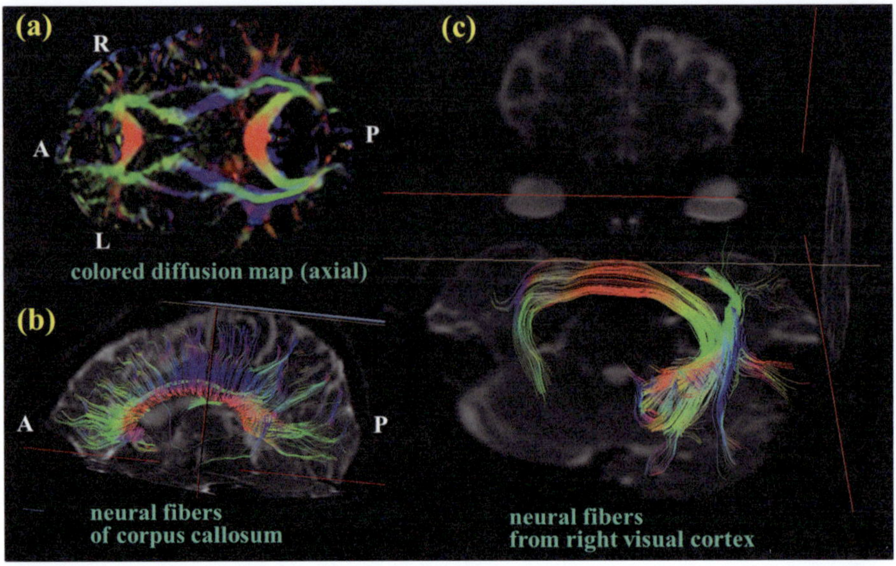


**Fig. 7.3** Diffusion anisotropy map and neural fiber tracking of the human brain (anatomical connectome). Tracking was performed using information on the anisotropic diffusivity of water molecules in neural fibers

necessary diagnostic method and has been used in many clinical and pre-clinical studies. High resolution sub-millimeter images are obtained with modern 3 T MRI scanners.

MRI signal intensity also depends on the flow of water, typically the flow of blood in the body. Flowing blood can be visualized by MRI, as shown in Fig. 7.2 (Haacke et al. 1999). Blood flows through blood vessels. An important point is that blood vessels can be visualized using MRI without contrast reagents. Blood vessel imaging is called "magnetic resonance angiography" (MRA). MRA has been employed in clinical and pre-clinical studies.

Diffusion weighted imaging (DWI) is a type of MRI that provides information on the diffusivity of water molecules (Brownian motion, not flow) from MRI signal intensity (Le Bihan et al. 1986). Since the diffusivity of water molecules changes markedly immediately after the onset of a disease, e.g., cerebral infarction (stroke), DWI provides an important way to assess acute stroke. The major cause of the change in the DWI signal during ischemia is cell swelling, i.e. the movement of water molecules from the interstitial space (high diffusivity) into cells (low diffusivity) (Moseley et al. 1990). Since water diffusion also significantly decreases in malignant tissues, DWI is considered to have the potential to detect malignant cancers.

Information on the three-dimensional diffusivity of water molecules can be added to DWI data (Basser et al. 1994). Since water molecule diffusion is anisotropic in muscle and neural fibers, the running directions of these fibers can be visualized by the information from anisotropy (Fig. 7.3a). The colors indicate the

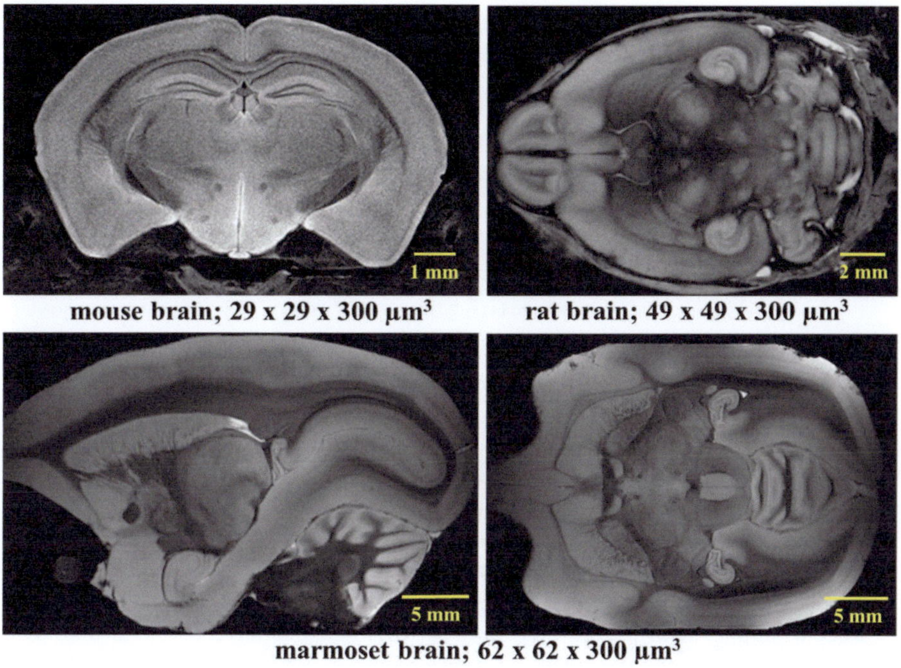


**Fig. 7.4** High resolution brain images of a mouse, rat, and marmoset. These fine images were obtained non-destructively using an 11.7 T MRI scanner

directions of neural fibers as follows: red: left-right, green: anterior-posterior, and blue: top-down. These fibers can also be tracked by tracing the most diffusive directions in the pixels (Mori and Barker 1999; Mori 2007). Figure 7.3b, c show the tracking of neural fibers in the human brain. The meaning of the colors is the same as in Fig. 7.3a.

One of the major limitations of the past MRI techniques was their spatial and temporal resolution. The resolution of MRI is determined by the signal to noise ratio (SNR), as in other imaging methods. Recently, ultra-high magnetic field MRI scanners (higher than 7 T) have been introduced, and they provide fine MR images at cellular resolution with animal scanners and at several hundred micrometers with human scanners. Ultra-high field scanners produce new tissue contrasts and fine vascular images. Tissues and organs of laboratory animals are sometimes fixed and stained in order to increase resolution and contrast (the scan time can be increased as well). Figure 7.4 shows MRI images of ex vivo animal brains provided by an 11.7 T MRI scanner.

These fine anatomical images are usable for histological assessments. In addition, since contrast agents of MRI are applicable in both in vivo and ex vivo studies, precise histological information can be obtained (Zhang et al. 2010). Magnetic resonance histology (or MRI histology) is one of the new applications of MRI for the study of tissues and organs.

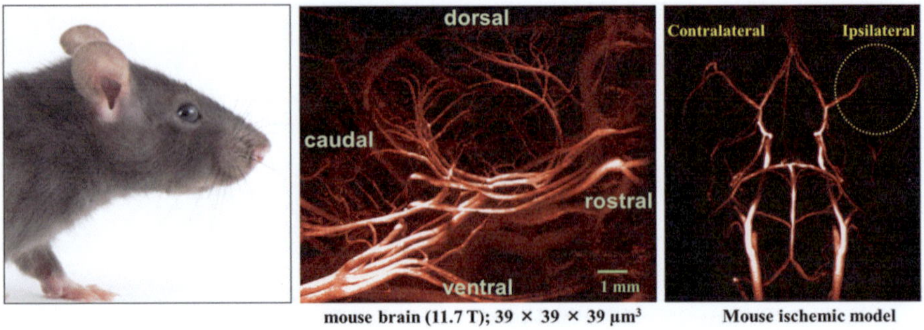


**Fig. 7.5** Visualization of blood vessels in a living mouse brain. *Center image*: side view of brain blood vessels (normal mouse). *Right image*: top-down view of brain blood vessels (the right MCA was occluded)

MR angiography (MRA) has been used in mice, as is shown in Fig. 7.5. Highly sensitive MRI scanners and optimized probe coils make it possible to visualize blood vessel trees in detail, even in the small mouse brain. They can be used to assist in the diagnosis of neovascularization and degenerative vascular changes following traumatic brain injury, ischemia, dementia, and so on.

Recent studies with a high-field animal MRI scanner have revealed that with specific contrast agents, the cells of a living mouse can be labeled and traced at the single cell level, and the distribution of labeled cells can be visualized (Mori et al. 2011; Mori and Yoshioka 2012).

Though it is a longstanding assumption that the central nervous system (CNS) is an "immune privileged site", some reports have demonstrated that immune cells in the CNS play an important role in neuroinflammatory diseases, such as multiple sclerosis and ischemic injury. However, the dynamic crosstalk between the CNS and immune cells has not been characterized by non-invasive imaging techniques. The non-invasive visualization of immune cell dynamics and immunological responses in vivo before/after neuroinflammatory conditions may lead to a greater understanding of the mechanistic underpinnings of CNS diseases and their repair. As a first step, it is important to elucidate immune cell dynamics in the CNS. MRI at 11.7 T provides a way to visualize immune cell dynamics and immune responses non-invasively. Figure 7.6 shows the accumulation of immune cells (black dots) in and around an inflammatory brain lesion (hyperintensity region). These images are obtained before and after brain injury. Labeled immune cells are detected in and around the injured region. Each small dark spot corresponds to a single cell.

In conclusion, the MRI techniques that are recently developed can adequately provide precise morphological information at the cellular and tissue levels, and such information is useful for pre-clinical studies. Many developments of MRI aim at overcoming its limitations in resolution, contrast and specificity, and MRI image quality and information precision are increasing day by day.

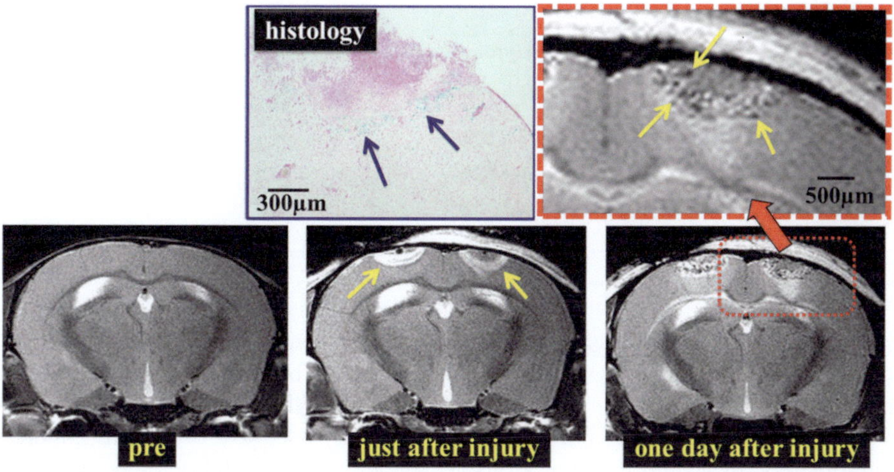


**Fig. 7.6** MRI image 1 day after injury shows the accumulation of labeled immune cells in and around the injured region

# 7.3 Magnetic Resonance Spectroscopy

## *7.3.1 Introduction*

Magnetic resonance imaging (MRI) and spectroscopy (MRS) are both based on the properties of spins in atomic nuclei. MRS is an analytical method used for the identification of chemical compounds and molecular interactions in physics and chemistry. MRS is also used in biological and biochemical studies to quantify and characterize biomolecules, and it can even be used in vivo. Proton MRS ($^{1}$H) is often used in biological studies because of the high natural abundance of hydrogen atoms in biological systems and the high sensitivity of the proton signal in this method (Table 7.1). While information provided by MRI is based on the water content, molecular information gathered by MRS is based on the chemical environment of nuclei.

MRS has some advantages over MRI. MRS can directly and specifically measure metabolite and neurotransmitter concentrations in the cells of brain and muscle tissues. Spectra detected by MRS have several peaks whose positions depend on the electron environment of chemical functional groups in molecules. This dependency is called a "chemical shift." MRS provides metabolic and physiological information, such as cell density, viability, energy metabolism, temperature, and pH, and such information can be helpful in disease diagnosis. On the other hand, MRI has some advantages over MRS. MRI is generally used to detect proton nuclear signals and the water content. When water is abundant in biological systems, MRI has high sensitivity and high spatial and temporal resolution. $^{1}$H MRS detects low concentrations of metabolites, ranging from 1 to 15 mM; this low as compared with water (~40 M) in biological systems. Since MRS has low sensitivity, it can only yield a

**Table 7.1** MRS properties of nuclei

| Isotope | Spin | Natural abundance (%) | Gyromagnetic ratio ($10^7$ rad/T/s) | NMR frequency at 2.35 T | Relative sensitivity |
|---|---|---|---|---|---|
| $^{1}H$ | 1/2 | 99.985 | 26.752 | 100.000 | 1.00 |
| $^{13}C$ | 1/2 | 1.108 | 6.728 | 25.145 | $1.76 \times 10^{-4}$ |
| $^{17}O$ | 5/2 | 0.037 | −3.628 | 13.562 | $1.08 \times 10^{-5}$ |
| $^{19}F$ | 1/2 | 100 | 25.181 | 94.094 | 0.834 |
| $^{23}Na$ | 3/2 | 100 | 7.080 | 26.466 | $9.27 \times 10^{-2}$ |
| $^{31}P$ | 1/2 | 100 | 10.841 | 40.481 | $6.65 \times 10^{-2}$ |

sufficient signal from a single voxel beyond several centimeters if used for several minutes; detection of a signal from multiple voxels beyond several centimeters takes several tens of minutes. For example, MRI could yield an anatomical image of a whole brain (roughly 20 cm in diameter, with a voxel size of several hundred micrometers) in several minutes. However, it can provide a highly temporal image of a whole brain (with a voxel size of several millimeters) in several hundredths of a millisecond. $^{1}H$ and $^{31}P$ MRS are used in clinical applications to measure metabolites *in vivo*.

Metabolite and neurotransmitter signals on MRS appear as peaks at different positions depending on the chemical shift. Since all the metabolic intermediate compounds contain hydrogen, carbon, and phosphorus nuclei, MRS in principle can detect all the metabolic compounds at different positions on the spectrum and measure their concentrations as signal peak areas. However, it is difficult to detect most metabolic intermediates if their concentrations are too low at a specific chemical shift or if there is overlap between peaks of different metabolites. The use of MRS is mostly limited to the brain, where there are few motion artifacts and low amounts of lipids.

### 7.3.2 Proton MRS

Proton MRS can simultaneously display the resonance peaks of metabolites and neurotransmitters in biological systems, such as the brain (Fig. 7.7). Since metabolites and neurotransmitters contain hydrogen nuclei, $^{1}H$ MRS is a powerful tool for detecting a large number of metabolites in vivo (Table 7.2). $^{1}H$ MRS of the brain detects a spectrum of metabolites, including *N*-acetyl aspartate, creatine and phosphocreatine, choline, myo-inositol, glutamate, and glutamine (Ross and Blüml 2001).

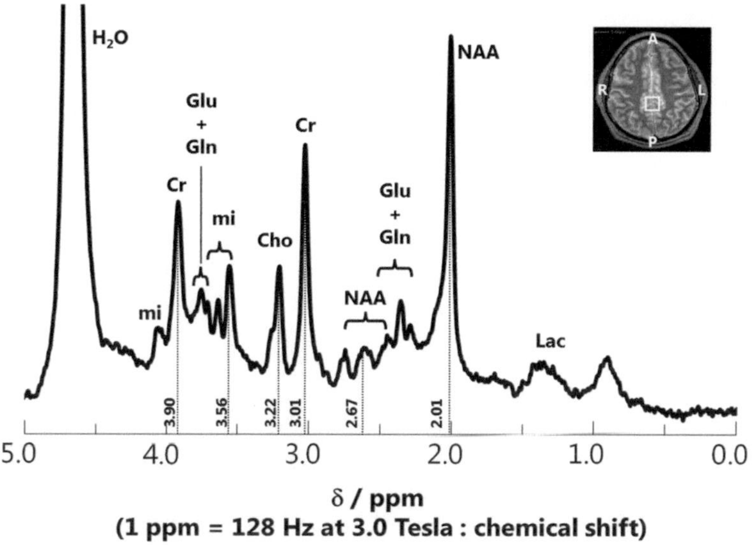


**Fig. 7.7** Proton MR spectrum obtained from human brain at 3.0 T (TR/TE = 3000 ms/30 ms; 2 cm × 2 cm × 2 cm) with partial water suppression. The voxel positions are indicated in the MR image. *Cr* total creatine, *mi* myo-inositol, *Cho* choline, *Glu* glutamate, *Gln* Glutamine, *NAA* N-acetyl aspartate, *Lac* lactate

**Table 7.2** Proton MRS – observed metabolites

| Metabolite | Main resonance (ppm) | Concentration in human brain (mM) | Significance |
|---|---|---|---|
| Lactate | 1.33 | 0.2–1.0 | Anaerobic glycolysis |
| NAA | 2.01 and 2.49 and 2.67 | 8–12 | Neuronal and axonal density and viability |
| Glutamate | 2.0–2.5 | 6–12 | Neurotransmitter and TCA cycle |
| Glutamine | 2.0–2.5 | 1.5–6.0 | Neurotransmitter and TCA cycle |
| GABA | 2.0–2.5 | 1.0–2.0 | Neurotransmitter and TCA cycle |
| Total Cr | 3.01 and 3.90 | 4.5–10 | Energy metabolism |
| Cho | 3.22 | 0.5–2.5 | Membrane metabolism and cell proliferation |
| Myo-inositol | 3.56 | 4.0–9.0 | Glial marker |

Hetherington et al. (1994), Pan et al. (1998), Pouwels and Frahm (1998), and Wang and Li (1998)

#### 7.3.2.1 *N*-Acetyl Aspartate

As Fig. 7.7 illustrates, *N*-acetyl aspartate (NAA) generates the most prominent peak at 2.01 ppm from a $CH_3$ group and smaller peaks at 2.49 and 2.67 ppm from a $CH_2$

group. NAA is synthesized from L-aspartate and acetyl-Co-A by the enzyme L-aspartate-N-acetyl-transferase. This enzyme, which is only detectable in the nervous system, is found in the mitochondria (Goldstein 1969) and cytoplasm of neurons. Since NAA is specifically and stably localized in neurons and axons of the adult brain, NAA is considered to be a marker of neuronal density and viability (Moffett et al. 2007; Signoretti et al. 2001). NAA is also found in the progenitor cells of oligodendrocytes and astrocytes (Urenjak et al. 1992) and increases in concentration during the maturation of the cerebellum, thalamus, and grey matter. Care is thus needed when NAA is used as a marker for neurons and axons during maturation. NAA is also used as a reference for other metabolite concentrations. Since NAA concentrations vary in different regions of the brain (Pouwels and Frahm 1998; Wang and Li 1998), care is needed when using NAA as a reference. A decrease in NAA concentration reflects degeneration of neurons and axons, and may indicate mitochondrial dysfunction. Mitochondrial dysfunction may occur in neurological disorders, such as epilepsy, hypoxia and multiple sclerosis (MS) (Signoretti et al. 2001). The physiological functions of NAA have not been fully elucidated, even though the localization of NAA to neurons is well known. Suggested physiological functions include (i) osmoregulation, i.e., removing water from neurons and counter anions in neurons (Taylor et al. 1995), and (ii) serving as a precursor for the production of N-acetyl aspartic glutamate (Moffett et al. 2007).

#### 7.3.2.2 Creatine and Phosphocreatine

The primary peaks at 3.01 ppm from a $CH_3$ group and at 3.9 ppm from a $CH_2$ group represent the combination of two compounds: phosphocreatine (PCr) and creatine (Cr); the combination is referred to as 'total Cr.' Since both compounds relate to ATP reserves, each of these peaks is regarded as a marker of energy metabolism (Ross and Blüml 1996). Since total Cr is relatively stable in the human brain at 6–8 mM (regardless of age), total Cr is often used as an internal reference for the relative concentrations of other brain metabolites. The ratio of metabolites to total Cr can be calculated (e.g., NAA/Cr) (Li et al. 2003). However, since the total Cr concentration varies with brain region (higher in the gray matter than in the white matter) (Pouwels and Frahm 1998), care should be taken when using total Cr concentrations as a reference. In addition, total Cr increases during trauma and decreases in hypoxia, stroke, and tumors.

#### 7.3.2.3 Choline

The peak at 3.22 ppm represents the combination of choline and choline-containing compounds, such as phosphocholine and glycerophosphocholine. Choline and its metabolites have three main physiological purposes: providing structural integrity, signaling in cell membranes, and facilitating cholinergic neurotransmission. Choline is a precursor of the neurotransmitter acetylcholine, which has many functions, including memory and muscle control. Choline is also a precursor of phosphatidylcholine, a major constituent of cell membranes (myelin sheaths in particular). Phospholipids are released during myelin breakdown, leading to an increase in choline concentration. Choline is thus a marker of cell membrane turnover. The concentration of choline is higher in the white matter than in the grey matter (Pouwels and Frahm 1998), and the concentration of choline in glial cells can be twice as high as that in neurons (Gill et al. 1989). The highest levels of choline are present at birth, and decrease with age. In adults, an increase in choline is associated with Alzheimer's disease, chronic hypoxia, and epilepsy.

#### 7.3.2.4 Myo-inositol

The main peak at 3.56 ppm and smaller peaks at 3.61 ppm and 4.05 ppm from a CH group include 80 % myo-inositol and 10–20 % inositol-monophosphate. Myo-inositol occurs primarily in glial cells and is incapable of crossing the blood-brain barrier; thus, this compound is a specific glial marker (Brand et al. 1993). Myo-inositol is abundant in the newborn brain and decreases during brain development. In the adult human brain, an increase in myo-inositol is regarded to indicate glial proliferation or an increase in glial cell size, both of which occur during inflammation (Soares and Law 2009). Myo-inositol also increases in Alzheimer's disease and diabetes mellitus, and decreases in stroke, tumor, and hypoxia encephalopathy.

#### 7.3.2.5 Glutamate, Glutamine, and γ-Aminobutyric Acid (GABA)

The peaks between 2.0 and 2.5 ppm are signals from the combination of protons in glutamate, glutamine, and GABA (β- and γ-$CH_2$ for glutamate and glutamine; β- and α-$CH_2$ for GABA). The peaks between 3.6 and 3.8 ppm are signals from the combination of protons (α-CH) in glutamate and glutamine. The peak at 3.00 ppm is GABA from γ-$CH_2$, but it is usually undetected because of the overlap with a large total Cr peak. Measurement at a higher magnetic field strength can improve the quantification of these compounds.

Glutamate is the most abundant excitatory neurotransmitter in the brain. It has been shown to increase in patients with multiple sclerosis (in both demyelinating lesions and apparently normal white matter) (Srinivasan et al. 2005). Glutamine is an intermediate molecule produced from glutamate in astrocytes; thus, this molecule serves as an astrocyte marker. Glutamine may increase in hepatic encephalopathy, hypoxic-ischemic events, and hyperosmolar states, and decrease in Alzheimer's disease and hyponatremia. GABA is the major inhibitory neurotransmitter in the brain. GABA is produced by the glutamate decarboxylase from glutamate in the mitochondria. GABA is associated with several neurological and psychiatric disorders, including depression and epilepsy. The glutamate-glutamine (and/or GABA) cycle exists in pre-synaptic and astroglial cells. Glutamate (or GABA) is released from the pre-synaptic into the cleft. After interaction with post-synaptic cells, glutamate is taken by the Na/K pump into the astroglia, where it is then converted to glutamine by glutamine synthetase. Glutamine is transported back to neurons, where it is converted to glutamate. This glutamate-glutamine cycle is similar to the GABA-glutamine cycle between neurons and astroglia (Magistretti and Pellerin 1999).

#### 7.3.2.6 Lactate

Lactate appears at 1.33 ppm but is usually not observed in normal brain spectra. Lactate is associated with glucose metabolism and measured immediately after stimulus onset in the visual cortex (Prichard et al. 1991; Frahm et al. 1996). Lactate is mainly detected in patients with stroke, tumor, hypoxia, and anoxia.

### *7.3.3 Phosphorous MRS*

Phosphorous ($^{31}$P) MRS is used to measure high energy phosphorus compounds, such as adenosine triphosphate (ATP), phosphocreatine (PCr), and inorganic phosphate ($P_i$). These compounds relate to energy metabolism in the brain, muscle, liver, and heart. $^{31}$P MRS provides high quality spectra in several minutes because of its relatively high sensitivity (approximately 7 % of $^{1}$H MRS) and high natural abundance of phosphorus in the body. Since $^{31}$P MRS has a relatively large chemical shift difference (e.g., 30 ppm compared to 10 ppm in $^{1}$H MRS), its spectral resolution is high even in the low magnetic fields used in clinical applications (Fig. 7.8).

The peaks of the main metabolites are measured by $^{31}$P MRS as follows: phosphomonoesters (PME) (6.5 ppm), $P_i$ (4.9 ppm), phosphodiesters (PDE) (2.6 ppm), PCr (0 ppm); ATP has three peaks: −2.5 (γ), −7.5 (α), and −16.3 (β) ppm. The peaks of adenosine diphosphate (ADP) at −7.1 (α) and −3.1 (β) ppm overlap with the ATP peaks. Nicotinamide adenine dinucleotide (NADH/NAD) has an additional peak at −8.3 ppm. Cellular energy metabolism is represented by ATP, PCr, and $P_i$, with PCr serving as a high energy phosphate storage compound in brain and muscle tissues. The PDE and PME compounds are related to membrane phospholipids. An increase in PME has been associated with rapid tissue growth or rapid membrane synthesis. PME has also been found to increase in the infant brain and in tumors. PCr resonance is conventionally used as an internal chemical shift reference between ATP and $P_i$. In addition, since the chemical shift of the peak in $P_i$ is pH dependent, the chemical shift between $P_i$ and α-ATP or PCr is commonly used to estimate intracellular acidity (pH) *in vivo*.

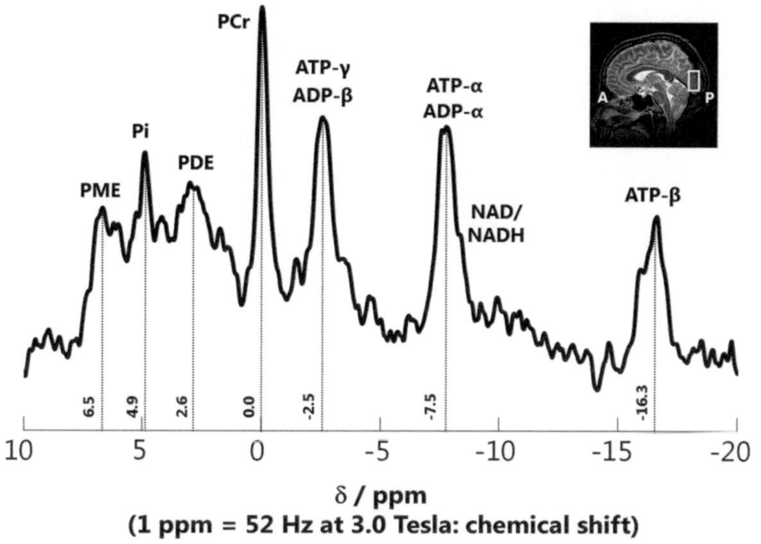


**Fig. 7.8** Phosphorous MR spectrum obtained from human brain at 3.0 T (TR/TE = 1000 ms/ 2.5 ms; 4 cm × 4 cm × 2 cm). The voxel positions are indicated in the MR image. *PME* phosphomonoesters, $P_i$ inorganic phosphate, *PDE* phosphodiester, *PCr* phosphocreatine, *ATP* adenosine triphosphate, *ADP* adenosine diphosphate, and *NAD/NADH* nicotinamide adenine dinucleotide

### *7.3.4 Carbon MRS*

If a biological system has a low natural abundance of $^{13}$C and the sensitivity of $^{13}$C detection (1.1 %; on the order of $10^{-4}$) is low, measurements can take several tens of minutes. For this reason, clinical application of $^{13}$C MRS is difficult. However, since the chemical shift difference is large (300 ppm, with specific peaks), the spectral resolution of $^{13}$C MRS is high and chemical shifts are easily assigned. Furthermore, in contrast to $^{31}$P MRS, $^{13}$C MRS can detect a large number of metabolites. Because of its low sensitivity, $^{13}$C MRS requires administered isotopes, such as $^{13}$C-glucose and $^{13}$C-acetate. $^{13}$C-glucose enters the normal metabolic pathway of $^{12}$C-glucose and has no toxic effects, allowing for the assessment of metabolic changes. $^{13}$C-glucose can be used to determine the rate of the tricarboxylic acid cycle in relation to the cerebral metabolic rates of oxygen and glucose.

### *7.3.5 Summary*

In the last few years, human MR systems with magnetic field strengths higher than 3.0 T have been developed. The development of coils and optimization of the RF pulse have led to greater sensitivity, faster acquisition, and higher spatial resolution for the clinical use of MR systems. Spectral resolution has been improved by stronger magnetic fields (better spectral separation between peaks). Concerns have been raised with regard to increased magnetic fields: inhomogeneity, structure-activity relationship limitations, and the problem of shimming. Practical and precise localization techniques are crucial for achieving the best field homogeneity and for effective water suppression. Despite these disadvantages, MRS is a powerful and promising tool for providing metabolic and viability data of biological systems. MRS uses the same system as MRI and provides structural and anatomical information.

## 7.4 Functional Magnetic Resonance Imaging (fMRI)

### *7.4.1 BOLD fMRI*

Magnetic resonance imaging (MRI) methods can clearly depict the morphological information both in clinical and in pre-clinical studies (Fig. 7.1). MRI is a useful and innovative imaging modality for non-invasive and simultaneous imaging of both morphological changes and cellular migration, even if the regions of interest are located in deep tissues and organs.

Functional magnetic resonance imaging (fMRI) of the brain is a non-invasive technique for measuring brain activity using MRI signal changes associated with brain neuronal activities. An example of fMRI during visual stimulation is shown in Fig. 7.9.

The most widely used method is based on the mechanism of BOLD (blood oxygenation level-dependent) contrast. It works by detecting changes in the blood oxygenation level that occur in response to local neural activities (Ogawa et al. 1990; Ogawa and Sung 2007). Since the electronic spin state of heme iron in deoxyhemoglobin is 2 (high spin state), deoxyhemoglobin is paramagnetic (Pauling and Coryell 1936) and has an effect on the MRI signal. Deoxyhemoglobin is a naturally occurring contrast agent for MRI. Oxyhemoglobin, on the other hand, has no effect on MRI parameters because the spin of heme iron is zero (low spin state). Therefore, the MRI signal depends on the blood oxygenation level. This magnetic property of blood was reported by Thulborn et al. in 1982 in an in vitro blood experiment. The blood oxygenation level changes during brain activation due to an increase in oxygen usage and a change in blood flow. The regional cerebral blood flow is changed by neurovascular coupling and is affected by the metabolites produced by neurons and glial cells during activation. In general, since the oxygen supply by blood flow overcompensates the oxygen demand of activated neurons, the blood oxygenation level increases during brain activation, and the MRI signal intensity increases. BOLD-fMRI was originally designed to detect activated brain areas using the rest-task protocol shown in Fig. 7.9b. fMRI can be used to produce activation maps that show which parts of the brain are involved in a particular

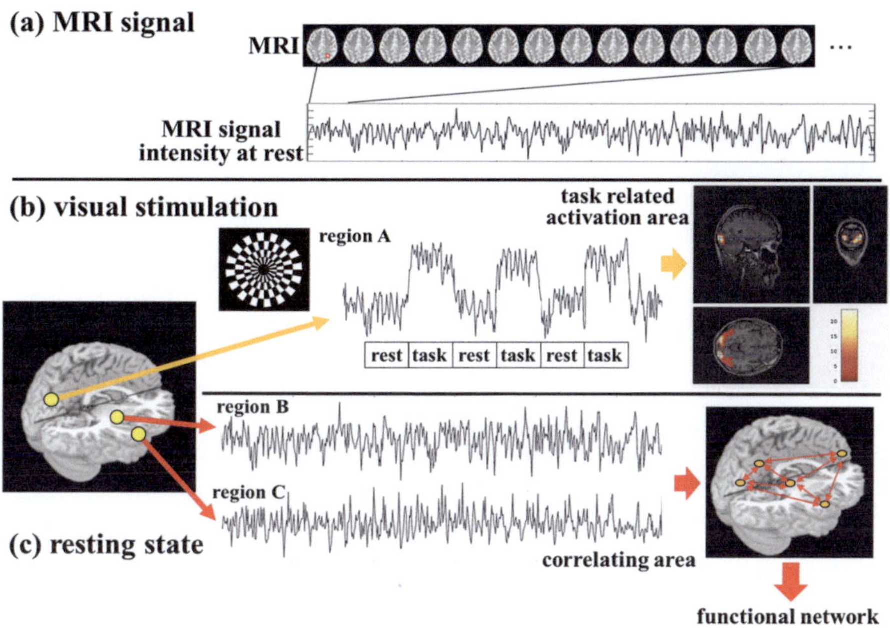


**Fig. 7.9** (**a**) MRI and MRI signal fluctuation with time. This fluctuation contains physiological modulations as well as artificial noise. (**b**) Task-related fMRI. (**c**) Resting state fMRI. Correlated regions are picked up

mental process. fMRI is an important application of MRI and used widely in neurosciences, human sciences, and economics, as well as in medical sciences (Greene and Haidt 2002; Bandettini 2012).

The main disadvantage of BOLD-fMRI over magnetoencephalography (MEG) and electroencephalography (EEG) is slow response time, and it is common to all measurements based on vascular changes. Vascular changes (hemodynamic responses) appear a few seconds after the start of neural activation. If events occur in short time scales relative to the fMRI response time, it is not easy to resolve individual events by fMRI.

The development of various non-invasive fMRI measurements other than BOLD-fMRI has been challenged. Cerebral blood flow (CBF) and cerebral blood volume (CBV) measurements, both of which are based on the vascular response like as BOLD-fMRI, have good specificity and sensitivity for activated blood vessels (Ogawa and Sung 2007). It has been shown that diffusion MRI signals respond to neuronal activation and their signal change has good spatial resolution (Le Bihan et al. 2006). This diffusion weighted MRI appears to reflect a change in the apparent diffusion constant (ADC) of water in the brain. The response could be explained by the swelling of neurons during neuronal activation. Direct detection of electromagnetic events caused by brain activity has also been tried using MRI. However, the direct in vivo detection of electromagnetic events is still challenging.

### 7.4.2 New Wave of fMRI – Resting State fMRI, Functional Connectivity (Connectome)

Figure 7.9b shows a typical block paradigm of event-related fMRI and Fig. 7.9c a resting-state fMRI experiment. Resting-state fMRI is a relatively new method to study spontaneous fluctuations in brain activity during the resting state (the subject is not performing an explicit task). A long MRI time series data analysis shows that the spontaneous low-frequency (<0.1 Hz) BOLD signal fluctuations under resting conditions often have strong correlations even in distant brain regions (Biswal et al. 1995). The distributed patterns of resting-state fMRI correlations are relatively consistent over several brain states; these patterns are so-called resting state networks (Fox and Raichle 2007). The biological significance of these spontaneous, correlated activities in the resting brain remains unclear; however, the study of resting state networks has already been shown to have potential clinical value, providing rich and sensitive markers of disease (Fox and Greicius 2010).

## 7.5 Introduction to Compressed Sensing (Faster Imaging Technology)

### 7.5.1 Introduction

Magnetic Resonance (MR) has many applications in biological science and industry. MR Imaging has been widely used in medical diagnosis because of its non-invasive manner and exquisite detail in images of soft tissues. The major restriction to the use of MRI is its long scan time.

The physical processes of excitation and reception of MR signals are inherently quite slow. To measure a MR signal takes milliseconds per pulse. The hardware of MRI has been developed and improved dramatically, but hardware development cannot shorten the physical processes required to acquire MR signals. MRI raw data are collected first in the frequency domain (called the 'k-space'). The data acquisition speed (k-space traversal speed), which depends on the gradient switching speed, is limited because of the current produced by the high power and high speed switching of gradient coils stimulate peripheral nerves. The gradient switching speed is restricted so as not to go over the limits of peripheral nerve stimulation. Further acceleration of the MR system requires alternative techniques. Many researchers are seeking for a new method of acceleration.

The reduction of scan time is an ongoing challenge. Many methods, such as parallel imaging and other undersampling strategies, have been invented to enable imaging of fast biological processes. Fundamentally, these methods utilize the redundancy of data (e.g., multiple receiver coils are designed to generate redundant data per acquisition). These techniques are mostly governed by the Nyquist sampling rate, and hence they cannot yield acceleration greater than the Nyquist limit due to the resulting aliasing artifacts. The Nyquist limit is the minimum sampling rate required to avoid aliasing; the sampling rate for any signal must be at twice the highest frequency contained within the signal.

On the other hand, compressed sensing (CS) (also known as compressive sampling) can recover the signal below the Nyquist limit. CS is a fundamentally new approach to data acquisition. CS can reconstruct sparse signals and images from what was previously believed to be incomplete information. Hence, CS has gained attention in signal processing, statistics, and computer science.

### 7.5.2 *Compressed Sensing*

Every day, we use data compression to process images and audio signals. Conventional data compression techniques take the full data set at the beginning. For example, after one takes a picture (full information) and acquires an entire image, compression tools encode a few significant data (most of the information is thrown away) and store it. One can decode and reconstruct the original entire image from the stored data. As this example shows, there is unnecessary data in original images. If there is a wise way to measure compressed data directly from the real world, only a small number of measurements are needed. This is a valuable line of approach for MRI because MRI measurements are expensive and slow. CS opens the door to this approach. The theory of compressed sensing was put forward by Candes et al. and Donoho (Candès et al. 2006; Donoho 2006). They pointed out that signals can be reconstructed from a very limited number of samples, provided that the measurements satisfy an incoherence property.

Since the CS acquisition process corresponds to the measurement of compressed data of images, CS requires images to be compressible on some basis. The image property that lies behind compressibility is sparsity: under transformation, many of the data values in the transform domain can be set to zero and the image can be reconstructed from the rest of the data without an appreciable effect. In other words, if a significant majority of the data or vectors have coefficients equal to zero and a few coefficients of it contain all the information, the data or vector is sparse.

The CS approach requires the following:

1. Random sampling (incoherence of undersampling artifacts)
2. Sparse signals in some basis (transform sparsity)
3. Non-linear reconstruction

Random sampling ensures the incoherence of undersampling artifacts; this means that aliasing is produced by sampling below the Nyquist limit. Sparsity of data defines the important coefficients that are used for reconstruction of the original signal. How many coefficients are required for reconstruction is decided by a threshold. The measured data used in the CS acquisition process is already compressed, and hence reconstruction of the original data functions as a decoding process.

### 7.5.3 Applying Compressed Sensing to MRI

The application of CS to MRI potentially provides significant scan time reductions, leading to benefits for patients and health care economics. It is also beneficial for the measurements of high speed biological processes, such as cardiac and rapid angiography.

The first point to be discussed is under what conditions CS can be applied to MRI. The application of CS in MRI has the following restrictions:

1. MRI data should be collected by random sampling
2. MRI images should be sparse (or transform domain)
3. Images should be reconstructed by nonlinear methods

The MR system acquires MR signals as encoded samples, and this allows CS to be successfully applied to MRI. MRI measures not direct pixel samples but rather the Fourier coefficients of an image. A common approach is to use undersampling to reduce the number of readout lines in the grid of the k-space (this undersampling pattern corresponds to sparse sampling in phase direction). Other sampling patterns, e.g., radial sampling, are also available.

Most MR images are compressible by sparse coding in an appropriate transform domain (e.g., by wavelet transform). For example, the images of MR angiography—consisting of contrast enhanced blood vessels with a black background—are already sparse in the pixel representation and also sparse in the finite differences domain. Other examples are cardiac cine images. They are sparse in the spatio-temporal Fourier domain. More complex images, such as brain images, can be sparsified in the wavelet transform domain.

It is important for the reconstruction process to impose a nonlinear regularization, such as the $L_1$ norm. The non-linear reconstruction process has consistency with acquired data, as a subset of the k-space data.

This raises the questions as to how to decide on the minimum number of measurements (coefficients) required to get acceptable SNR and how to ensure high data fidelity. This question is an optimization problem. The $L_0$ norm of the signal $x$ ($\|x\|_0$) corresponds to the sparsity of the signal $x$. This is the number of non-zero elements in the vector. In terms of the computational process, the $L_0$ norm problem is not convex in nature, and it is difficult and impractical to solve the $L_0$ norm problem. The solution of the $L_1$ problem is the closest convex approximation to the solution of the $L_0$ norm problem. The $L_1$ norm provides the absolute sum of the elements of the vector. The $L_1$ norm penalizes the presence of a large number of coefficients. In other words, solving the $L_1$ norm problem corresponds to seeking the transform sparsity. As for the question of how to ensure data fidelity, the $L_2$ norm penalizes small errors less and large errors more significantly, by the quadratic nature of the norm. The $L_2$ norm is an iterative approximation process that accounts for data consistency.

The constrained optimization problem in reconstruction can be described by the following equation (Lustig et al. 2007):

$$minimize\|\psi m\|_1$$
$$s.t.\|F_u\ m - y\|_2 < \varepsilon$$

where $m$ is the desired image represented by a complex vector, $y$ is the measured k-space data from MRI, $\psi$ is the linear operator that transforms from pixel representation into a sparse representation and the threshold parameter $\varepsilon$ is the parameter for error tolerance which controls the fidelity of the reconstruction to the measured data. To solve the problem, many strategies and algorithms (e.g., the non-linear conjugate gradient) have been studied and applied.

### *7.5.4 Intuitive Example: Incoherence*

One of the most important requirements of CS in MRI is incoherence. Let us examine the importance of incoherence and the feasibility of CS in more detail. A simple 1D example with 480 data points of acquisition is shown in Fig. 7.10.

An example of undersampling patterns in the k-space is in Fig. 7.10b. Equispaced undersampling in the k-space results in coherent aliasing (Fig. 7.10c) and brings about an inherent ambiguity: it is impossible to distinguish between the original and shifted copies of a signal. Incoherent undersampling seems to incur random noise (Fig. 7.10d). The artifact that appears to be a noise is not “noise” per

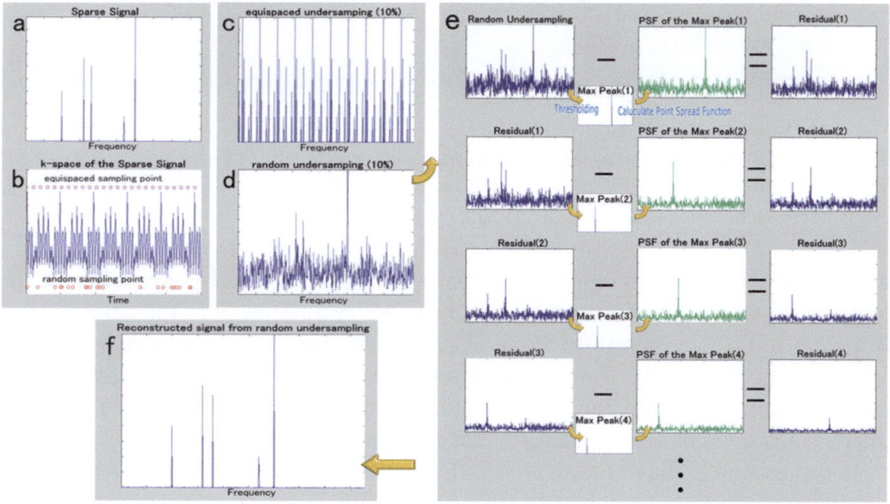


**Fig. 7.10** Reconstruction from undersampled data. (**a**) Sparse signal. (**b**) Undersampling pattern in the k-space. The *circles* denote the sampling timings. (**c**) Equispaced undersampling results in coherent signal aliasing. (**d**) Pseudo-random undersampling results in incoherent aliasing. (**e**) Iterative reconstruction process from undersampled data. (**f**) Recovered signal by iterative reconstruction

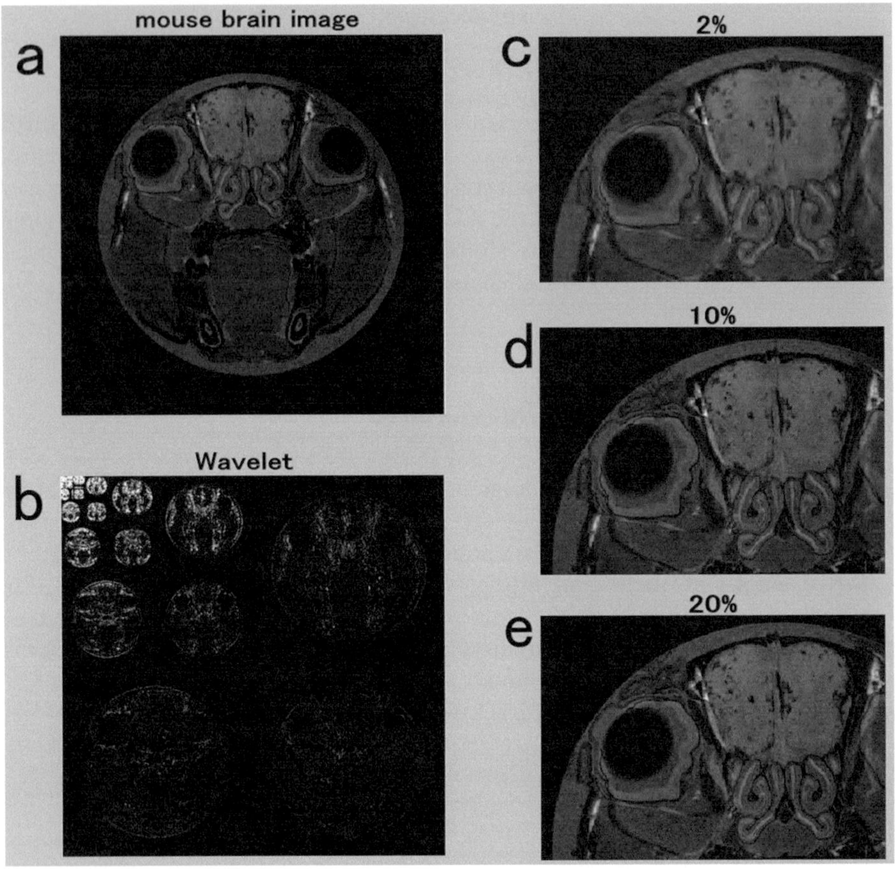


**Fig. 7.11** Sparse approximation of a compressible image. The image is approximated by its largest wavelet coefficients. (**a**) Axial T1 mouse brain image. (**b**) Wavelet coefficients of the image. (**c–e**) Brain images are reconstructed from a subset of 2, 10, and 20 % of the largest coefficients

se but rather the energy spreading from each individual non-zero value of the original signal. Incoherent undersampling removes ambiguity in the original data (Fig. 7.10e, f). The pattern of Point Spread Function (PSF), a calculation of the spreading of the energy from the original signal, is dependent on the pattern of random sampling. If the random sampling pattern is not incoherent enough, the interference level increases and prevents accurate reconstruction.

In this example, the iteration procedure is repeated until all the significant signal components are recovered. The process corresponds to a sort of iterative algorithm for solving the optimization problem, and in effect performs threshold and interference cancellation at each iteration. Therefore, there is a close connection between this simple example of interference cancellation and formal and practical applications of CS in MRI.

The next example illustrates that sparse approximation can be available even with complicated images of the brain of the mouse (Fig. 7.11). Signal approximation with 10 % of the largest coefficients results in good image quality. Nearly equivalent images can be gained using a significantly small number of coefficients. For CS in MRI to achieve the quality of image reconstruction that is shown in Fig. 7.11, it is necessary to develop a method to sample coefficients that correspond to those of the transform domain. In reality, it is not true that any sparse sampling pattern is possible; a practical sampling pattern is constrained by the hardware of MRI and physiological limitations.

### 7.5.5 *Future of CS in MRI*

The three requirements for the successful application of CS in MRI (as noted in 7.5.3) will be satisfied in many applications by ongoing developments. Potential applications of CS in MRI are based on some domain of sparsity: for example, sparsity in the temporal domain (e.g. dynamic imaging), sparsity in space (e.g. angiography), and sparsity as additive redundancy (e.g. resolution enhanced imaging). Even though different applications face different constraints, sampling trajectories and sparsifying transforms play key roles in matching the constraints. The ongoing development of CS in MRI, including the development of reconstruction algorithms, will bring new, fascinating applications and substantially benefit both the research and clinical usage of MRI.

## 7.6 Summary

In summary, MRI and MRS are proving to be extremely useful in neuroscientific research and neurological clinical practice. These powerful non-invasive tools are likely to continue to grow in importance in these fields and to gain ever more important clinical applications.

## Exercises

1. What is nuclear magnetic moment?
2. What is resonance?
3. What is the resonance frequency of proton at 1.5 T?
4. What is the chemical shift?
5. What is the longitudinal relaxation time?
6. What is the transverse relaxation time?
7. What limits the resolution of MRI?
8. Why is the gradient coil necessary for MRI?
9. What is the contrast agent for MRI?

10. What factors influence water diffusion?
11. What is the Nyquist frequency?
12. What is the wavelet transformation?

## References


Arnold, J.T., Dharmatti, S.S., Packard, M.E.: Chemical effects on nuclear induction signals from organic compounds. J. Chem. Phys. **19**, 507 (1951)

Bandettini, P.A.: Twenty years of functional MRI: the science and the stories. Neuroimage **62**, 575–588 (2012)

Basser, P.J., Mattiello, J., Le Bihan, D.: MR diffusion tensor spectroscopy and imaging. Biophys. J. **66**, 259–267 (1994)

Biswal, B., Yetkin, F.Z., Haughton, V.M., Hyde, J.S.: Functional connectivity in the motor cortex of resting human brain using echo-planar MRI. Magn. Reson. Med. **34**, 537–541 (1995)

Bloch, F., Hansen, W.W., Packard, M.: Nuclear induction. Phys. Rev. **69**, 127 (1946)

Brand, A., Richter-Landsberg, C., Leibfritz, D.: Multinuclear NMR studies on the energy metabolism of glial and neuronal cells. Dev. Neurosci. **15**, 289–298 (1993)

Candès, E.J., Romberg, J., Tao, T.: Robust uncertainty principles: exact signal reconstruction from highly incomplete frequency information. IEEE Trans. Inf. Theory **52**, 489–509 (2006)

Damadian, R.: Tumor detection by nuclear magnetic resonance. Science **171**, 1151–1153 (1971)

Donoho, D.L.: Compressed sensing. IEEE Trans. Inf. Theory **52**, 1289–1306 (2006)

Fox, M.D., Greicius, M.: Clinical applications of resting state functional connectivity. Front. Syst. Neurosci. **4**, 19 (2010)

Fox, M.D., Raichle, M.E.: Spontaneous fluctuations in brain activity observed with functional magnetic resonance imaging. Nat. Rev. Neurosci. **8**, 700–711 (2007)

Frahm, J., Krüger, G., Merboldt, K.D., Kleinschmidt, A.: Dynamic uncoupling and recoupling of perfusion and oxidative metabolism during focal brain activation in man. Magn. Reson. Med. **35**, 143–148 (1996)

Gill, S.S., Small, R.K., Thomas, D.G., Patel, P., Porteous, R., Van Bruggen, N., Gadian, D.G., Kauppinen, R.A., Williams, S.R.: Brain metabolites as $^1$H NMR markers of neuronal and glial disorders. NMR Biomed. **2**, 196–200 (1989)

Goldstein, F.B.: The enzymatic synthesis of N-acetyl-L-aspartic acid by subcellular preparations of rat brain. J. Biol. Chem. **244**, 4257–4260 (1969)

Greene, J., Haidt, J.: How (and where) does moral judgment work? Trends Cogn. Sci. **6**, 517–523 (2002)

Haacke, E.M., Brown, R.F., Thompson, M., Venkatesan, R.: Magnetic Resonance Imaging: Physical Principles and Sequence Design. Wiley, New York (1999)

Hetherington, H.P., Mason, G.F., Pan, J.W., Ponder, S.L., Vaughan, J.T., Twieg, D.B., Pohost, G. M.: Evaluation of cerebral gray and white matter metabolite differences by spectroscopic imaging at 4.1 T. Magn. Reson. Med. **32**, 565–571 (1994)

Lauterbur, P.C.: Image formation by induced local interactions: examples employing nuclear magnetic resonance. Nature **242**, 190–191 (1973)

Le Bihan, D., Breton, E., Lallemand, D., Grenier, P., Cabanis, E., Laval-Jeantet, M.: MR imaging of intravoxel incoherent motions: application to diffusion and perfusion in neurologic disorders. Radiology **161**(2), 401–407 (1986)

Le Bihan, D., Urayama, S., Aso, T., Hanakawa, T., Fukuyama, H.: Direct and fast detection of neuronal activation in the human brain with diffusion MRI. Proc. Natl. Acad. Sci. U. S. A. **103**, 8263–8268 (2006)

Li, B.S., Wang, H., Gonen, O.: Metabolite ratios to assumed stable creatine level may confound the quantification of proton brain MR spectroscopy. Magn. Reson. Imaging **21**, 923–928 (2003)

Lustig, M., Donoho, D., Pauly, J.: Sparse mri: the application of compressed sensing for rapid mr imaging. Magn. Reson. Med. **58**, 1182–1195 (2007)
Magistretti, P.J., Pellerin, L.: Cellular mechanisms of brain energy metabolism and their relevance to functional brain imaging. Philos. Trans. R. Soc. Lond. B Biol. Sci. **354**, 1155–1163 (1999)
Moffett, J.R., Ross, B., Arun, P., Madhavarao, C.N., Namboodiri, A.M.: N-Acetylaspartate in the CNS: from neurodiagnostics to neurobiology. Prog. Neurobiol. **81**, 89–131 (2007)
Mori, S.: Introduction to Diffusion Tensor Imaging. Elsevier, Amsterdam/Boston (2007)
Mori, S., Barker, P.B.: Diffusion magnetic resonance imaging: its principle and applications. Anat. Rec. **257**, 102–109 (1999)
Mori, Y., Yoshioka, Y.: Visualization of immune cell dynamics in mouse brain with 11.7 T MRI. Proc. Intl. Soc. Magn. Reson. Med. **20**, 911 (2012)
Mori, Y., Umeda, M., Fukunaga, M., Ogasawara, K., Yoshioka, Y.: MR contrast in mouse lymph nodes with subcutaneous administration of iron oxide particles: size dependency. Magn. Reson. Med. Sci. **10**, 219–227 (2011)
Moseley, M.E., Cohen, Y., Mintorovitch, J.: Early detection of regional cerebral ischemic injury in cats: evaluation of diffusion and T2-weighted MRI and spectroscopy. Magn. Reson. Med. **14**, 330–346 (1990)
Ogawa, S., Sung, Y.W.: Functional magnetic resonance imaging. Scholarpedia **2**(10), 3105 (2007)
Ogawa, S., Lee, T.M., Kay, A.R., Tank, D.W.: Brain magnetic resonance imaging with contrast dependent on blood oxygenation. Proc. Natl. Acad. Sci. U. S. A. **87**, 9868–9872 (1990)
Pan, J.W., Twieg, D.B., Hetherington, H.P.: Quantitative spectroscopic imaging of the human brain. Magn. Reson. Med. **40**, 363–369 (1998)
Pauling, L., Coryell, C.D.: The magnetic properties and structure of hemoglobin, oxyhemoglobin and carbonmonoxy hemoglobin. Proc. Natl. Acad. Sci. U. S. A. **22**, 210–216 (1936)
Pouwels, P.J., Frahm, J.: Regional metabolite concentrations in human brain as determined by quantitative localized proton MRS. Magn. Reson. Med. **39**, 53–60 (1998)
Prichard, J., Rothman, D., Novotny, E., Petroff, O., Kuwabara, T., Avison, M., Howseman, A., Hanstock, C., Shulman, R.: Lactate rise detected by 1H NMR in human visual cortex during physiologic stimulation. Proc. Natl. Acad. Sci. U. S. A. **88**, 5829–5831 (1991)
Purcell, E.M., Torrey, H.C., Pound, R.V.: Resonance absorption by nuclear magnetic moments in a solid. Phys. Rev. **69**, 37–38 (1946)
Rabi, I.I., Zacharias, J.R., Millman, S., Kusch, P.: A new method of measuring magnetic moments. Phys. Rev. **53**, 318 (1938)
Ross, B.D., Blüml, S.: New aspects of brain physiology. NMR Biomed. **9**, 279–296 (1996)
Ross, B., Blüml, S.: Magnetic resonance spectroscopy of the human brain. Anat. Rec. **265**, 54–84 (2001)
Signoretti, S., Marmarou, A., Tavazzi, B., Lazzarino, G., Beaumont, A., Vagnozzi, R.: N-Acetylaspartate reduction as a measure of injury severity and mitochondrial dysfunction following diffuse traumatic brain injury. J. Neurotrauma **18**, 977–991 (2001)
Soares, D.P., Law, M.: Magnetic resonance spectroscopy of the brain: review of metabolites and clinical applications. Clin. Radiol. **64**, 12–21 (2009)
Srinivasan, R., Sailasuta, N., Hurd, R., Nelson, S., Pelletier, D.: Evidence of elevated glutamate in multiple sclerosis using magnetic resonance spectroscopy at 3 T. Brain **128**, 1016–1025 (2005)
Taylor, D.L., Davies, S.E., Obrenovitch, T.P., Doheny, M.H., Patsalos, P.N., Clark, J.B., Symon, L.: Investigation into the role of N-acetylaspartate in cerebral osmoregulation. J. Neurochem. **65**, 275–281 (1995)
Thulborn, K.R., Warterton, C.J., Matthews, P.M., Radda, G.K.: Oxygenation dependence of the transverse relaxation time of water protons in whole blood at high field. Biochim. Biophys. Acta **714**, 265–270 (1982)
Urenjak, J., Williams, S.R., Gadian, D.G., Noble, M.: Specific expression of N-acetylaspartate in neurons, oligodendrocyte-type-2 astrocyte progenitors, and immature oligodendrocytes in vitro. J. Neurochem. **59**, 55–61 (1992)

Wang, Y., Li, S.J.: Differentiation of metabolic concentrations between gray matter and white matter of human brain by in vivo $^{1}$H magnetic resonance spectroscopy. Magn. Reson. Med. **39**, 28–33 (1998)

Zhang, X., Bearer, E.L., Perles-Barbacaru, A.T., Jacobs, R.E.: Increased anatomical detail by in vitro MR microscopy with a modified Golgi impregnation method. Magn. Reson. Med. **63**, 1391–1397 (2010)

# Chapter 8
# Advances in Neuroimaging Techniques with PET

**Eku Shimosegawa**

**Abstract** Using a radioisotope, positron emission tomography (PET) can obtain images of natural circulation through the body (from blood vessels to organs and from organs to blood vessels) as molecular images, which is naturally present *in vivo*. A distinctive feature of PET molecular imaging is to use radioisotope compounds as tracers. This chapter describes the latest status of imaging research on neurological functions, to the extent to which they are related to PET. Although PET has limitations, e.g., on spatial resolution, radiation exposure and observation period due to short half-life isotopes, it can visually and dynamically evaluate normal brain function and pathophysiology. Newly-developed imaging devices (e.g., semiconductor PET and PET/MR), in combination with improved imaging techniques and innovative analytical procedures, are expected to be powerful tools for studying the brain function in detail.

**Keywords** PET • Energy metabolism • CBF • Semiconductor PET • PET/MR

## 8.1 Introduction

In vivo imaging is largely classified into morphological, functional and molecular imaging. Morphological imaging represents spatial information, such as intraorgan structure and vasculature, based on anatomy, while functional imaging physiologically and biochemically visualizes circulation at the tissue level and intracellular changes. Molecular imaging visualizes the functions and dynamics of biofunctional molecules based on molecular biology. Specifically, gadolinium contrast agents used in magnetic resonance imaging (MRI) are artificial compounds that are not usually present in the body, and functional imaging similar to organ perfusion imaging can be performed by administrating these agents intravenously and obtaining their signals. However, the images thus obtained are confined only to circulation in blood vessels. Positron emission tomography (PET) can obtain

E. Shimosegawa (✉)
Graduate School of Medicine, Osaka University, Suita, Osaka, Japan
e-mail: eku@mi.med.osaka-u.ac.jp

 
M. Kasaki et al. (eds.), *Cognitive Neuroscience Robotics B*,
DOI 10.1007/978-4-431-54598-9_8

natural circulation through the body, from blood vessels to organs and from organs to blood vessels, as molecular images following labeling water, which is naturally present *in vivo*, with a radioisotope.

A problem in visualizing brain neuronal activity is the difficulty of comprehensively representing it, because the functions of the brain cover metaphysical areas, such as mind, consciousness and intention, and individual activities cross each other in a complicated manner, unlike in other organs. Therefore, also in PET molecular imaging with radioisotope compounds as tracers, it should be supposed that brain functions are localized, that each nerve type differentiates separately and that functional molecules come into and go out of the tissue in a quantitative manner. Furthermore, their quantity should be analyzed by collecting signals, *i.e.*, radiation released from the body, and converting them to a parameters set based on a virtual mathematical model. There are limitations in elucidating the whole picture of neurological functions by this method. However, tracers used in PET could be powerful tools in neuroimaging research, in that they are based on natural molecules constituting organisms, thereby making it possible to obtain specific molecular images of circulation, metabolism, neuroreceptor functions and activity of each cell.

This chapter describes the latest status of imaging research on neurological functions, that are mainly related to PET.

## 8.2 Brain Neuronal Activity and Cell-Specific Energy Metabolism

The brain is an organ with active energy metabolism, and oxygen consumption in the brain reaches approximately 20 % of the total oxygen consumption just to maintain basal metabolism. In the brain, most energy is consumed for active transport of ions involved in maintaining the neuronal membrane potential. Although the activity of brain tissue is maintained by a large amount of adenosine triphosphate (ATP) produced by oxidative phosphorylation of glucose, brain tissue has no efficient storage system such as glycogen and requires a constant supply of glucose and oxygen. If cerebral ischemia occurs due to the occlusion of cerebral blood vessels, reduced supply of glucose and oxygen to the brain induces impairment of the neuronal membrane potential, which suppresses neuronal firing and directly leads to a decrease in neurotransmission functions.

On the other hand, if brain neuronal activity increases by thinking and speaking, neuronal energy metabolism increases in the relevant local brain centers, and cerebral blood flow (CBF) also increases, leading to an increased supply of glucose and oxygen. Therefore, if the results of a relatively simple task are summarized as a certain brain function or thinking process, brain neuronal activity after receiving the task from the outside can be visualized as images of areas with an increase in CBF.

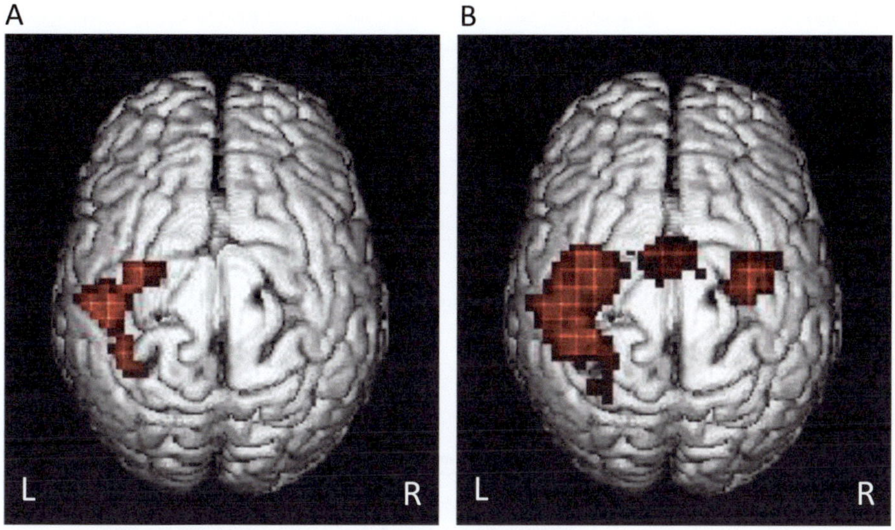


**Fig. 8.1** Brain activation study with moving right hand. (**a**) Simple motor task by finger tapping. (**b**) Complex motor task by finger tapping and position changing (Courtesy Dr. Atsushi Umetsu.)

This is called localized brain activation (stimulation) study. Figure 8.1 shows the visualization of differences in locally activated brain areas in response to different tasks, a simple or somewhat complex movement, using functional magnetic resonance imaging (f-MRI), in the task of moving one hand.

Up to the present, brain energy metabolism has been studied focusing on neurons. However, in recent years, new metabolic images are being visualized from new points of view. Although it had long been considered that the only role for glial cells was as a supporting tissue of the brain, these cells were revealed to play an important role in supplying lactic acid necessary for neuronal energy metabolism. Astrocytes, which are glial cells present in the central nervous system, incorporate more than twice as much glucose as neurons do and generate lactic acid from pyruvic acid through the glycolytic pathway. Lactic acid is a compound from which acetyl coenzyme A (acetyl-CoA), an important substrate in the tricarboxylic acid (TCA) cycle, is generated, and astrocytes are a major source of supply of lactic acid necessary for neurons through the astrocyte-neuron-lactate shuttle (Dienel and Cruz 2006; Dienel and Hertz 2001) (Fig. 8.2). In addition, glutamate generated by the TCA cycle metabolism in neurons is released as a neurotransmitter from the termini of presynaptic neurons to the synaptic cleft and binds to receptors on postsynaptic neurons, and thus neurotransmission is achieved. In this process, excitatory glutamate released in excess to the synaptic cleft acts as a neurotoxin, if its concentration is high. Astrocytes are also involved in glutamate-glutamine cycling, in which excitatory glutamate in the synaptic cleft is recovered, detoxified to glutamine through gamma-aminobutyric acid (GABA) and reused in the termini

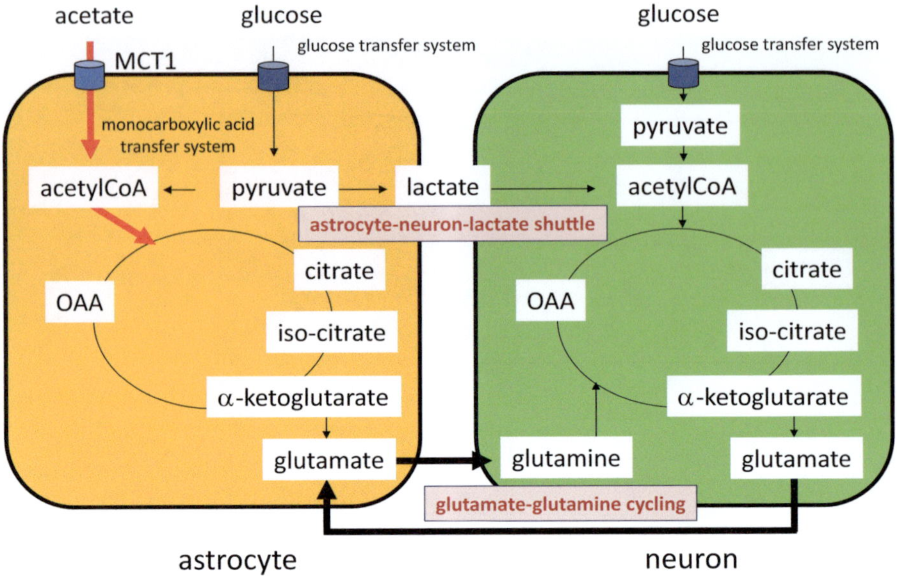


**Fig. 8.2** Model for energy cycling of neuronal and glial activity with glucose utilization. *MCT1* monocarboxylic acid transporter-1, *OAA* oxaloacetate. MCT1 is a specific transporter for acetate intake into astrocyte

of presynaptic neurons (Friston et al. 1991) (Fig. 8.2). As described above, astrocytes have been revealed to play an important role in supporting neuronal energy metabolism and neurotransmission.

Energy metabolism in astrocytes can be visualized by the radioactivation of energy substrates that are specifically incorporated into astrocytes but not into neurons. Acetic acid is specifically incorporated into astrocytes via monocarboxylate transporters and contributes to the turnover of the TCA cycle. Carboxylic acid in this compound can be radiolabeled with $^{11}C$ to prepare a PET agent. Figure 8.3 shows a normal brain PET image using $^{11}C$-acetic acid, in which glial cell-specific energy metabolism was visualized. Hertz et al. visualized changes in acetate metabolic activity in rats given auditory stimulation using $^{14}C$-labled acetic acid, compared them with areas where glucose metabolism was activated and proved that these metabolisms increase in the same areas (Heiss et al. 2004).

It has not been elucidated yet how the brain obtains and consumes energy required for higher brain function activity or how differences in activity among central nervous system cell types affect higher brain function. Research on energy metabolism imaging using PET may provide new clues in this field.

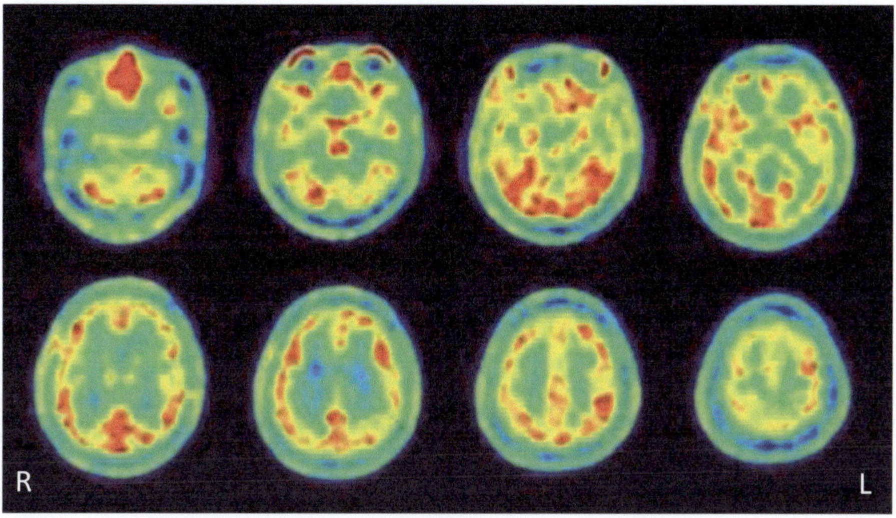


**Fig. 8.3** $^{11}C$-acetate brain distribution in a healthy volunteer measured by PET

## 8.3 Cognitive Impairment and Neuroimmunity

Population aging has advanced, and early detection of pathological cognitive impairment attracts high attention. Alzheimer's disease is a representative disease, and according to a famous hypothesis explaining its development, aggregation and accumulation of amyloid protein (Aβ) due to the cleavage of amyloid precursor protein (Fig. 8.4) and neurofibrillary tangles due to the expression and phosphorylation of the tau protein occur in a cascade fashion (Hertz et al. 2007). It is clinically important to establish an early screening system for middle-aged and elderly patients who have developed mild cognitive impairment (MCI) within the allowable range as a normal age-related change to assess the potential future development of Alzheimer's disease, to provide them with treatment and to improve their quality of life through enhanced activities of daily living.

Studies using PET employ procedures in which compounds that specifically bind to aggregated Aβ are labeled to perform imaging, and many agents have been developed, represented by [$^{11}$C]Pittsburgh compound B ($^{11}$C-PIB), [$^{11}$C]2-(2-[2-dimethylaminothiazol-5-yl]ethenyl)-6-(2-[fluoro]ethoxy)benzoxazole ($^{11}$C-BF227), 2-(1-6-[(2-[$^{18}$F]fluoroethyl)(methyl)amino]-2-naphthylethylidene) malononitrile ($^{18}$F-FDDNP), $^{18}$F-BAY 1008472 and $^{11}$C-AZD2995. In particular, $^{11}$C-PIB, which was developed in the early days, was tested in a large-scale international clinical trial, and it was reported that patients with low accumulation of $^{11}$C-PIB in the brain did not develop Alzheimer's disease (Kato et al. 2008). However, its specificity for predicting conversion to Alzheimer disease is low in MCI patients with $^{11}$C-PIB accumulation. In addition, its accumulation is low in disease types other than Alzheimer's disease, such as frontotemporal dementia, and

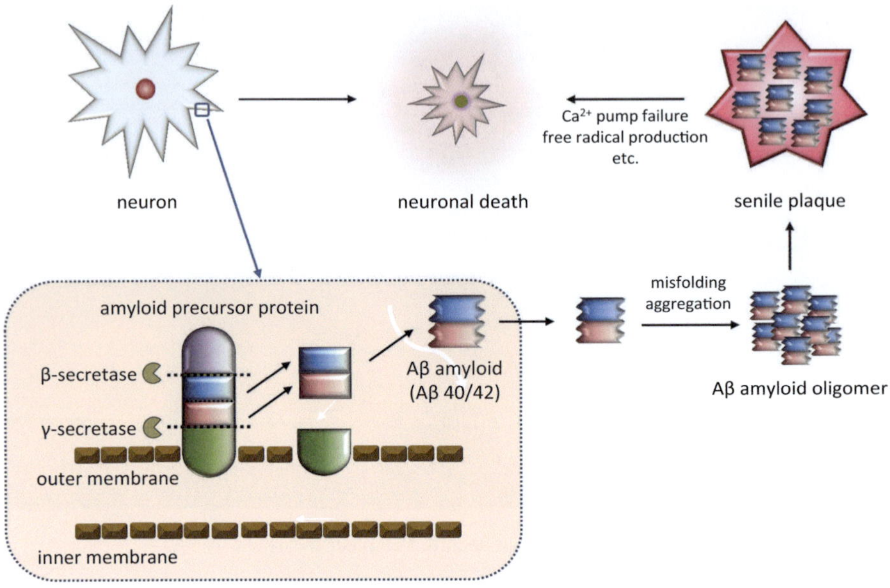


**Fig. 8.4** Schema of senile plaque formation based amyloid hypothesis in Alzheimer's disease

it is also difficult to differentiate between such disease types and normal conditions. Therefore, agents that bind to abnormally accumulated substances in the brain with higher pathological specificity, such as the tau protein, are being searched for. Furthermore, new PET agents are being developed for diseases that have molecular pathologically distinct accumulated substances, such as α-synuclein and TAR DNA-binding protein 43 (TDP-43), and are also pathologically distinguished from Alzheimer's disease, such as Parkinson's disease and frontotemporal dementia.

Attention has not often been paid up till now to immune reactions in the field of neuroscience. However, it was recently observed that activated microglia induce immune responses against aggregated Aβ leading to neuroinflammation, in the pathogenesis of Alzheimer's disease (Matsuda et al. 2007). It is now possible to perform imaging of abnormally expressed activated microglia using a PET agent prepared by labeling a protein that binds to receptors on the mitochondrial outer membrane of activated microglial cells.

## 8.4 Improvement in Resolution and Sensitivity of PET Devices

There are several points that should be kept in mind when localized brain activation studies are conducted using PET. One is that, because of the influence of the positron range and internal absorption/scattering of radiation, PET images have poor spatial resolution compared to MRI and CT images, making it difficult to identify anatomical sites, and it takes time to collect data. To overcome these disadvantages, devices with highly sensitive detectors and new data collection methods have been developed. Detectors can be thinned and high-resolution images can be obtained by adopting semiconductors in detectors (Fig. 8.5). Local brain anatomy can also be identified in small animal imaging by using semiconductor PET devices (Fig. 8.6). In addition, sensitivity in the detection of radioactive concentration has improved and data acquisition time and exposure dose can be reduced by the three-dimensional data collection of radiation generated in the field of view of devices (Fig. 8.7). The development of the above latest devices has made it possible to analyze the energy metabolism in small structure of the brain, which could not have been visualized by the conventional PET scanner. In particular, the brainstem contains many nuclei from which various nervous systems originate, but has not been studied in the field of higher brain function activity. High glucose metabolic activity can be visualized in areas where there are aggregates of brain

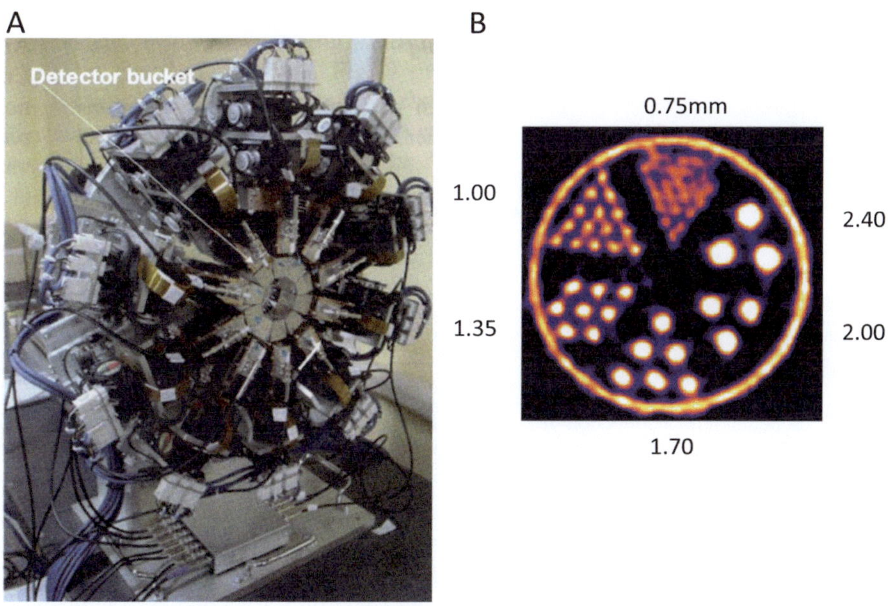


**Fig. 8.5** High resolution and high sensitivity PET system made by semiconductor detectors. (**a**) A semiconductor PET system without gantry cover. (**b**) Images of the Derenzo phantom obtained by a semiconductor PET. Axial spatial resolution of this PET system is 1.00 mm

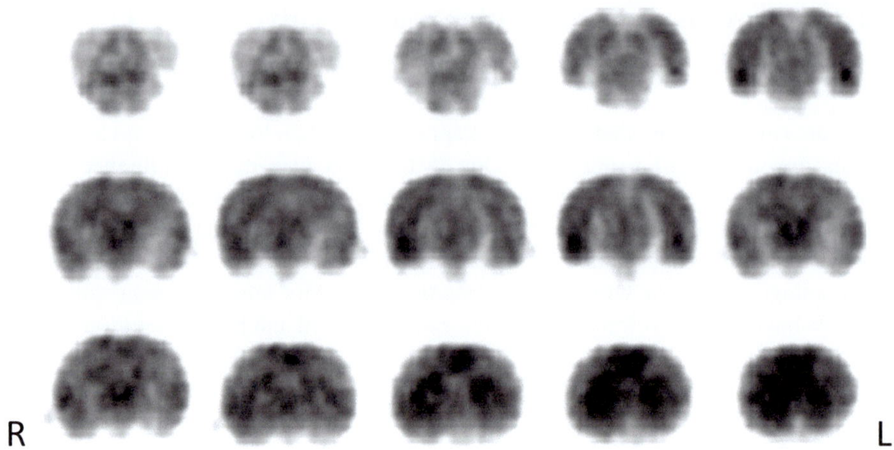


**Fig. 8.6** Quantitative image of cerebral metabolic rate of glucose utilization in a normal rat measured by a high-resolution semiconductor PET system

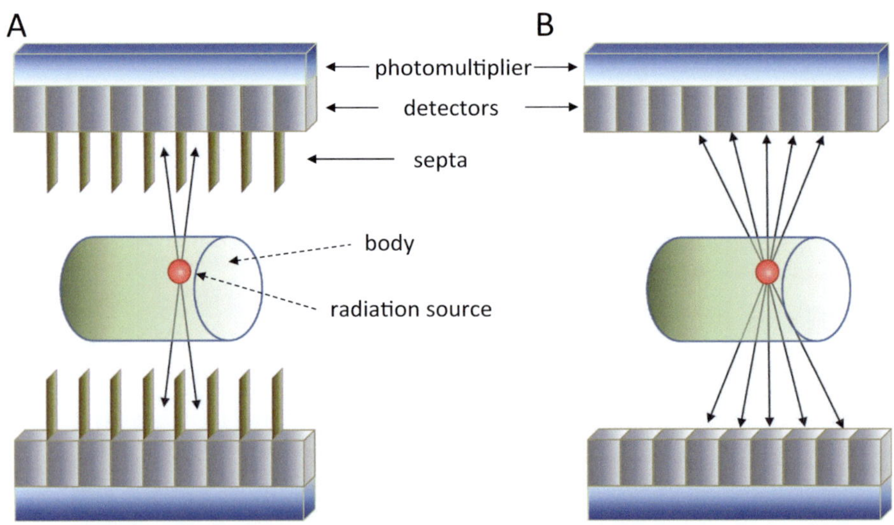


**Fig. 8.7** Schema of data collection of 2-dimensional (2-D) and 3-dimensional (3-D) PET system. (**a**) 2-D PET has septa between each detector to eliminate scatter radiation. (**b**) 3-D PET system has high image sensitivity because it can collect many radiations from the radiation source without being disturbed by septa

nuclei, such as the brainstem reticular formation involved in consciousness and perception, by using high-resolution and high-sensitivity PET devices (Minoshima et al. 1995) (Fig. 8.8). Interrelation and differences in activity between these

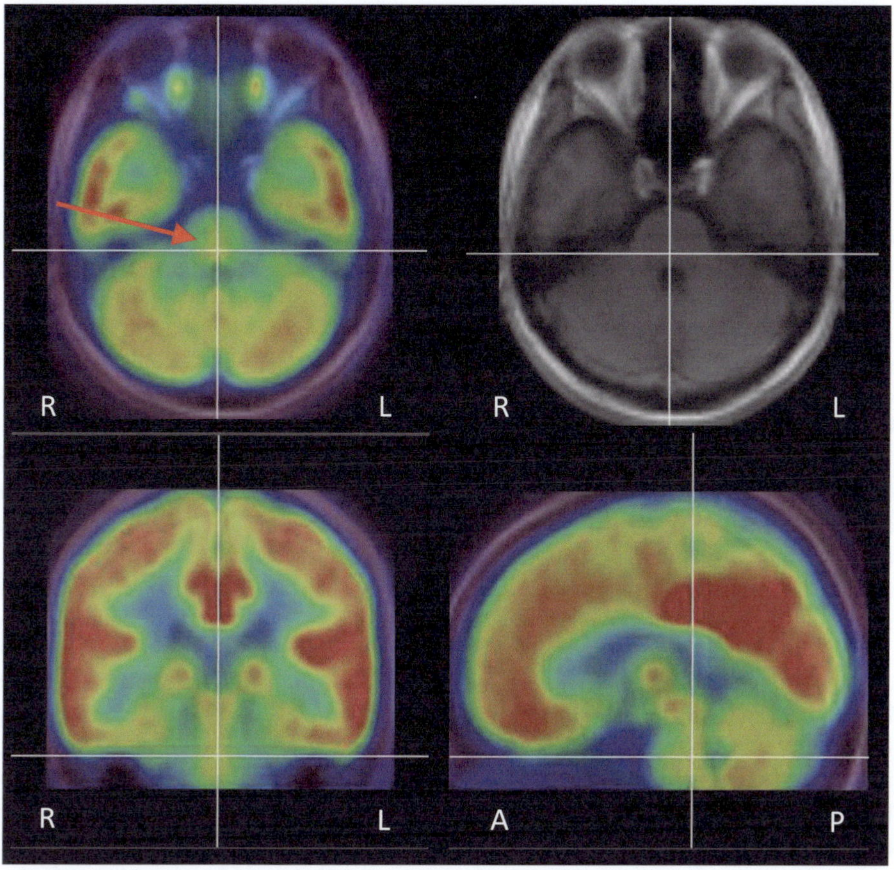


**Fig. 8.8** Normal brainstem glucose metabolism measured by $^{18}$F-FDG PET. Averaged PET image of healthy volunteers (*upper left*) shows high glucose metabolism in the dorsal area of the pons (*red arrow*), corresponding to the brainstem reticular formation

brainstem nuclei and the cerebral neocortex are unexplored fields in localized brain activation studies.

## 8.5 Automated Image Analysis and Its Statistical Evaluation

Methods conventionally used for analysis of PET images are visual evaluation and analysis with a regions-of-interest (ROI) setting. Visual evaluation, however, does not allow quantitative analysis, thus making it impossible for PET image readers to enjoy the advantage of analyzing the linear relationship between biofunctions and radioactivity level. Analysis with the ROI setting involves the step of setting an ROI of any form in freely selected regions of the image and the step of quantitatively measuring the radioactivity level in the pixels within the ROI. Problems with this approach lay in that the selection of the place and size of the ROI is affected by arbitrary elements and that it is difficult to set an ROI covering the entire brain.

In recent years, statistical image analysis has been advancing as a means allowing the simple and objective analysis of images. A representative method of this category is Statistical Parametric Mapping (SPM) devised by Friston et al. as a method of inter-group comparison (Okello et al. 2009). In the field of medicine, the neurologic statistical image analysis program (NEUROSSTAT; University of Washington) and the easy Z-scores imaging system (e-ZIS; Fiji Film RI Pharma Co.) are used extensively as techniques for comparison between group (healthy group or sick group) and individuals (Pellerin et al. 2007; Talairach and Tournoux 1988).

With basic methods of statistical image analysis such as SPM, measurement is done on the PET images collected from the subjects to be compared, and the continuous probability distribution of brain tissue radioactivity level in each pixel is determined and fed into a database. From the continuous probability distribution data for each pixel from each group collected in this manner, pixels with a statistically significant elevation or reduction of probability are extracted from each group and depicted as a map of the statistical values of the pixels concerned on brain images. With this method, several steps of image processing are needed before the statistical analysis. First, the brain images of varying forms from individual subjects must undergo linear or nonlinear transformation and are fit into a single template, i.e., the standard brain (anatomical normalization). Through this step, the pixels constituting the brain images of subjects are all located on the same axis of the coordinate within the three-dimensional visual space. Next, the radioactivity levels are standardized by setting of a threshold level for the radioactivity level of individual pixels or dividing the radioactivity level in individual pixels by the whole brain radioactivity level or the radioactivity in a reference area. Subsequently, using a prescribed statistical method, the continuous probability distribution of individual pixels is compared between groups, to extract pixels with significant inter-group differences. The pixels extracted following this procedure are developed on the Talairach brain atlas (Tanzi and Bertram 2005) or the standard brain image, to identify the anatomical site concerned (Fig. 8.9). In addition to these steps, some other steps such as realignment and smoothing are incorporated into this method to reduce statistical noise from the images.

Several requirements need to be satisfied to enable adoption of this analytical method. First, the brain of the subject to be analyzed must be free of marked deformation, defect or atrophy as compared to the standard brain. If the brain form differs markedly from that of the standard brain, pixel dislocation may occur during the process of anatomical normalization, and an accurate analysis is not possible due to an admixture with the radioactivity in tissue other than the target site. Furthermore, to

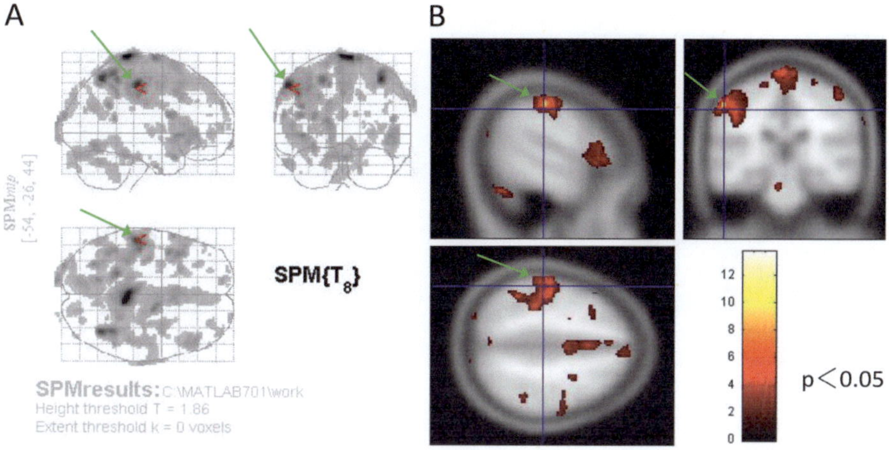


**Fig. 8.9** Image analysis for the CBF difference between normal male and female using statistical parametric mapping (SPM). (**a**) Areas with significantly higher CBF in female than in male demonstrated on the Talairach atlas. Coordinate-54, -26, and 44 (*green arrow*) corresponded to the Brodmann area 2 (postcentral gyrus). (**b**) Overlayed image on three intersecting (sagittal, coronal, transaxial) slices of MRI-$T_1$ template image

enable automated anatomical normalization, it is necessary to use a PET agent whose radioactivity can be distributed extensively within the brain, like the distribution on brain perfusion images or brain glucose metabolism images. In addition, all subjects covered by a single analysis need to undergo imaging with the same PET device combined with the same PET agent, followed by image reconstruction with the same technique. This is necessary because different PET agents have different characteristics in local brain distribution and because the image quality can be altered by the method of image collection or reconstruction, causing statistical noise. Regarding the composition of the groups of subjects, it is necessary to cover a large population so that the precision of the statistical detective power can be improved, and it is also necessary to ensure uniformity of all groups in terms of background elements (e.g. age group and gender) possibly affecting the analytical results.

In any event, statistical image analysis is a noninvasive method for objective evaluation of radioactivity distribution, not necessitating measurement of the input function by arterial blood sampling (needed for quantitative test) but enabling simple analysis if data from a reliable database are collected with the same device and the same method. SPM is a computer program of the "freeware" category (http://www.fil.ion.ucl.ac.uk/spm/) which has been used extensively not only for PET image analysis but also for analysis of functional MRI data in regional brain activation testing. In addition, this program can be used for interim processes (anatomical normalization, smoothing, etc.) in its unique way. Because of these

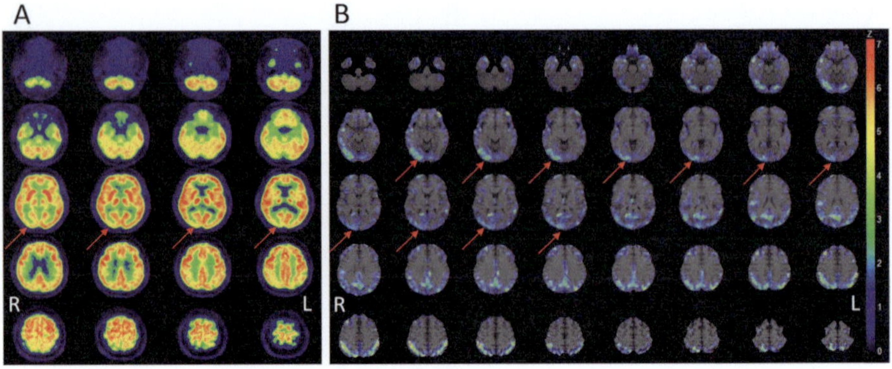


**Fig. 8.10** Statistical image analysis of the brain glucose metabolism in a patient with probable diffuse Lewy body disease. (**a**) Original $^{18}$F-FDG-PET images showed subtle decrease of glucose metabolism in the right primary visual cortex (*red arrow*). (**b**) Statistical image analysis by NEUROSTAT software demonstrated areas with significantly decrease of glucose metabolism compared with normal database

features, the program has been applied to diverse studies involving images. NEUROSTAT and e-ZIS, which enable group vs. individual comparison, are often used for identification of brain sites with reduced perfusion or glucose metabolism through neuronal degeneration in the diagnosis of dementia and for disease typing, thus providing a useful tool in clinical medicine (Fig. 8.10). They are also used for the analysis of neuroreceptor images to identify the epileptic focus.

## 8.6 Integration of Different Imaging Devices and Advances in Small Animal Imaging

To improve the diagnostic ability of PET imaging with poor spatial resolution, PET/CT devices in which PET and CT are combined are currently generally used (Fig. 8.11). The acquisition of CT images before and after acquisition of PET images and fusion of these images using an integrated CT-PET device have made it possible to more clearly identify the visual biodistribution of radioisotope compounds (Fig. 8.12). In addition, the measurement of the internal permeability of X-ray from the outside using a CT device has also made it possible to estimate the internal absorption of radioactivity to correct measured values obtained with a PET device.

Devices involving combinations of different technologies, as described above, represent a major trend in the development of the latest devices. In recent years, attention has also been paid to the combination of PET and MR devices, and devices offering practical indications in the clinical field are now commercially

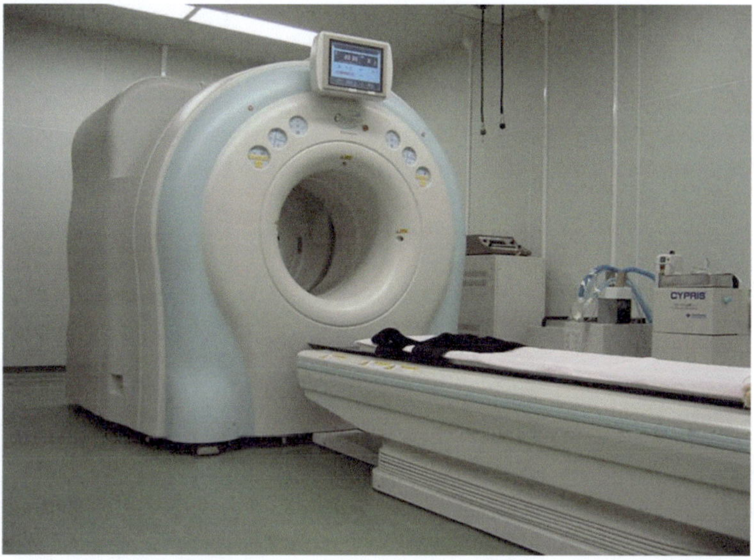

**Fig. 8.11** Integrated PET/CT scanner for clinical measurement. The scanner can collect radiations from field-of-view with 3-D data acquisition system, which can reduce radiation dose without disturbing image quality

available. Particularly in the field of neuroscience, studies using integrated PET/MR devices are prospective. In higher brain function tests, it is possible to simultaneously collect the distribution of various radioactive tracers using PET and functional images, such as MRI tractography and functional MRI images, and to compare them. In addition, PET images with low resolution have to be normalized to the anatomical standard brain by software processing after higher brain function tests, and activated areas of local brain function have to be evaluated in comparison with the brain atlas. However, simultaneous acquisition of PET and MR images in the same subjects using PET/MR devices makes it easier to evaluate active areas in the deformed brain due to developmental abnormalities and in the pathologically damaged brain that cannot be processed by software. Furthermore, a method has been devised to correct the underestimation of radioactivity due to the presence of tissues other than the brain, air, etc. in image voxels (partial volume effect) by collecting PET and MR images simultaneously (Villemagne et al. 2011) (Fig. 8.13).

The combination of PET and MR devices also greatly benefits small animal imaging (Fig. 8.14). Naturally, it is even more difficult to anatomically evaluate radioactivity distribution in the imaging of small animals using PET with poor resolution, but the fusion of PET and MRI, the latter of which has excellent density resolution, makes it easier to analyze radioactivity distribution in organs such as the

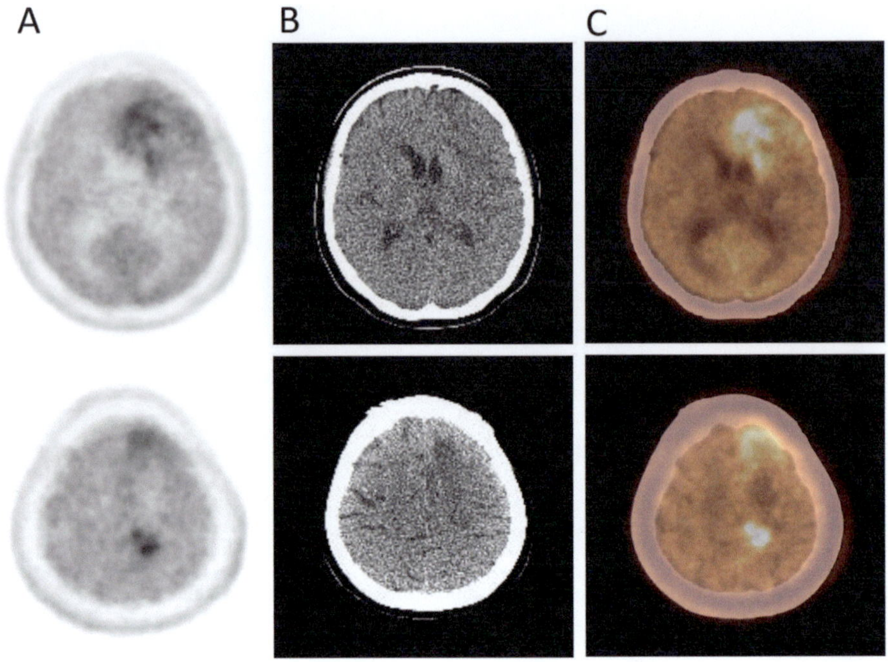


**Fig. 8.12** Diagnostic utility of fused image between PET and CT in a patient with glioblastoma. (**a**) $^{11}$C-methionine PET image (amino acid tracer) for the detection of tumor infiltration in the brain, (**b**) CT image at the same level of $^{11}$C-methionine PET obtained by integrated PET/CT system, and (**c**) fused image between PET and CT. Functional brain damage examined by PET can be clearly identified by morphological information by CT

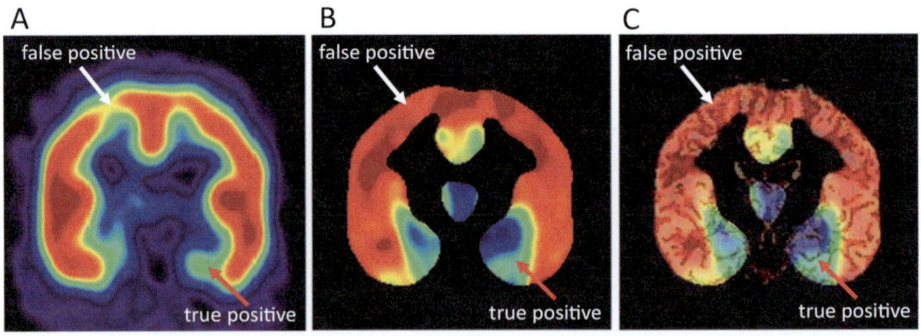


**Fig. 8.13** Improvement of diagnosis in a patient with epilepsy by partial volume effect (PVE) correction. (**a**) Uncorrected PET or single photon computed tomography (SPECT) image, (**b**) PVE-corrected PET or SPECT image, and (**c**) fused image between PVE-corrected image and gray matter MRI map. False positive are (*white arrow*) and true positive are (*red arrow*) precisely diagnosed by the PVE-corrected image

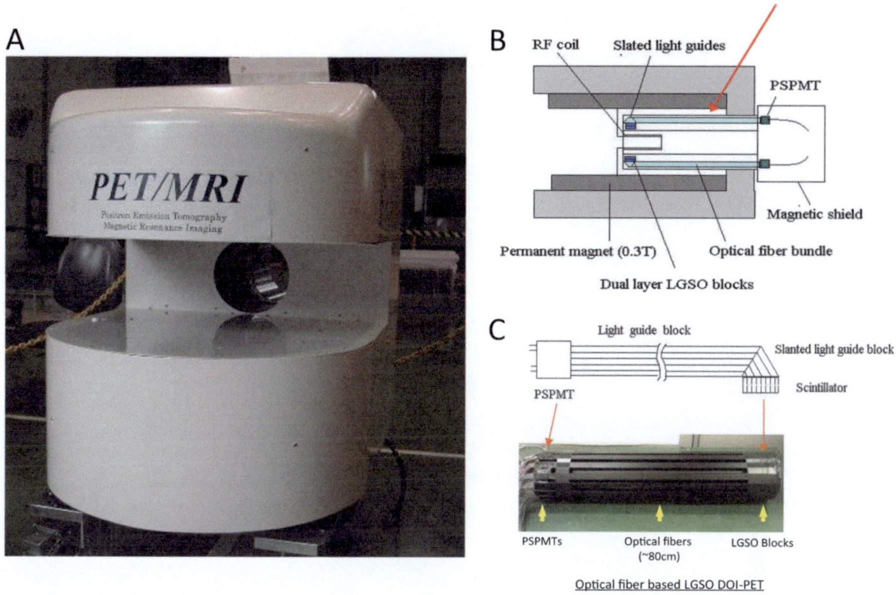


**Fig. 8.14** (**a**) Appearance of an integrated PET/MRI scanner for animal experiment, (**b**) its vertically-sectioned diagram, and (**c**) the detector system in PET portion. The scanner combined PET portion with an open-type MRI portion with a 0.3 T permanent magnet. The PET portion (*red arrow*), consisted of detectors and optical fibers, which was placed between the radiofrequency (RF) coil and permanent magnet for MRI. Position-sensitive photo-multiplier tubes (PSPMT), covered with magnetic shield, are located behind the yoke

brain and spinal cord (Fig. 8.15). In addition, in pathological studies of cerebrovascular disease, such as those using transient cerebral ischemia models, brain functional images and the state of cerebrovascular occlusion and recanalization can be examined by simultaneously obtaining cerebral blood flow and metabolic images using PET and magnetic resonance angiography (MRA) images. Although a large number of animals have been used so far in such studies, and autopsied to obtain ideal experimental models, based on these results leading to the subsequent development of innovative combined technology systems, such tasks can now be replaced by the in vivo imaging of a small number of animals. The development of the above-described innovative experimental devices has made it possible to obtain molecular images in living animals using protocols similar to those used for humans.

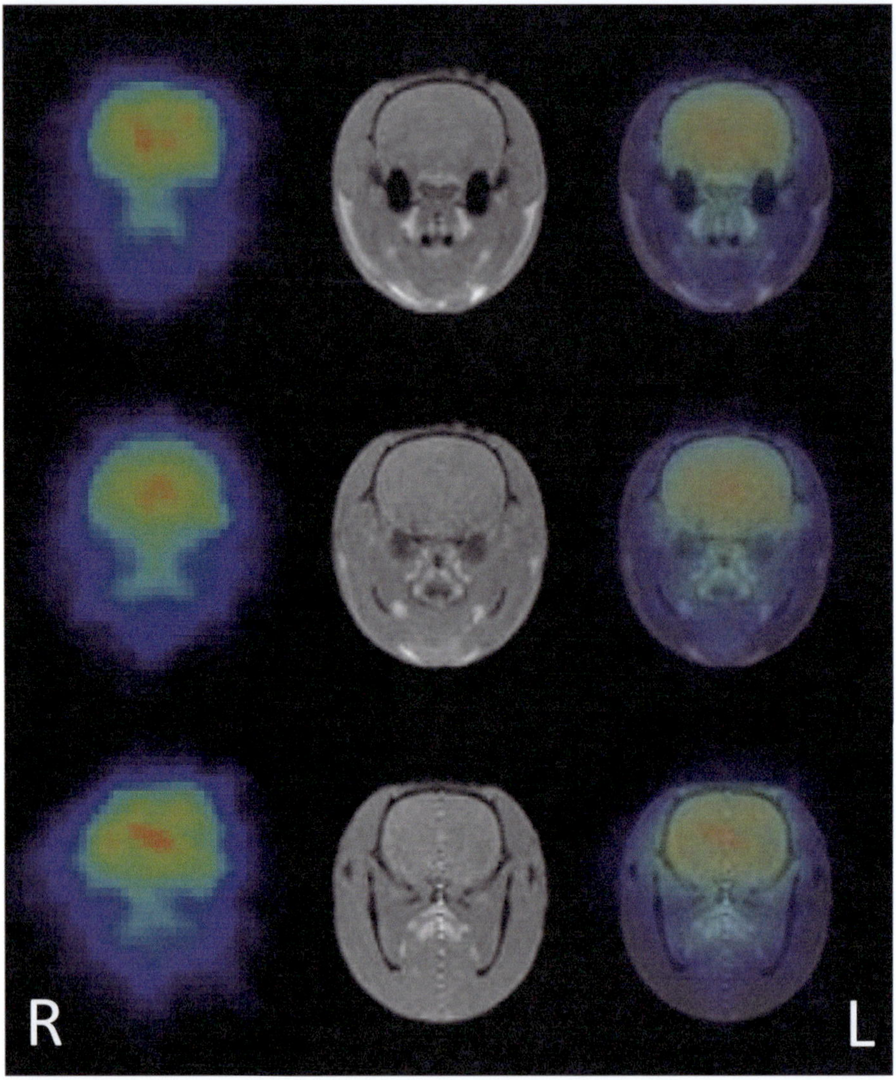


**Fig. 8.15** Image fusion between PET and MRI in a normal rat obtained by the integrated PET/MR scanner. $^{18}$F-FDG PET image (*left column*), MRI (*middle column*), and fused PET and MR image (*left column*)

## 8.7 Concluding Remarks

The brain is an organ that controls the total organism, including what is referred to as the mind. Although various means, including PET, have been devised to observe the activities and movement of the mind from the outside, it is difficult to comprehensively investigate its profound areas. However, organisms are consisted of molecules, and if the functions and activities of the brain can be converted to signals, digitized and analyzed as molecular actions, it is possible to perform their imaging. Although PET has limitations, such as spatial resolution, radiation exposure and limited observation periods due to short half-life isotopes, efforts are being continuously made to elucidate more minute brain functions by introducing new radiopharmaceuticals, developing devices, improving imaging techniques and innovating analytical procedures.

## Exercise

What significance do the fusion of a functional image and the morphological image have for neurologic image analysis?

## References

Dienel, G.A., Cruz, N.F.: Astrocyte activation in working brain: energy supplied by minor substrates. Neurochem. Int. **48**, 586–595 (2006)

Dienel, G.A., Hertz, L.: Glucose and lactate metabolism during brain activation. J. Neurosci. Res. **66**, 824–838 (2001)

Friston, K.J., Frith, C.D., Liddle, P.F., et al.: Comparing functional (PET) images: the assessment of significant change. J. Cereb. Blood Flow Metab. **11**, 690–699 (1991)

Heiss, W.D., Habedank, B., Klein, J.C., et al.: Metabolic rates in small brain nuclei determined by high-resolution PET. J. Nucl. Med. **45**, 1811–1815 (2004)

Hertz, L., Peng, L., Dienel, G.A.: Energy metabolism in astrocytes: high rate of oxidative metabolism and spatiotemporal dependence on glycolysis/glycogenolysis. J. Cereb. Blood Flow Metab. **27**, 219–249 (2007)

Kato, H., Shimosegawa, E., Oku, N., et al.: MRI-based correction for partial-volume effect improves detectability of intractable epileptogenic foci on $^{123}$I-iomazenil brain SPECT images. J. Nucl. Med. **49**, 383–389 (2008)

Matsuda, H., Mizumura, S., Nagao, T., et al.: Automated discrimination between very early Alzheimer disease and controls using an easy Z-score imaging system for multicenter brain perfusion single-photon emission tomography. Am. J. Neuroradiol. **28**, 731–736 (2007)

Minoshima, S., Frey, K.A., Koeppe, R.A., et al.: A diagnostic approach in Alzheimer's disease using three-dimensional stereotactic surface projections of fluorine-18-FDG PET. J. Nucl. Med. **36**, 1238–1248 (1995)

Okello, A., Edison, P., Archer, H.A., et al.: Microglial activation and amyloid deposition in mild cognitive impairment: a PET study. Neurology **72**, 56–62 (2009)

Pellerin, L., Bouzier-Sore, A.K., Aubert, A., Serres, S., Merle, M., Costalat, R., Magistretti, P.J.: Activity-dependent regulation of energy metabolism by astrocytes: an update. Glia **55**, 1251–1262 (2007)

Talairach, J., Tournoux, P.: Co-Planar Stereotactic Atlas of the Human Brain. Theime Verlag, Stuttgart (1988)

Tanzi, R.E., Bertram, L.: Twenty years of the Alzheimer's disease amyloid hypothesis: a genetic perspective. Cell **120**, 545–555 (2005)

Villemagne, V.L., Pike, K.E., Chételat, G., et al.: Longitudinal assessment of Aβ and cognition in aging and Alzheimer disease. Ann. Neurol. **69**, 181–192 (2011)

# Chapter 9
# Movement Disorders and Motor Cortex Stimulation

**Naoki Tani and Youichi Saitoh**

**Abstract** Since the initial reports in the early 1990s, stimulation of the primary motor cortex (MCS) has been used to treat chronic intractable pain conditions and movement disorders. Corticobasal ganglia dysfunction is widely present in patients with movement disorders, but can be improved by methods of cortical stimulation, such as transcranial magnetic stimulation and chronic stimulation with implanted electrodes. Repetitive transcranial magnetic stimulation can lead to transient improvements in motor performance when the primary motor cortex or the supplementary motor area is targeted; and chronic electrical stimulation can improve symptoms of a variety of movement disorders, including Parkinson's disease, tremor and post-stroke dystonia. Immediate neuromodulation could be mediated by imposition of a specific pattern of activity and suppression of abnormal, disease-associated rhythmicity of oscillations in the corticobasal ganglia-cortical circuit. Additionally, delayed clinical benefit from MCS could come from synaptic plasticity, long-term potentiation, long-term depression, and expression of secondary messengers or polarization of brain tissue.

In this chapter, we will describe the advantages and shortcoming of MCS for the treatment of movement disorders, and discuss the mechanism of action of MCS in movement disorders.

**Keywords** Motor cortex • Transcranial magnetic stimulation • Parkinson's disease • Pain • Movement disorder • Motor cortex stimulation • Akinesia • Central sulcus • Electrode

## 9.1 Introduction

The groundwork for using chronic stimulation of the primary motor cortex (M1) to treat pain and movement disorders was laid 50 years ago. In 1985 and then 1991, Tsubokawa and colleagues reported the use of chronic motor cortex stimulation

---

This chapter was adopted from Tani and Saitoh (2011).

N. Tani • Y. Saitoh (✉)
Graduate School of Medicine, Osaka University, Suita, Osaka, Japan
e-mail: neurosaitoh@mbk.nifty.com

 
M. Kasaki et al. (eds.), *Cognitive Neuroscience Robotics B*,
DOI 10.1007/978-4-431-54598-9_9

(MCS) to treat post-thalamic stroke pain (Tsubokawa et al. 1991a, b). More recently, epidural MCS was found to concomitantly reduce motor impairments that were associated with pain, although it was initially studied with the aim of treating chronic pain (Nguyen et al. 1998; Katayama et al. 2002; Franzini et al. 2000).

Parkinson's disease (PD) is a movement disorder, and it is caused by loss of dopamine neurons in the substantia nigra that project to striatum. Its main symptoms are tremor, rigidity, bradykinesia and postural instability. These symptoms are thought to arise as a result of a complex rearrangement in the neuronal circuits that depend on the striatum, such as hyperactivity and a change in firing rate in the subthalamic nucleus (STN) and globus pallidus (Boraud et al. 1998; DeLong 1990; Filion and Tremblay 1991). There are several drugs available to effectively treat the initial symptoms of PD, but long-term medical management is often complicated by levodopa-induced motor fluctuations, which result in rapid changes between periods of severe akinesia and periods of mobility accompanied by troublesome dyskinesia (Goetz et al. 2005). Therefore, new treatment approaches have been developed, such as deep brain electrical stimulation (DBS). DBS is applied particularly to the STN and globus pallidus. DBS is highly effective at controlling motor symptoms of PD (Krack et al. 1998; Limousin et al. 1998). However, patients with poor response to levodopa or those with cognitive impairment, advanced age, considerable brain atrophy or cerebral ischemic foci in the white matter, are considered unsuitable for DBS, because of their increased surgical risk (Lopiano et al. 2002; Pahwa et al. 2005).

MCS might represent an alternative treatment for patients with PD who do not fulfill all DBS inclusion criteria. There is some evidence that MCS may relieve motor symptoms of PD (Canavero and Bonicalzi 2002; Pagni et al. 2005). However, there are also a number of contradictory reports regarding the efficacy of MCS for patients with PD. The apparent discrepancies among these reports can be accounted for by a number of factors: different selection criteria, different surgical (i.e., epidural vs. subdural electrodes) or methodological (i.e., stimulation frequencies) approaches, and importantly, different methods of assessing motor performance. This chapter summarizes current evidence on these topics.

## 9.2 Transcranial Magnetic Stimulation of the Primary Motor Cortex

Transcranial magnetic stimulation (TMS) is a non-invasive technique that delivers electrical currents into the cortex. TMS has been widely used in neuroscience over the last two decades because of its ability to modulate brain activity and function by depolarizing neurons or their axons (Siebner et al. 2009). The magnetic coil can deliver pulses in a repeated fashion where frequency, number of pulses and intensity of stimulation can be controlled with different consequences on the cortex (Siebner et al. 2009).

Several studies report that repetitive transcranial magnetic stimulation (rTMS) shows short-term benefits on symptoms of PD patients, such as akinesia, tremor and rigidity. The benefits varies depending on the stimulus frequency and the site of stimulation. On the one hand, high frequency (5–25 Hz) rTMS of M1 improves motor impairments (Siebner et al. 2000; de Groot et al. 2001; Bornke et al. 2004; Lefaucheur et al. 2004), but the antiparkinsonian effects are rather weak (10–25 % improvement in motor scores when compared with a control group of patients) and short-term (lasting a few hours). rTMS of M1 also improves speech (Dias et al. 2006) and bladder dysfunction (Brusa et al. 2009). On the other hand, low-frequency rTMS is reported to have inconsistent effects on motor performance (Lefaucheur et al. 2004; Okabe et al. 2003). The inconsistencies may reflect differences in coil type or patient selection.

Studies on rTMS of) the supplementary motor area (SMA) are conflicting. rTMS of SMA reduces levodopa-induced dyskinesias (Koch et al. 2005; Brusa et al. 2006) when delivered at a low frequency (1 Hz); but it has no clinical impact (Koch et al. 2004, 2005) or even worsens motor performance (Boylan et al. 2001) when delivered at a high frequency (5–10 Hz). However, a recent large Japanese multicenter trial has reported that repeated sessions of 5 Hz rTMS over SMA for 8 weeks reduces Unified Parkinson's Disease Rating Scale motor scores by 20 % (Hamada et al. 2008, 2009).

It has been proposed that the response to rTMS of M1 may predict the efficacy of MCS in patients with PD. For example, we had a patient with akinesia for whom high frequency rTMS of M1 and subsequent MCS showed similar beneficial effects on motor symptoms. In patients with intractable pain, Andre-Obadia demonstrated that the response to 20 Hz rTMS of M1 predicted the efficacy of subsequent MCS for pain reduction (Andre-Obadia et al. 2006) and Hosomi et al. reported that there was good correlation between the pain reduction with 5 Hz rTMS and that with MCS (Hosomi et al. 2008). Thus, it might be possible to use rTMS to better define the targets and parameters of subsequent chronic MCS.

## 9.3 Chronic Electrical Stimulation of the Motor Cortex

rTMS can produce transient improvement of motor performance when M1 or SMA is targeted. However, in order to produce permanent therapeutic effects, chronic stimulation with implanted electrodes is necessary. MCS has been used not only to relieve refractory pain but also to improve motor symptoms of a variety of movement disorders, including PD, tremor, and poststroke dystonia (Pagni et al. 2008; Arle and Shils 2008; Moro et al. 2011; Katayama et al. 2002; Canavero and Bonicalzi 2007; Franzini et al. 2000). The surgical procedure of MCS in patients with PD is the same as that of MCS in patients with intractable pain. However, because of the small number of reported case of MCS for PD, there are still controversies concerning details of the surgery, e.g., how to use unilateral or bilateral stimulation and epidural or subdural electrodes. Details of the surgical procedure of MCS in patients with PD will be discussed in this section (they have previously been reported in detail by Saitoh and Hosomi (2009)).

### 9.3.1 Pre-surgical Preparation

Prior to the surgical procedure, the hand motor area (the target of MCS) is identified using functional MRI (fMRI) or anatomical MRI. The lower limb motor area is located in the inter-hemispheric fissure, and it is difficult to implant an electrode into the lower limb motor area. fMRI (Fig. 9.1) is useful to precisely localize the site of M1, and some anatomical landmarks can help to identify the central sulcus. The central sulcus is characterized by the lack of sulcal branches, and lies just anterior to the pars marginalis of the cingulate sulcus on the interhemispheric surface (Naidich et al. 1995, 2001). The precentral knob sign corresponds to the hand motor area, and it is easily identifiable on MRI surface view (Fig. 9.2).

### 9.3.2 Neuronavigation

After localization of the targeted area, neuronavigation is used during surgery to precisely locate M1. Several kinds of navigation systems can be used to estimate the position of the central sulcus from above the skin surface (Tirakotai et al. 2007). Neuronavigation combined with fMRI data can help to identify the optimal location for craniotomy and placement of the stimulating paddle (Rasche et al. 2006). A drawback of neuronavigation is that it requires that the patient's head be fixed in a three-point pin holder or vacuum headrest (Tirakotai et al. 2007), which may not be tolerated by all patients under local anesthesia. For this reason, some surgeon prefers not to fix the patient's head but to operate without neuronavigation, i.e., to use Taylor-Haughton lines instead of neuronavigation lines (Fig. 9.3; Greenberg 2001).

### 9.3.3 Anesthesia

MCS electrodes are implanted under local or general anesthesia. General anesthesia is induced by a loading dose of Remifentanil 3–4 ng/ml in continuous infusion, being followed 5–8 min later by Propofol 5.5 μg/ml as an induction dose (total intravenous anesthesia). Endotracheal intubation is facilitated by vecuronium bromide 0.1 mg/kg and no further doses of muscle relaxants are administered throughout the surgery. The lungs are mechanically ventilated with a 50 % $O_2$-in-air

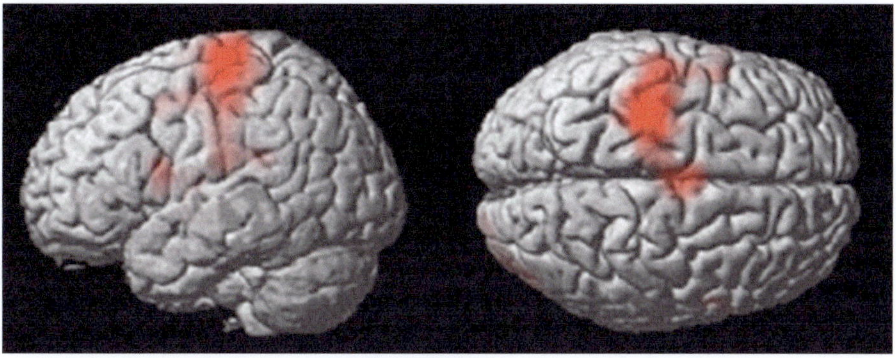

**Fig. 9.1** Motor cortex localization by fMRI. During the acquisition of fMRI data (echo planar imaging), the patient performed twelve 30-s epochs of right hand grasp with identical rest epochs. Data analysis was performed in MATLAB 2008a (Math Works, Inc., Natick, MA) using Statistical Parametric Mapping software (SPM@; Wellcome Department of Imaging Neuroscience, London, England)

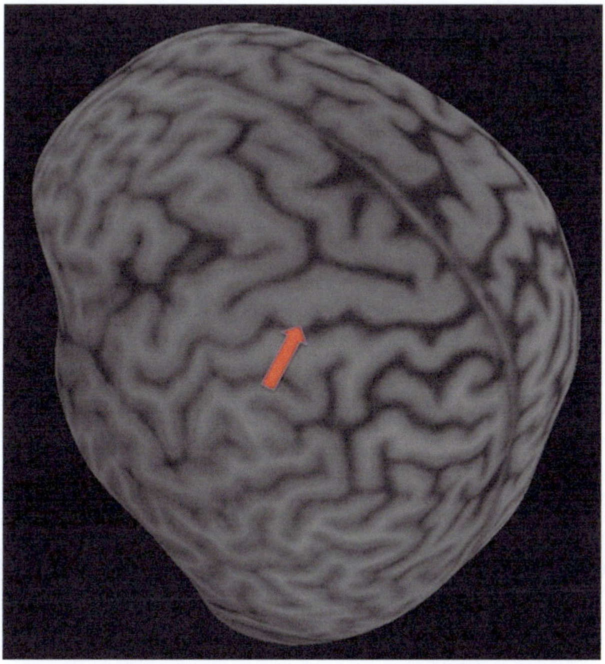

**Fig. 9.2** MRI brain surface view. *Red arrow* indicates the left M1 hand area

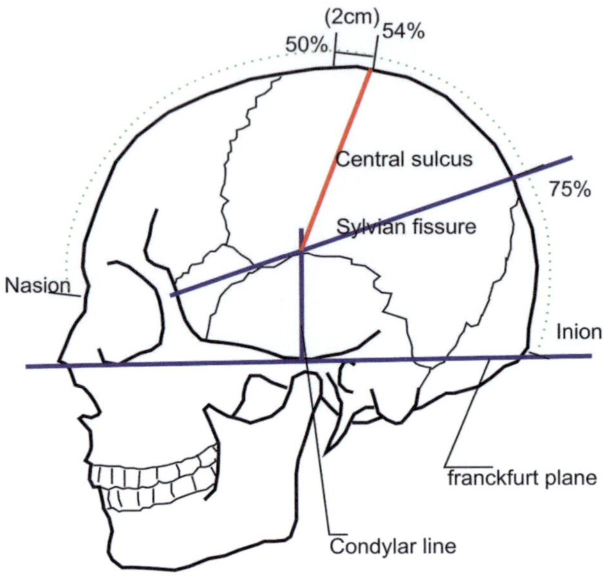


**Fig. 9.3** The Taylor-Haughton line (*red*) indicates the position of the central sulcus. The Frankfurt plane is the line from the inferior margin of the orbit through the upper margin of the external auditory meatus. The distance from the nasion to the inion is measured across the top of the calvaria and divided into quarters. The condylar line runs perpendicular to the baseline through the mandibular condyle (intersecting the line representing the Sylvian fissure). The Sylvian fissure is approximated by a line connecting the lateral canthus to a point three-quarters of the way posterior along the arc running over the convexity from nasion to inion. The central sulcus is drawn from the 54 % point on the naso-inoin line to the point where the Sylvian line intersects with the condylar line

mixture, in order to maintain end tidal concentration of $CO_2$ at 30–35 mmHg. Anesthesia is maintained with Remifentanil (5–6 ng/ml, up to 7–8 ng/ml if necessary) and Propofol (2.5–3.0 μg/ml). At the end of the surgical procedure, all patients are awakened within 15–30 min after cessation of total intravenous anesthesia.

### *9.3.4 Electrophysiological Localization of Hand Motor Area*

To localize the central sulcus, most neurosurgeons use the method of somatosensory evoked potentials (SSEPs) during contralateral median nerve stimulation. In this method, the phase reversal of the N20 (sensory cortex)/P20 (motor cortex) waves is used to confirm the location of the central sulcus (Wood et al. 1988), either using a multi-contact grid electrode and central scalp electroencephalogram (EEG) leads or directly using a definitive four-contact strip electrode overlying the dura matter (Fig. 9.4). Recently, an enlarged and displaced motor map for the hand area was found with regard to patients with PD. Map shifts were found in the majority of

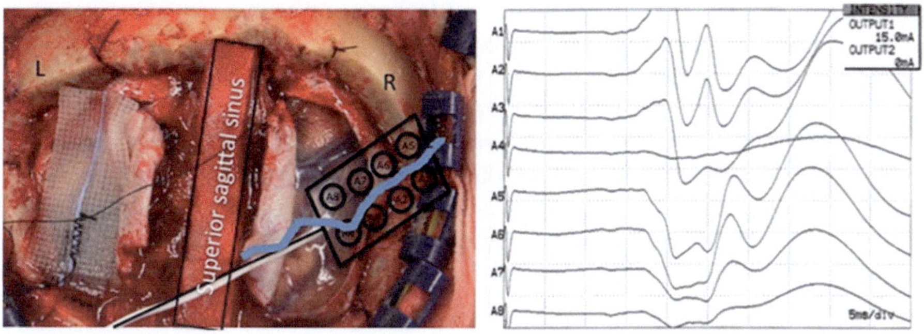


**Fig. 9.4** SSEP recording on the right central sulcus using an eight-contact subdural electrode. An eight-polar plate electrode (Specify Lead, 3998; Medtronic, Minneapolis, MN) was placed on the right central sulcus and the left median nerve was stimulated at the wrist with single pulses (0.5 ms, 4.7 Hz, 20 mA) to produce a small but consistent contraction of the thumb muscle. SSEPs were recorded from each cortical electrode referenced to the ipsilateral ear lobe. Individual SSEP signals were differentially amplified and filtered. The responses to 200 stimuli were averaged through a digital signal analyzer with sample interval of 100 ms. *Blue line* indicates the position of the central sulcus confirmed by SSEP. *SSPE* somatosensory evoked potential

the patients (Thickbroom et al. 2006). It is often inadequate or impossible to determine electrode placement solely by the method of SSEPs. According to Velasco et al. (2002), a simple and reliable way to determine electrode placement is by recording corticocortical evoked responses, and it is superior to the use of SSEPs. M1 stimulation elicits negative corticocortical evoked responses over the frontal scalp, whereas stimulation of the primary sensory cortex elicits positive responses over parietal and occipital scalp regions.

Most neurosurgeons attempt to use intraoperative test stimulation with quadripolar or grid electrodes. Bipolar test stimulation (210–1000 μsec (but generally 400–500-μsec), 1–5 Hz to 500 Hz, at voltages or intensities up to 50 mA) is applied anodally and cathodally, using contacts placed over M1. Muscle responses are recorded from muscle bellies of the contralateral upper limb, either with EMG (electromyography) electrodes or visually. In general, the stimulus intensity needed to produce motor responses is higher for epidural stimulation than for subdural stimulation. Motor responses can be elicited at relatively low intensities when general anesthesia is not used. The 1 Hz stimulation frequency is preferable to higher frequencies, as it does not habituate and has lower potential to trigger seizures.

### 9.3.5 Electrode Implantation

The great majority of investigators prefer epidural electrodes to subdural electrodes for MCS. This is mainly because subdual MCS has a greater risk of developing complications, such as cerebrospinal fluid leakage, difficulties in fixing the electrode, hemorrhage, and iatrogenic seizures. However, according to the physical model suggested by Holsheimer et al. (2007) and Manola et al. (2007), subdural MCS is more energy-efficient than epidural MCS is. In some patients with brain atrophy, the cortical surface and the dura mater are wide apart, and they may fail to respond to epidural stimulation. A subdural approach, therefore, may be considered in selected cases.

There is no study that has directly compared unilateral with bilateral MCS. However, some studies have reported that unilateral MSC has bilateral effects as clinical outcomes (Cilia et al. 2008; Pagni et al. 2005). The bilateral effects of unilateral MCS are probably due to bilateral afferent and efferent connections between cortical and subcortical structures (Leichnetz 1986).

A four-contact electrode array or two side-by-side four-contact electrode strips are placed on the hand motor area. Some surgeons place an electrode perpendicular to the central sulcus on the precentral (cathode) and postcentral (anode) gyri in the attempt to improve selectivity (Nguyen et al. 1999), whereas others place an electrode in a parallel fashion, i.e. with all contacts on M1 or S1 (Canavero and Bonicalzi 2002; Rasche et al. 2006). No polarity-related difference in pain relief is seen for most patients with epidural electrodes (Katayama et al. 1998).

#### 9.3.5.1 Epidural Electrodes

To implant epidural electrodes, Canavero et al. (2002, 2003) made an oblique linear skin incision (6–10 cm) parallel to and 1 cm ahead of or behind the projection of the central sulcus; and then, to accommodate a connector between the looping lead and the extension, they drilled two burr holes 2–4 cm apart and a bony groove parallel to a stimulating paddle. The stimulating paddle was inserted from the edge of one burr hole into the epidural space overlying the precentral gyrus contralateral to the side of the body most severely affected by the relevant movement disorders. The bony bridge between the two holes, then, held the plate in place and simultaneously reduced the durocortical gap (Fig. 9.5). This technique entails no risk of epidural hematoma, and accidental displacement of the electrode has never been observed (S. Canavero, personal communication).

#### 9.3.5.2 Subdural Electrodes

In patients with advanced cortical atrophy, epidural stimulation may not be effective because of the duro-cortical separation. The cortical surface and the interhemispheric surface may be selected as targets for stimulation. However, large bridging veins sometimes interfere with implantation on the interhemispheric surface. To implant subdural electrodes, we have performed a bilateral craniotomy over the superior sagittal sinus. After opening the dura mater, the location of the central sulcus is confirmed using the phase reverse of the N20 component of the

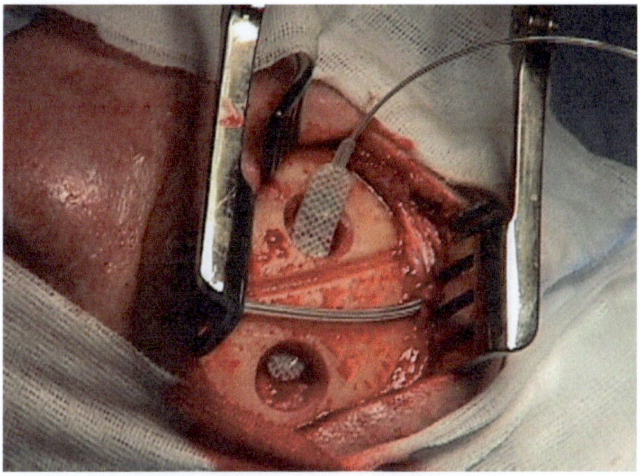

**Fig. 9.5** An image of two burr-hole surgery. The locations of burr holes are marked on the scalp according to anatomical landmarks (courtesy of Prof. Canavero)

SSEP elicited by stimulation of the median nerve. Subdural quadripolar electrodes (Resume II, model 3587A; Medtronic, Minneapolis, MN) are then placed bilaterally on the M1 adjacent to the superior sagittal sinus. At the end of surgery, in order to prevent dislocation, the lead extension is fixed to the dura or the border of the burr hole with a silk suture. However, migration of electrodes seems to be a bigger problem with the subdural approach than with the epidural approach. Meticulous, watertight dura closure is mandatory for minimizing the risk of cerebrospinal fluid leakage.

## *9.3.6 Test Stimulation*

Following closure of the craniotomy, the electrode cable(s) are connected to a trial stimulator. Based on over 3–14 days of test stimulation, the best stimulation parameters and electrodes are decided. There is considerable variation in the stimulation parameters; amplitudes range from 2 to 6 V, rates from 10 to 130 Hz, and pulse widths from 60 to 450 μsec (Canavero et al. 2002, 2003; Cilia et al. 2007, 2008; Fasano et al. 2008; Gutierrez et al. 2009; Pagni et al. 2005; Strafella et al. 2007; Tani et al. 2007). The optimal location and orientation of the electrode array are generally determined in such a way that it allows for bipolar stimulation with an appropriate pair of electrodes. Once the pulse width and frequency have been optimized, most investigators increase stimulus intensities using a percentage of the motor threshold as a guide. Many investigators begin by increasing the intensity by 20 % of the motor threshold and keep it at to 80 % of the motor threshold during the test stimulation.

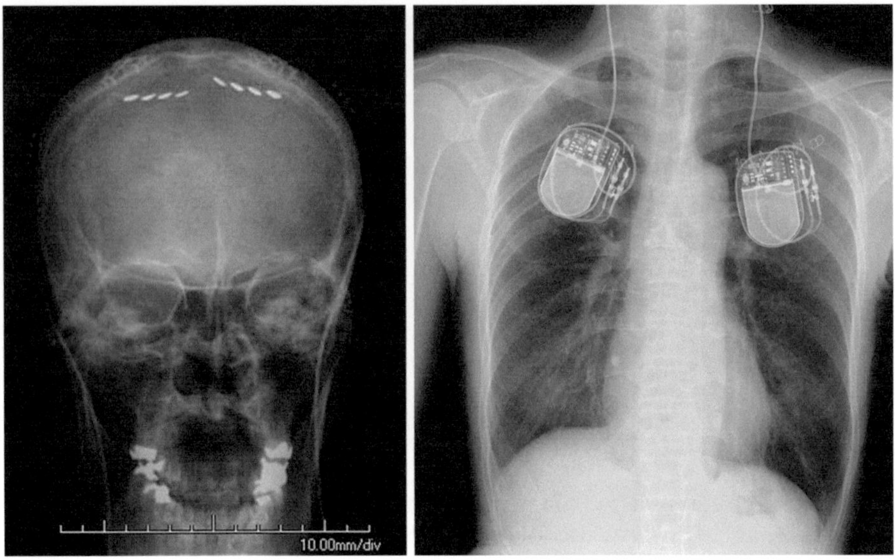


**Fig. 9.6** Subdural electrodes implanted bilaterally on the M1 (*left panel*) and connected pulse generators (*right panel*). The two Resume II subdural electrodes are connected to the pulse generators (Itrel III; Medtronic) that are implanted in the bilateral anterior chest

### 9.3.7 *Implantation of Pulse Generators*

If the test stimulation improves motor symptoms, patients are returned to the operating room and the electrode(s) are connected to implanted pulse generator (s) (Itrel III IPG; Medtronic Inc., Minneapolis, MN, USA), which are usually placed subcutaneously over the pectoralis muscle under general anesthesia (Fig. 9.6).

### 9.3.8 *Clinical Outcomes*

Table 9.1 shows clinical outcomes of MCS according to a summary of several studies. The summary includes single case reports and a multicenter retrospective clinical review; a small case series of patients with advanced PD have been reported to have variable clinical results. Some studies report a significant improvement in overall motor performance (De Rose et al. 2012; Arle and Shils 2008), whereas Cilia et al. (2007) report that epidural MCS has produced no motor benefit but subjective improvement in axial motor symptoms, such as gait, stooped posture and postural instability. Moro et al. (2011) report on the double-blind outcomes at 3 months and the unblinded outcomes at 1 year after unilateral subdural MCS in five patients with advanced PD, and conclude that no significant clinical benefit is evident for parkinsonian symptoms. Nevertheless, the scores of certain items of the

**Table 9.1** Summary of clinical outcomes of MCS from published studies

| Authors | Cases | Stimulation condition | Clinical outcome |
|---|---|---|---|
| De Rose | 10 | Epidural, unilateral, bipolar, 3.5–4.7 V, 40–80 Hz, 180 μsec | Moderate improvement in both total UPDRS and UPDRS part III in off-medication condition are observed in all patients, decreasing 12 months later. Improvements are only very slight in on-medication condition. Greater significant and long-lasting improvements are observed in axial symptoms in ff-medication condition. There is marked attenuation of L-dopa-induced dyskinesias and dystonia with significant reduction of UPDRS IV score for up to 18 months |
| Cilla | 6 | Epidural, unilateral, monopolar, 40–60 Hz, 3.0 ± 0.67 V, 180–210 μsec | No significant mean changes in UPDRS part II (40 % improvement), UPDRS part III (20 % improvement), as well as medication dose (15 % reduction) between baseline, 6-month ON stimulation, and 6-month OFF stimulation. Objective motor benefits are observed mainly in axial symptoms, such as gait, stooped posture and postural instability |
| Pagni | 6 | Epidural, unilateral, bipolar, 25–40 Hz, 2.5–6 V, 100–180 μsec | UPDRS: tremor; bilaterally: more than 80 % reduction in 3 of 4 cases; rigidity; more than 50 % reduction in 5 of 6 cases in contralateral side and in 4 of 6 cases in ipsilateral side;, activities of daily living:; more than 50 % reduction in 3 of 6 cases;, axial symptom:.; more than 50 % reduction in 3 of 6 cases;, dyskinesia:; more than 50 % reduction in 4 of 5 cases, 11–70 % reduction of L-dopa daily dose |
| Strafella | 4 | Subdural, unilateral, 50 or 130 Hz, 3–5 V, 60–90 μsec | No reduction of daily L-dopa dose; no significant difference in the UPDRS motor score between OFF stimulation (43.0 ± 7.9) and ON stimulation (39.5 ± 12.5) |
| Gutierrez | 6 | Epidural, unilateral, bipolar, 10–30 Hz, 3–4.5 V, 330–450 μsec | Mild improvement in UPDRS part III in 2 of 6 cases(14.7 % and 7.3 % improvement respectively) (ON stimulation on medication vs OFF stimulation on medication); 17.1 ± 11.1 % reduction of L-dopa daily dose |
| Canavero | 3 | Epidural, unilateral, bipolar, 20–31 Hz, 2–3.5 V, 90–330 μsec | Case 1: independent walk, absence of rigidity, trochlea, and tremor to all four limbs; 80 % reduction of L-dopa daily dose. Case 2: 50 % improvement in UPDRS part III, absence of freezing gait, tremor, improvement of bradykinesia, and 70 % reduction of L-dopa daily dose. Case 3: absence of tremor, rigidity, and improvement of gait and speech |

(continued)

**Table 9.1** (continued)

| Authors | Cases | Stimulation condition | Clinical outcome |
|---|---|---|---|
| Fasano | 1 | Epidural, bilateral, bipolar, 130 Hz, subthreshold intensity, 120 μsec | 20 % improvements of UPDRS part III, mainly in the Axial score (UPDRS items 27–30) |
| Tani | 1 | Subdural, bilateral, bipolar, 100 Hz, 1.8 V, 210 μsec | 50 % improvement in UPDRS part III and dramatic improvement in walking |

Unified Parkinson's Disease Rating Scale (UPDRS), such as gait and balance, show a slight improvement when patients with PD were off medication. The improvement predominantly involves axial symptoms, L-dopa-induced dyskinesias and dystonia, and quality of life. Overall, the degree of the clinical improvement obtained with MCS is lower than the reported clinical improvement with DBS (Cilia et al. 2008; Gutierrez et al. 2009).

Clinical improvements are produced by MCS several days to 4 weeks after stimulation parameter modifications (Cilia et al. 2008). Several investigators have reported that rigidity and tremor are abolished within several minutes of MCS, but the full effects of MCS on bradykinesia, axial symptoms, and especially gait disturbances are evident only after a longer period of time. After cessation of MCS, a slow worsening of the motor symptoms occurs over more than 2 days (Canavero and Bonicalzi 2002; Cilia et al. 2008; Pagni et al. 2003; Tani et al. 2007). Some studies have assessed the motor symptoms soon after modification of stimulation parameters, and so the improvement in motor symptom might have been underestimated.

### *9.3.9 Complications*

The only complication reported concerning MCS for PD was misplaced electrode (Pagni et al. 2005). Because the number of reported cases of MCS for PD is limited, this section focuses on reported complications of MCS for patients with intractable pain. Although the majority of studies have reported no adverse events of MCS for patients with intractable pain (Gharabaghi et al. 2005; Saitoh et al. 2000; Tsubokawa et al. 1991b), serious complications have been reported. Montes et al. (2002) analyzed event-related potentials and behavioral performance during an auditory target-detection task in 11 consecutive patients. They are tested both during MCS and 10 min after MCS was discontinued. Sensory responses were unaffected by MCS, but there was a significant delay of brain potentials reflecting target detection in older patients during MCS, which was rapidly reversed after MCS was discontinued. No effect was observed in patients aged younger than 50 years. Cognitive effects of MCS were mild and non-specific, and directly related to the stimulation period (i.e., with no post-effect), being similar to the cognitive effects reported during rTMS of M1. Thus, MCS may interfere with relatively simple cognitive processes, such as those underlying target detection, notably in the elderly and in the presence of preexisting cerebral lesions.

The occurrence of epileptic seizures has been reported during test stimulation in a minority of patients. The low rate of epileptic seizures during chronic stimulation (0.2 %) means that stimulation of M1 within an appropriate range of parameters is reasonably safe. The reported most serious complications are epidural and subdural hematomas. The occurrence of hematomas is exceptional in the epidural approach, and some surgeons have never observed them. Therefore, the risk of peri-operative hemorrhage is much lower than with DBS.

Wound infections have been reported by most neurosurgeons. If infection occurs, all devices, including paddles, extension leads, and pulse generators, must be removed temporally. The implanted pulse generator can accidentally turn off due to electromagnetic interference from household devices in a close (<10 cm) proximity, such as electric appliances, anti-theft devices, metal detectors, and magnets in loudspeakers. If impedance exceeds 2000 Ω, a connection problem, such as broken cable or lead fracture, must be suspected. The operator thus should measure impedance in a unipolar configuration and assign a value to a single contact.

## 9.4 Mechanisms of Action

The exact mechanisms of action of MCS are poorly understood. It is noteworthy that while the effects of MCS on rigidity and tremor are almost immediate (observed within the first minute of stimulation), the clinical benefit on akinesia and axial symptoms requires a longer stimulation time. It is known that the time required for high-frequency STN-DBS to have clinical effect varies, depending on the type of parkinsonian motor symptom;, it is fast (less than 1 min) for rigidity and tremor, and slow (from a few minutes to a few days) for other symptoms, such as bradykinesia and akinesia (Krack et al. 2002). As discussed by others, the delay in clinical benefit may be due to synaptic plasticity, long-term potentiation, long-term depression, expression of secondary messengers or polarization of brain tissue (Drouot et al. 2004; Krack et al. 2002; Priori and Lefaucheur 2007), and the immediate clinical benefit may result from imposing a specific pattern of activity and suppressing abnormal, disease-associated rhythmicity of oscillation in the corticobasal ganglia-cortical circuit (Brown 2006; Fasano et al. 2008; Garcia et al. 2003; Priori and Lefaucheur 2007).

Studies of rTMS of M1 reveal that PD is associated with excessive excitability or reduced inhibition in M1 (Cantello et al. 2002), and rigidity and tremor may be caused by the hyperactivity of M1 (Haslinger et al. 2001; Rodriguez-Oroz et al. 2009). In patients with PD, activation of M1 during production of a voluntary output is inadequately modulated, perhaps because intracortical inhibition mediated by γ-aminobutyric acid A (GABA) and GABA receptors is reduced (Cantello et al. 2002). Canavero et al. suggest that MCS increases cortical GABA in patients affected by central pain syndromes (Canavero and Bonicalzi 1995, 1998). MCS might reduce cortical hyperactivity by increasing GABA concentration and activating inhibitory neurons (Hanajima et al. 2002). Indeed, 99mTc-ECD SPECT (Single Photon Emission Computerized Tomography) data has demonstrated that resting-state neuronal activity decreases in the motor cortical areas during MCS (Cilia et al. 2008).

Functional neuroimaging studies have shown a significant increase in cerebral perfusion in the SMA and the DLPFC (dorsolateral prefrontal cortex) in ON-stimulation condition (Drouot et al. 2004; Fasano et al. 2008; Tani et al. 2007). The SMA and the DLPFC are under-active in patients with PD. This probably underlies bradykinesia (Haslinger et al. 2001; Jahanshahi et al. 1995; Rascol et al. 1992). These cortical metabolic abnormalities can be reversed by antiparkinsonian therapy, such as dopaminergic treatment (Jenkins et al. 1992), pallidotomy (Grafton et al. 1995), STN-DBS (Limousin et al. 1997), or GPi (Globus pallidus interna)-DBS (Fukuda et al. 2001). These types of therapies having similar results suggest that they may have similar therapeutic benefit in patients with PD, and might have some common mechanism (Table 9.2).

**Table 9.2** Summary of functional neuroimaging studies

| Authors | Cases | Stimulation | Modality | Neuroimaging |
|---|---|---|---|---|
| Cilla | 6 | Unilateral, 40–60 Hz | Tc-SPECT ON stimulation vs OFF stimulation under rest | Decrease in bilateral M1, bilateral premotor cortex, left DLPFC, right caudate nucleus, and left middle occipital gyrus |
| | | | | Increase in right cerebellar lobe, left inferior occipital gyrus, left cerevellar lobe, and vermis. |
| Strafella | 4 | Unilateral, 50 or 130 Hz | [$^{15}$O] $H_2O$ PET movement vs rest | Changes in rCBF were not significantly different if compared across different stimulation settings (OFF vs 50 Hz vs 130 Hz) |
| Canavero | 3 | Unilateral, 20–31 Hz | IBZM-SPECT | Before MCS, asymmetrical binding (right less than left) in basal ganglia |
| | | | | During MCS, renormalization of basal ganglia anomalies |
| | | | ECD-SPECT | Before MCS, bilateral parieto-temporal hypoperfusion |
| | | | | During MCS, renormalization of cortical metabolism on the side of stimulation but not contralaterally |
| Fasano | 1 | Bilateral, 130 Hz | ECD-SPECT OFF stimulation vs ON stimulation (on medication) | Increase in right frontal, right parietal and left frontal cortices |
| Tani | 1 | Bilateral, 100 Hz | [$^{15}$O] $H_2O$ PET STIM-OFF vs STIM-ON (on medication) | Increase in left SMA and right DLPFC |

## 9.5 Conclusion

MCS can provide benefit for patients with PD, and this benefit differs from the one observed with DBS. Although MSC produces a lower degree of clinical improvement than DBS does, the fact that MCS can improve axial symptoms of PD provides invaluable benefits for patients with PD; it is difficult for STN-DBSTMS to do this. In addition, MCS may have valuable applications in the treatment of movement disorders. But further work should be carried out to elucidate their mechanisms of action and therapeutic values.

## Exercise

Which is the best target of the direct electrical cortical stimulation for Parkinson's disease?

1. Supplementary motor cortex
2. Primary motor cortex of hand area
3. Primary motor cortex of leg area
4. Primary sensory cortex of hand area
5. Cerebellar cortex

## References

Andre-Obadia, N., Peyron, R., Mertens, P., Mauguiere, F., Laurent, B., Garcia-Larrea, L.: Transcranial magnetic stimulation for pain control. Double-blind study of different frequencies against placebo, and correlation with motor cortex stimulation efficacy. Clin. Neurophysiol. **117**, 1536–1544 (2006)

Arle, J.E., Shils, J.L.: Motor cortex stimulation for pain and movement disorders. Neurotherapeutics **5**, 37–49 (2008)

Boraud, T., Bezard, E., Guehl, D., Bioulac, B., Gross, C.: Effects of L-DOPA on neuronal activity of the globus pallidus externalis (GPe) and globus pallidus internalis (GPi) in the MPTP-treated monkey. Brain Res. **787**, 157–160 (1998)

Bornke, C., Schulte, T., Przuntek, H., Muller, T.: Clinical effects of repetitive transcranial magnetic stimulation versus acute levodopa challenge in Parkinson's disease. J. Neural Transm. Suppl. **68**, 61–67 (2004)

Boylan, L.S., Pullman, S.L., Lisanby, S.H., Spicknall, K.E., Sackeim, H.A.: Repetitive transcranial magnetic stimulation to SMA worsens complex movements in Parkinson's disease. Clin. Neurophysiol. **112**, 259–264 (2001)

Brown, P.: Bad oscillations in Parkinson's disease. J. Neural Transm. Suppl. **70**, 27–30 (2006)

Brusa, L., Versace, V., Koch, G., Iani, C., Stanzione, P., Bernardi, G., Centonze, D.: Low frequency rTMS of the SMA transiently ameliorates peak-dose LID in Parkinson's disease. Clin. Neurophysiol. **117**, 1917–1921 (2006)

Brusa, L., Finazzi Agro, E., Petta, F., Sciobica, F., Torriero, S., Lo Gerfo, E., Iani, C., Stanzione, P., Koch, G.: Effects of inhibitory rTMS on bladder function in Parkinson's disease patients. Mov. Disord. **24**, 445–448 (2009)

Canavero, S., Bonicalzi, V.: Cortical stimulation for central pain. J. Neurosurg. **83**, 1117 (1995)
Canavero, S., Bonicalzi, V.: The neurochemistry of central pain: evidence from clinical studies, hypothesis and therapeutic implications. Pain **74**, 109–114 (1998)
Canavero, S., Bonicalzi, V.: Therapeutic extradural cortical stimulation for central and neuropathic pain: a review. Clin. J. Pain **18**, 48–55 (2002)
Canavero, S., Bonicalzi, V.: Extradural cortical stimulation for movement disorders. Acta Neurochir. Suppl. **97**, 223–232 (2007)
Canavero, S., Paolotti, R., Bonicalzi, V., Castellano, G., Greco-Crasto, S., Rizzo, L., Davini, O., Zenga, F., Ragazzi, P.: Extradural motor cortex stimulation for advanced Parkinson disease. Report of two cases. J. Neurosurg. **97**, 1208–1211 (2002)
Canavero, S., Bonicalzi, V., Paolotti, R., Castellano, G., Greco-Crasto, S., Rizzo, L., Davini, O., Maina, R.: Therapeutic extradural cortical stimulation for movement disorders: a review. Neurol. Res. **25**, 118–122 (2003)
Cantello, R., Tarletti, R., Civardi, C.: Transcranial magnetic stimulation and Parkinson's disease. Brain Res. Brain Res. Rev. **38**, 309–327 (2002)
Cilia, R., Landi, A., Vergani, F., Sganzerla, E., Pezzoli, G., Antonini, A.: Extradural motor cortex stimulation in Parkinson's disease. Mov. Disord. **22**, 111–114 (2007)
Cilia, R., Marotta, G., Landi, A., Isaias, I.U., Vergani, F., Benti, R., Sganzerla, E., Gerundini, P., Pezzoli, G., Antonini, A.: Cerebral activity modulation by extradural motor cortex stimulation in Parkinson's disease: a perfusion SPECT study. Eur. J. Neurol. **15**, 22–28 (2008)
De Groot, M., Hermann, W., Steffen, J., Wagner, A., Grahmann, F.: Contralateral and ipsilateral repetitive transcranial magnetic stimulation in Parkinson patients. Nervenarzt **72**, 932–938 (2001)
de Rose, M., Guzzi, G., Bosco, D., Romano, M., Lavano, S.M., Plastino, M., Volpentesta, G., Marotta, R., Lavano, A.: Motor cortex stimulation in Parkinson's disease. Neurol. Res. Int. **2012**, 502096 (2012)
Delong, M.R.: Primate models of movement disorders of basal ganglia origin. Trends Neurosci. **13**, 281–285 (1990)
Dias, A.E., Barbosa, E.R., Coracini, K., Maia, F., Marcolin, M.A., Fregni, F.: Effects of repetitive transcranial magnetic stimulation on voice and speech in Parkinson's disease. Acta Neurol. Scand. **113**, 92–99 (2006)
Drouot, X., Oshino, S., Jarraya, B., Besret, L., Kishima, H., Remy, P., Dauguet, J., Lefaucheur, J. P., Dolle, F., Conde, F., Bottlaender, M., Peschanski, M., Keravel, Y., Hantraye, P., Palfi, S.: Functional recovery in a primate model of Parkinson's disease following motor cortex stimulation. Neuron **44**, 769–778 (2004)
Fasano, A., Piano, C., de Simone, C., Cioni, B., di Giuda, D., Zinno, M., Daniele, A., Meglio, M., Giordano, A., Bentivoglio, A.R.: High frequency extradural motor cortex stimulation transiently improves axial symptoms in a patient with Parkinson's disease. Mov. Disord. **23**, 1916–1919 (2008)
Filion, M., Tremblay, L.: Abnormal spontaneous activity of globus pallidus neurons in monkeys with MPTP-induced parkinsonism. Brain Res. **547**, 142–151 (1991)
Franzini, A., Ferroli, P., Servello, D., Broggi, G.: Reversal of thalamic hand syndrome by long-term motor cortex stimulation. J. Neurosurg. **93**, 873–875 (2000)
Fukuda, M., Mentis, M., Ghilardi, M.F., Dhawan, V., Antonini, A., Hammerstad, J., Lozano, A.M., Lang, A., Lyons, K., Koller, W., Ghez, C., Eidelberg, D.: Functional correlates of pallidal stimulation for Parkinson's disease. Ann. Neurol. **49**, 155–164 (2001)
Garcia, L., Audin, J., D'alessandro, G., Bioulac, B., Hammond, C.: Dual effect of high-frequency stimulation on subthalamic neuron activity. J. Neurosci. **23**, 8743–8751 (2003)
Gharabaghi, A., Hellwig, D., Rosahl, S.K., Shahidi, R., Schrader, C., Freund, H.J., Samii, M.: Volumetric image guidance for motor cortex stimulation: integration of three-dimensional cortical anatomy and functional imaging. Neurosurgery **57**, 114–120 (2005); discussion 114-20

Goetz, C.G., Poewe, W., Rascol, O., Sampaio, C.: Evidence-based medical review update: pharmacological and surgical treatments of Parkinson's disease: 2001 to 2004. Mov. Disord. **20**, 523–539 (2005)

Grafton, S.T., Waters, C., Sutton, J., Lew, M.F., Couldwell, W.: Pallidotomy increases activity of motor association cortex in Parkinson's disease: a positron emission tomographic study. Ann. Neurol. **37**, 776–783 (1995)

Greenberg, M.: Handbook of Neurosurgery. Thieme Verlag, New York (2001)

Gutierrez, J.C., Seijo, F.J., Alvarez Vega, M.A., Fernandez Gonzalez, F., Lozano Aragoneses, B., Blazquez, M.: Therapeutic extradural cortical stimulation for Parkinson's disease: report of six cases and review of the literature. Clin. Neurol. Neurosurg. **111**, 703–707 (2009)

Hamada, M., Ugawa, Y., Tsuji, S.: High-frequency rTMS over the supplementary motor area for treatment of Parkinson's disease. Mov. Disord. **23**, 1524–1531 (2008)

Hamada, M., Ugawa, Y., Tsuji, S.: High-frequency rTMS over the supplementary motor area improves bradykinesia in Parkinson's disease: subanalysis of double-blind sham-controlled study. J. Neurol. Sci. **287**, 143–146 (2009)

Hanajima, R., Ashby, P., Lang, A.E., Lozano, A.M.: Effects of acute stimulation through contacts placed on the motor cortex for chronic stimulation. Clin. Neurophysiol. **113**, 635–641 (2002)

Haslinger, B., Erhard, P., Kampfe, N., Boecker, H., Rummeny, E., Schwaiger, M., Conrad, B., Ceballos-Baumann, A.O.: Event-related functional magnetic resonance imaging in Parkinson's disease before and after levodopa. Brain **124**, 558–570 (2001)

Holsheimer, J., Nguyen, J.P., Lefaucheur, J.P., Manola, L.: Cathodal, anodal or bifocal stimulation of the motor cortex in the management of chronic pain? Acta Neurochir. Suppl. **97**, 57–66 (2007)

Hosomi, K., Saitoh, Y., Kishima, H., Oshino, S., Hirata, M., Tani, N., Shimokawa, T., Yoshimine, T.: Electrical stimulation of primary motor cortex within the central sulcus for intractable neuropathic pain. Clin. Neurophysiol. **119**, 993–1001 (2008)

Jahanshahi, M., Jenkins, I.H., Brown, R.G., Marsden, C.D., Passingham, R.E., Brooks, D.J.: Self-initiated versus externally triggered movements. I. An investigation using measurement of regional cerebral blood flow with PET and movement-related potentials in normal and Parkinson's disease subjects. Brain **118**(Pt 4), 913–933 (1995)

Jenkins, I.H., Fernandez, W., Playford, E.D., Lees, A.J., Frackowiak, R.S., Passingham, R.E., Brooks, D.J.: Impaired activation of the supplementary motor area in Parkinson's disease is reversed when akinesia is treated with apomorphine. Ann. Neurol. **32**, 749–757 (1992)

Katayama, Y., Fukaya, C., Yamamoto, T.: Poststroke pain control by chronic motor cortex stimulation: neurological characteristics predicting a favorable response. J. Neurosurg. **89**, 585–591 (1998)

Katayama, Y., Oshima, H., Fukaya, C., Kawamata, T., Yamamaoto, T.: Control of post-stroke movement disorders using chronic motor cortex stimulation. Acta Neurochir. Suppl. **79**, 89–92 (2002)

Koch, G., Oliveri, M., Brusa, L., Stanzione, P., Torriero, S., Caltagirone, C.: High-frequency rTMS improves time perception in Parkinson disease. Neurology **63**, 2405–2406 (2004)

Koch, G., Brusa, L., Caltagirone, C., Peppe, A., Oliveri, M., Stanzione, P., Centonze, D.: rTMS of supplementary motor area modulates therapy-induced dyskinesias in Parkinson disease. Neurology **65**, 623–625 (2005)

Krack, P., Pollak, P., Limousin, P., Hoffmann, D., Xie, J., Benazzouz, A., Benabid, A.L.: Subthalamic nucleus or internal pallidal stimulation in young onset Parkinson's disease. Brain **121**(Pt 3), 451–457 (1998)

Krack, P., Fraix, V., Mendes, A., Benabid, A.L., Pollak, P.: Postoperative management of subthalamic nucleus stimulation for Parkinson's disease. Mov. Disord. **17**(Suppl 3), S188–S197 (2002)

Lefaucheur, J.P., Drouot, X., von Raison, F., Menard-Lefaucheur, I., Cesaro, P., Nguyen, J.P.: Improvement of motor performance and modulation of cortical excitability by repetitive transcranial magnetic stimulation of the motor cortex in Parkinson's disease. Clin. Neurophysiol. **115**, 2530–2541 (2004)

Leichnetz, G.R.: Afferent and efferent connections of the dorsolateral precentral gyrus (area 4, hand/arm region) in the macaque monkey, with comparisons to area 8. J. Comp. Neurol. **254**, 460–492 (1986)

Limousin, P., Greene, J., Pollak, P., Rothwell, J., Benabid, A.L., Frackowiak, R.: Changes in cerebral activity pattern due to subthalamic nucleus or internal pallidum stimulation in Parkinson's disease. Ann. Neurol. **42**, 283–291 (1997)

Limousin, P., Krack, P., Pollak, P., Benazzouz, A., Ardouin, C., Hoffmann, D., Benabid, A.L.: Electrical stimulation of the subthalamic nucleus in advanced Parkinson's disease. N. Engl. J. Med. **339**, 1105–1111 (1998)

Lopiano, L., Rizzone, M., Bergamasco, B., Tavella, A., Torre, E., Perozzo, P., Lanotte, M.: Deep brain stimulation of the subthalamic nucleus in PD: an analysis of the exclusion causes. J. Neurol. Sci. **195**, 167–170 (2002)

Manola, L., Holsheimer, J., Veltink, P., Buitenweg, J.R.: Anodal vs cathodal stimulation of motor cortex: a modeling study. Clin. Neurophysiol. **118**, 464–474 (2007)

Montes, C., Mertens, P., Convers, P., Peyron, R., Sindou, M., Laurent, B., Mauguiere, F., Garcia-Larrea, L.: Cognitive effects of precentral cortical stimulation for pain control: an ERP study. Neurophysiol. Clin. **32**, 313–325 (2002)

Moro, E., Schwalb, J.M., Piboolnurak, P., Poon, Y.Y., Hamani, C., Hung, S.W., Arenovich, T., Lang, A.E., Chen, R., Lozano, A.M.: Unilateral subdural motor cortex stimulation improves essential tremor but not Parkinson's disease. Brain **134**, 2096–2105 (2011)

Naidich, T.P., Valavanis, A.G., Kubik, S.: Anatomic relationships along the low-middle convexity: part I – Normal specimens and magnetic resonance imaging. Neurosurgery **36**, 517–532 (1995)

Naidich, T.P., Blum, J.T., Firestone, M.I.: The parasagittal line: an anatomic landmark for axial imaging. AJNR Am. J. Neuroradiol. **22**, 885–895 (2001)

Nguyen, J.P., Pollin, B., Feve, A., Geny, C., Cesaro, P.: Improvement of action tremor by chronic cortical stimulation. Mov. Disord. **13**, 84–88 (1998)

Nguyen, J.P., Lefaucheur, J.P., Decq, P., Uchiyama, T., Carpentier, A., Fontaine, D., Brugieres, P., Pollin, B., Feve, A., Rostaing, S., Cesaro, P., Keravel, Y.: Chronic motor cortex stimulation in the treatment of central and neuropathic pain. Correlations between clinical, electrophysiological and anatomical data. Pain **82**, 245–251 (1999)

Okabe, S., Ugawa, Y., Kanazawa, I.: 0.2-Hz repetitive transcranial magnetic stimulation has no add-on effects as compared to a realistic sham stimulation in Parkinson's disease. Mov. Disord. **18**, 382–388 (2003)

Pagni, C.A., Zeme, S., Zenga, F.: Further experience with extradural motor cortex stimulation for treatment of advanced Parkinson's disease. Report of 3 new cases. J. Neurosurg. Sci. **47**, 189–193 (2003)

Pagni, C.A., Zeme, S., Zenga, F., Maina, R.: Extradural motor cortex stimulation in advanced Parkinson's disease: the Turin experience: technical case report. Neurosurgery **57**, E402 (2005); discussion E402

Pagni, C.A., Albanese, A., Bentivoglio, A., Broggi, G., Canavero, S., Cioni, B., Rose, M.D., Simone, C.D., Franzini, A., Lavano, A., Landi, A., Meglio, M., Modugno, M., Romanelli, L., Romito, L.M., Sturiale, C., Valzania, F., Zeme, S., Zenga, F.: Results by motor cortex stimulation in treatment of focal dystonia, Parkinson's disease and post-ictal spasticity. The experience of the Italian Study Group of the Italian Neurosurgical Society. Acta Neurochir. Suppl. **101**, 13–21 (2008)

Pahwa, R., Wilkinson, S.B., Overman, J., Lyons, K.E.: Preoperative clinical predictors of response to bilateral subthalamic stimulation in patients with Parkinson's disease. Stereotact. Funct. Neurosurg. **83**, 80–83 (2005)

Priori, A., Lefaucheur, J.P.: Chronic epidural motor cortical stimulation for movement disorders. Lancet Neurol. **6**, 279–286 (2007)

Rasche, D., Ruppolt, M., Stippich, C., Unterberg, A., Tronnier, V.M.: Motor cortex stimulation for long-term relief of chronic neuropathic pain: a 10 year experience. Pain **121**, 43–52 (2006)

Rascol, O., Sabatini, U., Chollet, F., Celsis, P., Montastruc, J.L., Marc-Vergnes, J.P., Rascol, A.: Supplementary and primary sensory motor area activity in Parkinson's disease. Regional cerebral blood flow changes during finger movements and effects of apomorphine. Arch. Neurol. **49**, 144–148 (1992)

Rodriguez-Oroz, M.C., Jahanshahi, M., Krack, P., Litvan, I., Macias, R., Bezard, E., Obeso, J.A.: Initial clinical manifestations of Parkinson's disease: features and pathophysiological mechanisms. Lancet Neurol. **8**, 1128–1139 (2009)

Saitoh, Y., Hosomi, K.: From localization to surgical implantation. In: Canavero, S. (ed.) Therapeutic Cortical Stimulation. Nova Science Publishers, Inc., New York (2009)

Saitoh, Y., Shibata, M., Hirano, S., Hirata, M., Mashimo, T., Yoshimine, T.: Motor cortex stimulation for central and peripheral deafferentation pain. Report of eight cases. J. Neurosurg. **92**, 150–155 (2000)

Siebner, H.R., Rossmeier, C., Mentschel, C., Peinemann, A., Conrad, B.: Short-term motor improvement after sub-threshold 5-Hz repetitive transcranial magnetic stimulation of the primary motor hand area in Parkinson's disease. J. Neurol. Sci. **178**, 91–94 (2000)

Siebner, H.R., Bergmann, T.O., Bestmann, S., Massimini, M., Johansen-Berg, H., Mochizuki, H., Bohning, D.E., Boorman, E.D., Groppa, S., Miniussi, C., Pascual-Leone, A., Huber, R., Taylor, P. C., Ilmoniemi, R.J., DE Gennaro, L., Strafella, A.P., Kahkonen, S., Kloppel, S., Frisoni, G.B., George, M.S., Hallett, M., Brandt, S.A., Rushworth, M.F., Ziemann, U., Rothwell, J.C., Ward, N., Cohen, L.G., Baudewig, J., Paus, T., Ugawa, Y., Rossini, P.M.: Consensus paper: combining transcranial stimulation with neuroimaging. Brain Stimul. **2**, 58–80 (2009)

Strafella, A.P., Lozano, A.M., Lang, A.E., Ko, J.H., Poon, Y.Y., Moro, E.: Subdural motor cortex stimulation in Parkinson's disease does not modify movement-related rCBF pattern. Mov. Disord. **22**, 2113–2116 (2007)

Tani, N., Saitoh, Y.: Electrical stimulation of primary motor cortex for Parkinson's syndrome. In: Dushanova, J. (eds.) Diagnostics and Rehabilitation of Parkinson's Disease, pp. 493–508. InTech (2011)

Tani, N., Saitoh, Y., Kishima, H., Oshino, S., Hatazawa, J., Hashikawa, K., Yoshimine, T.: Motor cortex stimulation for levodopa-resistant akinesia: case report. Mov. Disord. **22**, 1645–1649 (2007)

Thickbroom, G.W., Byrnes, M.L., Walters, S., Stell, R., Mastaglia, F.L.: Motor cortex reorganisation in Parkinson's disease. J. Clin. Neurosci. **13**, 639–642 (2006)

Tirakotai, W., Hellwig, D., Bertalanffy, H., Riegel, T.: Localization of precentral gyrus in image-guided surgery for motor cortex stimulation. Acta Neurochir. Suppl. **97**, 75–79 (2007)

Tsubokawa, T., Katayama, Y., Yamamoto, T., Hirayama, T., Koyama, S.: Chronic motor cortex stimulation for the treatment of central pain. Acta Neurochir. Suppl. (Wien) **52**, 137–139 (1991a)

Tsubokawa, T., Katayama, Y., Yamamoto, T., Hirayama, T., Koyama, S.: Treatment of thalamic pain by chronic motor cortex stimulation. Pacing Clin. Electrophysiol. **14**, 131–134 (1991b)

Velasco, M., Velasco, F., Brito, F., Velasco, A.L., Nguyen, J.P., Marquez, I., Boleaga, B., Keravel, Y.: Motor cortex stimulation in the treatment of deafferentation pain. I. Localization of the motor cortex. Stereotact. Funct. Neurosurg. **79**, 146–167 (2002)

Wood, C.C., Spencer, D.D., Allison, T., Mccarthy, G., Williamson, P.D., Goff, W.R.: Localization of human sensorimotor cortex during surgery by cortical surface recording of somatosensory evoked potentials. J. Neurosurg. **68**, 99–111 (1988)

# Chapter 10
# Brain Machine-Interfaces for Sensory Systems

**Takashi Fujikado**

**Abstract** A brain–machine interface (BMI) is a system that provides direct communication between the brain and an external device. It is designed to assist or repair human cognitive or sensory-motor functions. BMI research has studied sensory systems with a focus on the use of neuro-prosthetic devices to restore impaired hearing or vision.

Because of cortical plasticity, signals from implanted prostheses can be processed by the brain after rehabilitation.

The most widely used neuro-prosthetic devices are cochlear implants. A cochlear implant is activated by sound waves, and signals generated from it are filtered and decomposed into an envelope and a temporal structure, which are then converted into electrical energy. The electrical energy then stimulates the auditory nerve. Speech perception scores typically continue to improve during the first 3–12 months of implant use, suggesting that plastic changes occur in the brain to use sparse inputs better.

In retinal prosthesis, a camera takes an image, and a computer processes and transmits it wirelessly to the implant. The implant maps the image across an array of electrodes, and it stimulates neurons in the retina and sends neural signals to the visual cortex. The recipients then perceive a monochromatic pattern of dots.

Several approaches exist in rental prosthesis; epiretinal prosthesis in which electrode array is inserted under the retina, subretinal prosthesis in which electrode array is fixed on the retina, suprachoroidal prosthesis in which electrode array is inserted in the suprachoroidal space or in the sclera pocket.

Currently, the best decimal visual acuity achieved by retinal prosthesis is 0.037 by subretinal prosthesis. Even though the visual acuity is still poor, patients with implants can read large letters. To achieve an improvement of the quality of life for blind patients implanted with a retinal prosthesis, not only an improvement of surgical procedures and engineering advancement but also a development of effective rehabilitation are necessary.

**Keywords** Brain-machine interface • Cochlear implant • Retinal prosthesis • Visual prosthesis

---

T. Fujikado (✉)
Graduate School of Medicine, Osaka University, Suita, Osaka, Japan
e-mail: fujikado@ophthal.med.osaka-u.ac.jp



M. Kasaki et al. (eds.), *Cognitive Neuroscience Robotics B*,
DOI 10.1007/978-4-431-54598-9_10

## 10.1 Introduction

A brain–machine interface (BMI) (also called a 'brain–computer interface' (BCI) recently) is a system that provides direct communication between the brain and an external device. A BMI is designed to assist or repair human cognitive or sensory-motor functions. BMI research studies sensory systems with a focus on neural prostheses, and aims at restoring impaired hearing or vision. Because of cortical plasticity, signals from implanted prostheses can be processed by the brain after rehabilitation. The most widely used neuro-prosthetic devices are )cochlear implants (auditory prostheses), and they have been implanted in approximately 220,000 people worldwide until 2011 (Wilson et al. 2011).

## 10.2 Cochlear Implants

### *10.2.1 General*

Cochlear implants (CIs) contain an electrode array, and it is implanted into the cochlea. CIs detect and process sounds and generate electrical signals to stimulate the auditory nerve. The microphone of the CI system receives sound waves from the external environment and sends sound wave signals to a processor. The processor digitizes sound waves and splits them into separate frequency bands.

Digitized sound waves are sent to appropriate tonotopic regions in the cochlea. For example, activation near the basal end of the cochlea indicates the presence of high-frequency sounds, and activation in the regions closer to the apical end indicates the presence of lower-frequency sounds (Fig. 10.1).

### *10.2.2 Speech Perception After Implantation of Cochlear Implant*

The number of electrodes in a CI is up to 22, and the number of auditory neurons is 30,000. So the spatial selectivity of CI is poor (Rubinstein 2004).

Once a cochlear implant is implanted in patients, they generally tend to improve their speech perception scores gradually after they have auditory experience with their implants. The scores typically continue to improve during the first 3–12 months of implant use (Helms et al. 1997). This time span reflects plastic changes in brain function that are necessary to utilize the sparse input of the implant. These results generally apply to patients who have lost their hearing after they acquire language (postlingually deafened patients). If children are implanted at the age of 18 months or younger, then the results of implantation are as good as those of postlingually deafened patients (Niparko et al. 2010). If children

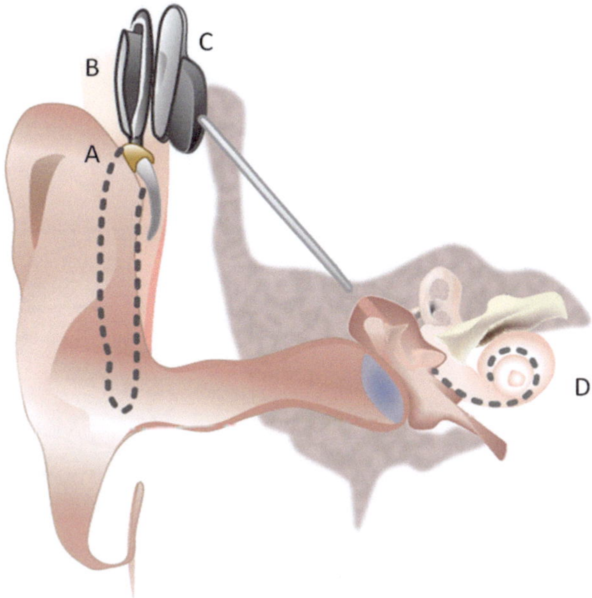


**Fig. 10.1** Schematic diagram of a cochlear implant. The microphone (*A*) receives sound from the external environment and sends it to the processor (*B*). The processor digitizes the sound and filters it into separate frequency bands. They are sent to the internal devices (*C*), and then to the stimulating electrodes inserted in the tonotopic region of the cochlea (*D*)

are implanted at 2 or 3 years of age, the results of implantation are usually worse than they are for postlingual patients. The results of implantation are very poor for children implanted after 4–6 years of age.

The improvement of performance of cochlear implants depends on the understanding of the plasticity of the brain and sensitive periods of brain development. The combination of electrical engineering, signal processing, and cognitive neuroscience is necessary to maximize the performance of cochlear implant.

## 10.3 Visual Prosthesis

### *10.3.1 General*

A visual prosthesis creates a visual image by electrically stimulating neurons in the visual system. A camera takes an image, and a computer processes and transmits it wirelessly to the implant. The implant maps the image across an array of electrodes, and it stimulates neurons in the retina and sends neural signals to the visual cortex. This process creates an image of visual objects. Any point along the optic pathway can be stimulated, such as the optic nerve or visual cortex (da Cruz et al. 2013).

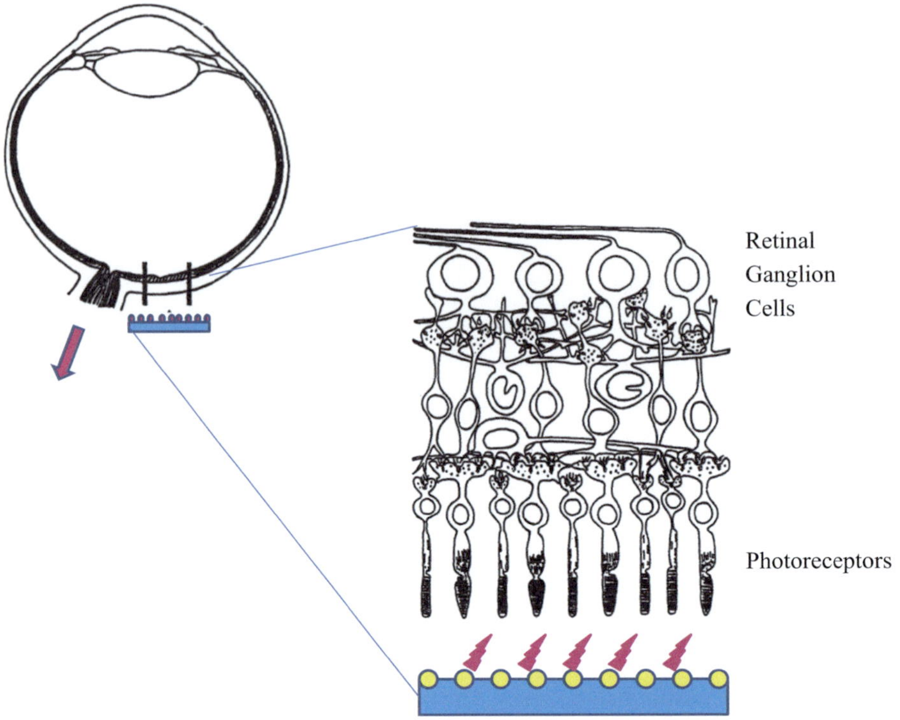


**Fig. 10.2** Schematic diagram of retinal structure and the effect of electrical stimulation by a retinal implant. Photoreceptors are neurons that specialize in converting photons into electrical signals. Bioelectrical signals sent from the photoreceptors of the retina are processed by the retinal neurons and sent to the visual cortex via retinal ganglion cells. Electrical currents stimulate the inner retinal neurons and activate the retinal ganglion cells in eyes of patients who have photoreceptor degeneration and are implanted with a retinal prosthesis

### *10.3.2 Retinal Structures*

When light enters the eye, light stimuli are received by the photoreceptors, i.e., the rods and cones of the retina. Photoreceptors are neurons that specialize in converting photons into electrical signals. Bioelectrical signals are sent from the photoreceptors of the retina, and then they are processed by the retinal neurons and sent to the visual cortex through the axons of the retinal ganglion cells in the optic nerve (Fig. 10.2).

Damage to any part of this pathway can cause impairment of vision or even blindness. Age-related macular degeneration (AMD) and retinitis pigmentosa (RP) are two common degenerative diseases that affect the outer retina. A retinal prosthesis stimulates residual retinal neurons by electrical pulses and produces phosphenes in eyes with photoreceptor degeneration.

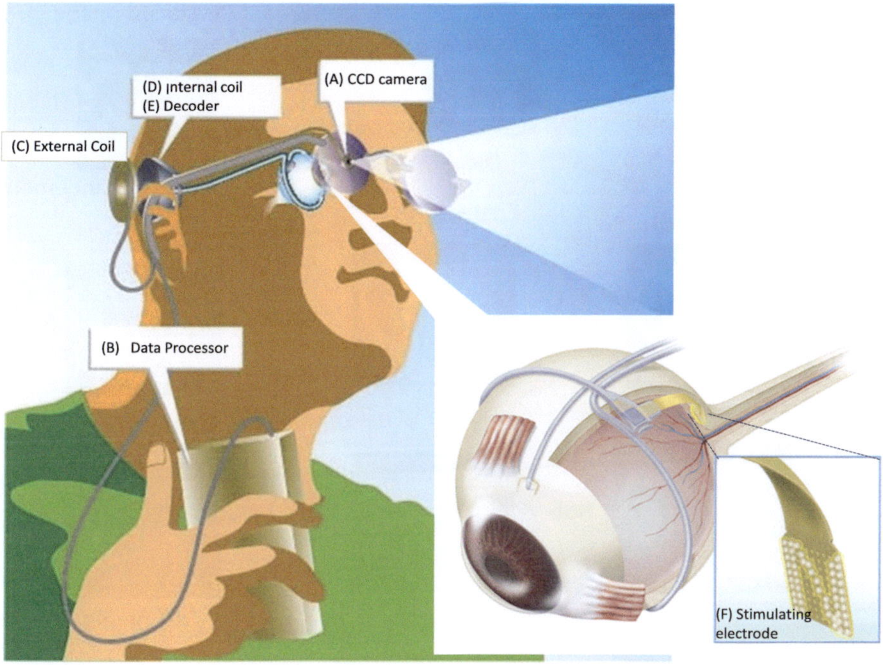


**Fig. 10.3** Schematic diagram of the entire visual prosthesis system. The external visual prosthesis system consists of a CCD camera (*A*), a data processor (*B*), and an external coil (*C*). If an implanted imaging system is used, the external devices are not required. Power and data are transmitted to the internal system wirelessly from the external unit. The internal system consists of an internal coil (*D*), a decoder (*E*), and a stimulating electrode array (*F*). The decoder receives power and data and converts data to a signal. The signal is delivered to the nervous system through stimulating electrodes. Stimulating electrodes can be placed on the retina, optic nerve, or visual cortex

### *10.3.3 Systems of Visual Prosthesis*

A visual prosthesis is generally composed of an external and an internal system. The CCD camera in the external system captures images of the outside world. The information of images is transmitted wirelessly to the internal system. The electric power is also transmitted by the same wireless system. The internal system decodes the transmitted information to an analogue signal, and sends electrical currents to individual electrodes (Fig. 10.3). A visual prosthesis placed under the retina uses an implantable microelectrode, instead of a CCD camera, which captures images.

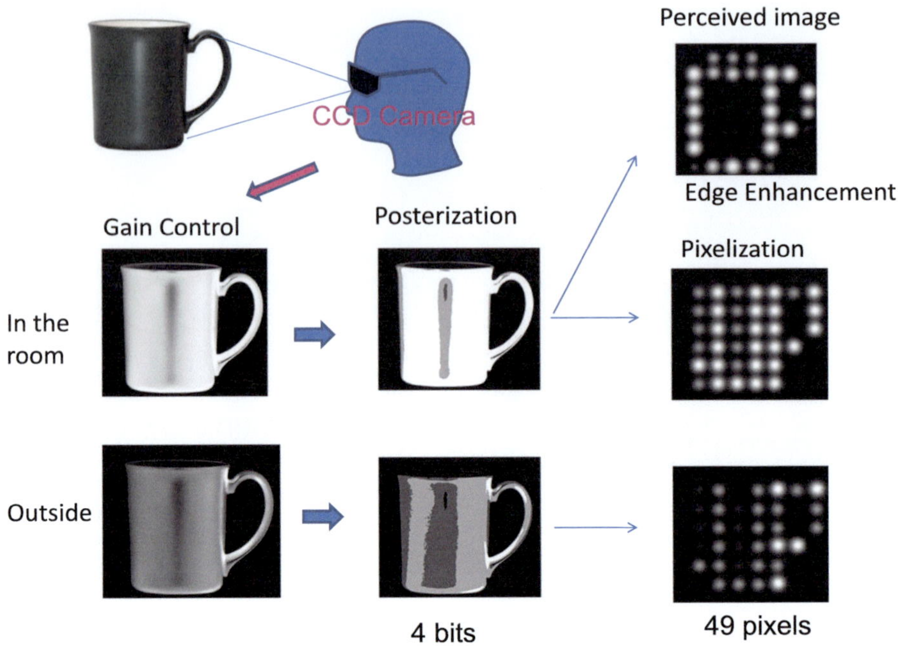


**Fig. 10.4** Simulated images of a cup perceived by a patient with a visual prosthesis. The CCD camera acquires an image of the cup and the gain control system in the data processor adjusts the level of brightness of the image. The gain is adjusted to a higher level in a dark room and to a lower level in the outdoors. After posterization and pixelization, stimulus pulses are generated (in some cases, an edge enhancement mode is used). Stimulus currents are generated in the decoder and delivered to each electrode to generate artificial visual sensation

## *10.3.4 Image Perception with Visual Prosthesis*

A patient with a visual prosthesis can perceive images as patterns of monochromatic dots. The quality of perceived images depends on the number of electrodes and the grades of tones. If the patient sees a cup, the CCD camera captures an image of the cup and the gain is adjusted according to the brightness of the background. The grey tone scale is similarly adjusted and then pixelized according to the number of electrodes. The edge of an image can be enhanced if the patient wants to see the details of the cup more clearly (Fig. 10.4). A simulated image of fingers perceived by the patient with a visual prosthesis is shown in Fig. 10.5. As pixels and gray-tone levels increase, the patient can obtain more information; 4 gray-scale levels and 100 pixels might be sufficient to identify the number of fingers.

When the receptive field of a stimulated neuron does not match the visual field captured by the CCD camera, patients may be misdirected and fail to reach the object with a proper positioning. In the case that the CCD camera attached to an eyeglass frame and the electrode array placed on the visual cortex, the retinotopic map would not be preserved. In normal subjects, images of the outside world are

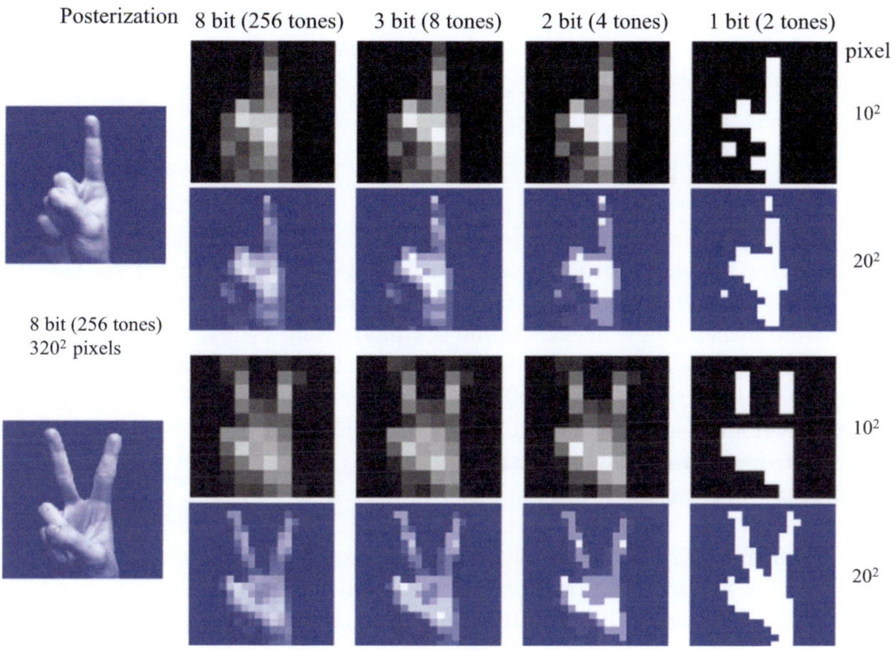


**Fig. 10.5** Simulated images of fingers perceived by a patient with a visual prosthesis. Even after reducing the number of pixels to 100, the number of fingers can still be discriminated. The number of tones is useful in representing the details of objects when the number of pixels is large

captured by retinal photoreceptor cells, and visual information is transmitted to the corresponding cortical area.

In patients implanted with a visual prosthesis, the retina-based coordinate system and CCD-based coordinate system are different when their eyes do not fixate the target. If a patient has an epiretinal prosthesis and the retinal nerve fibers are stimulated, retinotopic map is disturbed because retinal ganglion cells (RGCs) are placed differently from stimulated nerve fibers, i.e., the axons of the RGCs. If this happens, visual rehabilitation with the aid of cortical plasticity may help to adjust for the discrepancy.

## *10.3.5 Types of Prosthesis*

### 10.3.5.1 General

Visual prostheses can be classified into cortical and non-cortical prostheses. The cortical prosthesis is designed for the retina or optic nerve pathologies in which most of RGCs are damaged, such as glaucoma or optic neuritis. The non-cortical prosthesis is designed for the eye pathologies in which most of photoreceptors are damaged but a considerable number of retinal bipolar cells and/or RGCs are preserved. The retinal prosthesis and optic nerve prosthesis are classified as non-cortical prosthesis.

#### 10.3.5.2 Cortical Prosthesis

The cortical prosthesis was first developed in the 1970s. The development was pioneered by Brindley (Brindley and Rushton 1974) and followed by Dobelle (Dobelle and Mladejovsky 1974). Dobelle reported that 64 channel cortical electrodes were placed on the surface of the occipital visual cortex, and blind patients with them succeeded at recognizing characters at 5 ft; they had approximately 20/1200 visual acuity (Dobelle 2000). There were still a number of challenges to the further development of the cortical prosthesis: pain was induced by currents from large electrodes during meningeal stimulation, and a development of epileptic foci was induced by focal electrical stimulation (Schmidt et al. 1996).

In addition, the major part of the visual cortex lies deep within the calcarine fissure and away from surface electrodes. To reach those inner neurons, it is necessary for surface electrodes to produce high currents. Even though this is necessary, it has many collateral effects and damages the underlying tissues.

To overcome these problems, intracortical stimulation has been introduced. Intracortical stimulation uses intracortical penetrating electrodes, and they produce high fidelity images with low electrical energy and reduce damage to nervous tissues.

The Utah Electrode Array is a typical intracortical prosthesis. This array consists of multiple silicon spikes each of which has a platinum electrode at the tip (Maynard et al. 1997). Currently, the intracortical prosthesis has been examined in primates, and it is demonstrated that micro stimulation of the primary visual cortex evokes visual sensations.

The cortical visual prosthesis has an advantage over non-cortical prostheses in that it bypasses all the damaged neurons in the visual pathway peripheral to the primary visual cortex. As such, this approach has the potential to restore vision for a large number of blind patients.

#### 10.3.5.3 Non-cortical Prosthesis

##### 10.3.5.3.1 Optic Nerve Prosthesis

Veraart et al. have developed a spiral nerve-cuff electrode array for the optic nerve as a visual prosthesis (Veraart et al. 1998). They have implanted cuff electrodes circumferentially on the surface of the optic nerve. By controlling the pulse frequency of stimulation, the position of phosphenes can change and this change supports the identification of the positions of targets, such as knives (Brelen et al. 2010).

#### 10.3.5.3.2 Retinal Approaches

Retinal approaches are divided into three kinds of prosthesis: the epiretinal prosthesis in which electrodes are placed on the retina, the subretinal prosthesis in which electrodes are placed beneath the retina, and the suprachoroidal prosthesis in which electrodes are placed in the suprachoroidal space or in the scleral pocket.

#### 10.3.5.3.3 Epiretinal Prostheses

When an epiretinal prosthesis is implanted, an electrode array is placed on the nerve fiber layer of the retina. Images captured by a CCD camera are processed by a microprocessor, and it sends image signals wirelessly to the internal devises. Signals captured by a secondary coil are processed by a decoder, and it sends pulses to the electrode array.

Humayun et al. developed an intraocular retinal prosthesis and implanted it in patients (Humayun et al. 1996). Humayun and Second Sight have recently developed the 60 channel stimulating epiretinal microelectrode array named ARGUS II System. This system is the only commercially available retinal implant in the world, and its use is approved by the FDA of the USA (Humayun et al. 2012).

Humayun et al. did a multi-center study to test the Second Sight ARGUS II System in 2012. This system was implanted in 30 patients, and they performed several tasks statistically better in the "system-on" condition than in the "system-off" condition: for object localization, 96 % of the subjects did better with the system on; for motion discrimination, 57 %of the subjects did better with the system on; and for discrimination of grating orientation, 23 % of the subjects did better with the system on.

The mean performance on the orientation and mobility tasks was significantly better when the system was on than it was off. 70 % of the patients did not have any serious adverse events (SAEs). The most commonly reported SAEs were conjunctival erosion and dehiscence over the extraocular implant. They were treated successfully in all the patients except one who required explantation of the system (Humayun et al. 2012).

Similarly, da Cruz et al. asked 21 subjects to perform a forced-choice letter identification task in the system-on and system-off conditions in 2013. The mean $\pm$ standard deviation (SD) percentages of correct letter identification were the following: for letters L, T, E, J, F, H, I, and U, $72.3 \pm 24.6$ % with the system on and $17.7 \pm 12.9$ % with the system off; for letters A, Z, Q, V, N, W, O, C, D, and M, $55.0 \pm 27.4$ % with the system on and $11.8 \pm 10.7$ % with the system off; and for letters K, R, G, X, B, Y, S, and P, $51.7 \pm 28.9$ % with the system on and $15.3 \pm 7.4$ % with the system off ($P < 0.001$ for all groups). The average implant duration was 19.9 months (da Cruz et al. 2013).

The advantage of the epiretinal prosthesis is that it has a long history of clinical trials and is developed to be stable for a long time. The shortcoming of the recent

Argus II system is the relatively high percentage of side effects on the eye, such as endophthalmitis and conjunctival erosion.

#### 10.3.5.3.4 Subretinal Prosthesis

A subretinal prosthesis is placed between the photoreceptor layer and the retinal pigment epithelium to replace the function of degenerated photoreceptors. A microphotodiode array captures lights, and the electrical current stimulates the retinal ganglion cells and/or bipolar cells. The merit of this system is that a stimulation electrode is placed adjacent to each microphotodiode, and so it can stimulate the retina in the same region as light is received (Zrenner 2002).

Zrenner et al. developed a subretinal prosthesis that was powered via transdermal inductive transmission and carried 1500 independent microphotodiode-amplifier-electrode elements on a 9 $mm^2$ chip. The chip was implanted subretinally in 9 blind patients. Light perception was present in 8 of the 9 patients, light localization in 7 of them, motion detection with angular speed up to 35 deg/s in 5 of them, and grating acuity of 3.3 cycles/deg in 6 of them. Decimal acuity of 0.037 was achieved in 2 of the 9 patients.

In addition, 8 of the 9 patients, in repeated tests over a 9-month period, significantly improved the abilities to identify, localize, and discriminate objects ($P < 0.05$ for each subtest). Among the 9 patients, 3 patients were able to read letters spontaneously, and a patient was able to read letters after training; 5 patients used implant-mediated visual perceptions in daily life within a field of 15° of visual angle.

These results showed that the subretinal implant can restore visual functions and they are useful for daily life. In several patients, however, the implant was removed during the follow-up period because of its technical instability (Stingl et al. 2013).

#### 10.3.5.3.5 Suprachoroidal Prosthesis

A Japanese group has developed a new approach of implanting a prosthesis, and named the new approach ‘suprachoroidal-transretinal stimulation’ (STS). They have hypothesized that placing a stimulating electrode in the suprachoroidal space or in a sclera pocket with a ground electrode in the vitreous cavity (Fig. 10.6) is a less-invasive method to stimulate the retina (Fujikado et al. 2007) than other ways of implantation.

Fujikado et al. reported on the results of a semi-chronic suprachoroidal implantation in 2 patients with RP (Patient 1 and Patient 2) (Fujikado et al. 2011). The visual acuity of the patients before implantation was light perception. A 49 electrode array was placed in the scleral pocket without retinal detachment or vitreous hemorrhage being caused, and internal devices were implanted under the skin on the temporal side of the head (Fig. 10.7). The array of electrodes was placed under the lower temporal side of macula (Fig. 10.8). Biphasic pulses were delivered

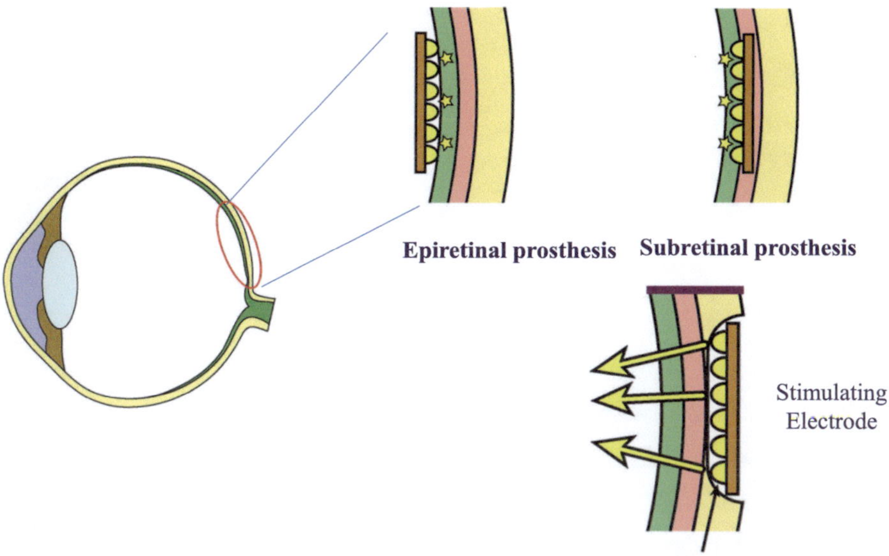


**Fig. 10.6** Three types of retinal prostheses. The epiretinal prosthesis uses electrodes placed on the retina. The subretinal prosthesis uses electrodes placed subretinally, and the suprachoroidal prosthesis uses electrodes placed in the suprachoroidal space or in a scleral pocket

sequentially to 9 active electrodes (Fig. 10.9). The implants remained functional during the 4 weeks of the study.

Phosphenes were elicited by currents delivered through 6 electrodes in Patient 1 and through 4 electrodes in Patient 2. Phosphenes were perceived in the upper-nasal visual field in both patients, and these phosphenes were consistent with the activation of the retina in the lower temporal area (Fig. 10.10). The positions of the perceived phosphenes elicited by simultaneously stimulating two electrodes roughly matched the positions of the electrodes (Fig. 10.11).

The success rate of discriminating 2 bars was higher than chance in both patients. In Patient 2, the success rate of a grasping task was higher than chance, and the success rate of identifying a white bar on a touch panel was increased with repeated testing (Fig. 10.12) (Fujikado et al. 2011).

An Australian group has investigated the feasibility of suprachoroidal placement of stimulating microelectrode array (Villalobos et al. 2013). There are several advantages to the suprachoroidal prosthesis over epiretinal and subretinal prostheses. First, the surgery for suprachoroidal placement is relatively uncomplicated. Second, suprachoroidally placed electrodes are relatively non-invasive to the retina and stably fixed. Third, it is relatively easy to remove or replace suprachoroidally placed electrodes. Fourth, the visual field of patients with a suprachoroidal prosthesis is relatively wide because it is implanted outside the choroid. And fifth, retinal photoreceptors in the eyes of patients with inherited photoreceptor

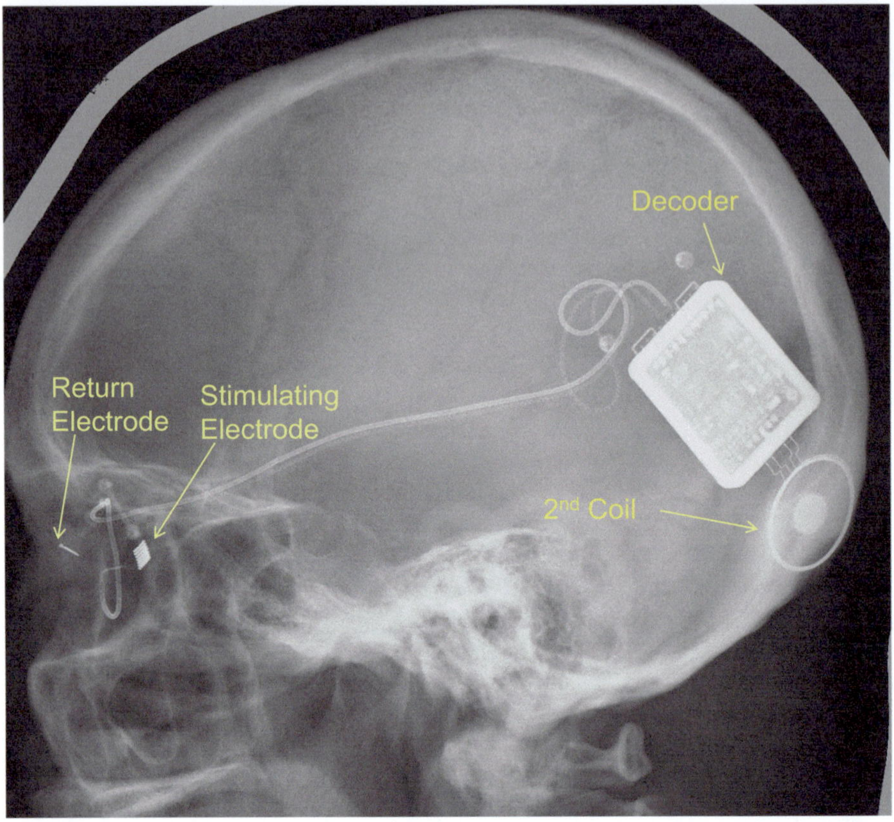


**Fig. 10.7** X-ray image after the surgery for implanting a suprachoroidal retinal prosthesis. The internal devices are implanted on the temporal side of the head. The stimulating electrode array is placed around the posterior pole of the eye, and the return electrode is placed at the anterior portion of the eye (Modified from (Fujikado et al. 2011), with kind permission from Association for Research in Vision and Ophthalmology)

degeneration can be protected by extraocular electrical stimulation (this has been proved in animals (Morimoto et al. 2012) and in humans (Schatz et al. 2011)).

There are possible disadvantages to the suprachoroidal prosthesis. First, suprachoroidally placed electrodes are relatively farther from the retina, and so high currents are required. Second, the current spread could be high, and so the resolution of perceived images might be limited.

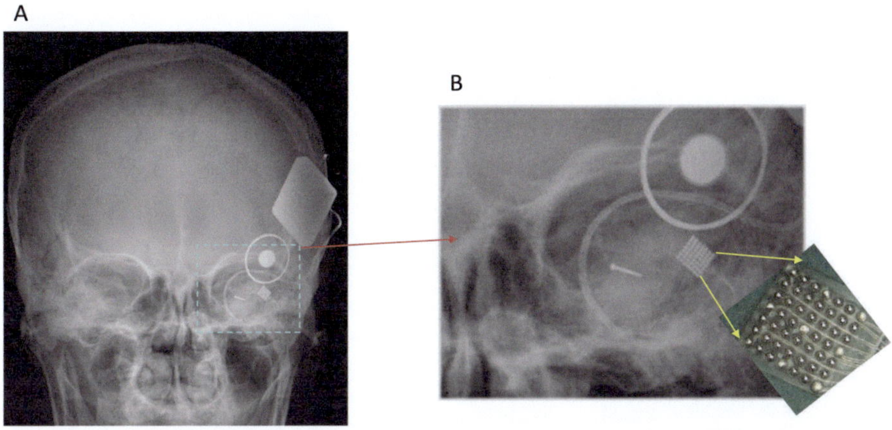


**Fig. 10.8** Position of the electrode array in relation to the retina. (**a**) Anterior-posterior x-ray image. (**b**) Magnified image of **a.** The x-ray image shows the electrode array placed in the lower temporal area of the left eye (Adapted from (Fujikado et al. 2011), with kind permission from Association for Research in Vision and Ophthalmology)

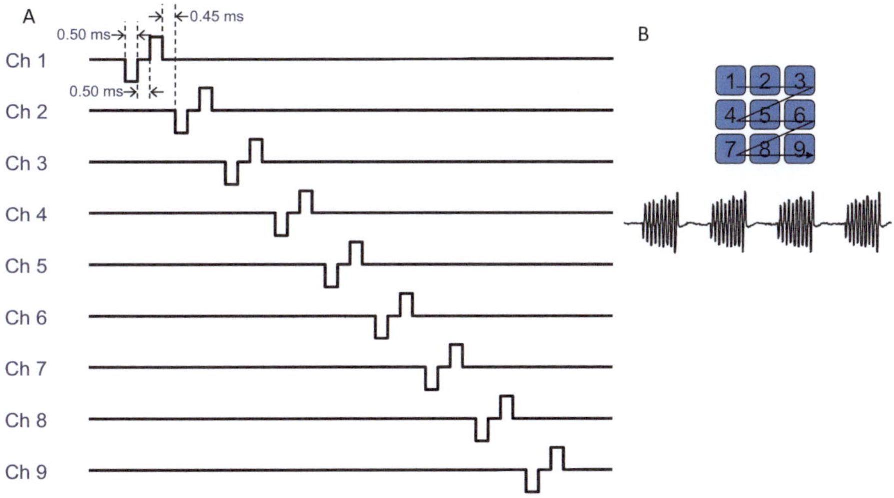


**Fig. 10.9** Time sequence of stimulating electrical current pulses. (**a**) The first pulse is a cathodic current and the second pulse is an anodic current to balance the charge. (**b**) A pair of pulses are delivered sequentially from the Chs 1–9 electrodes (Modified from (Fujikado et al. 2011), with kind permission from Association for Research in Vision and Ophthalmology)

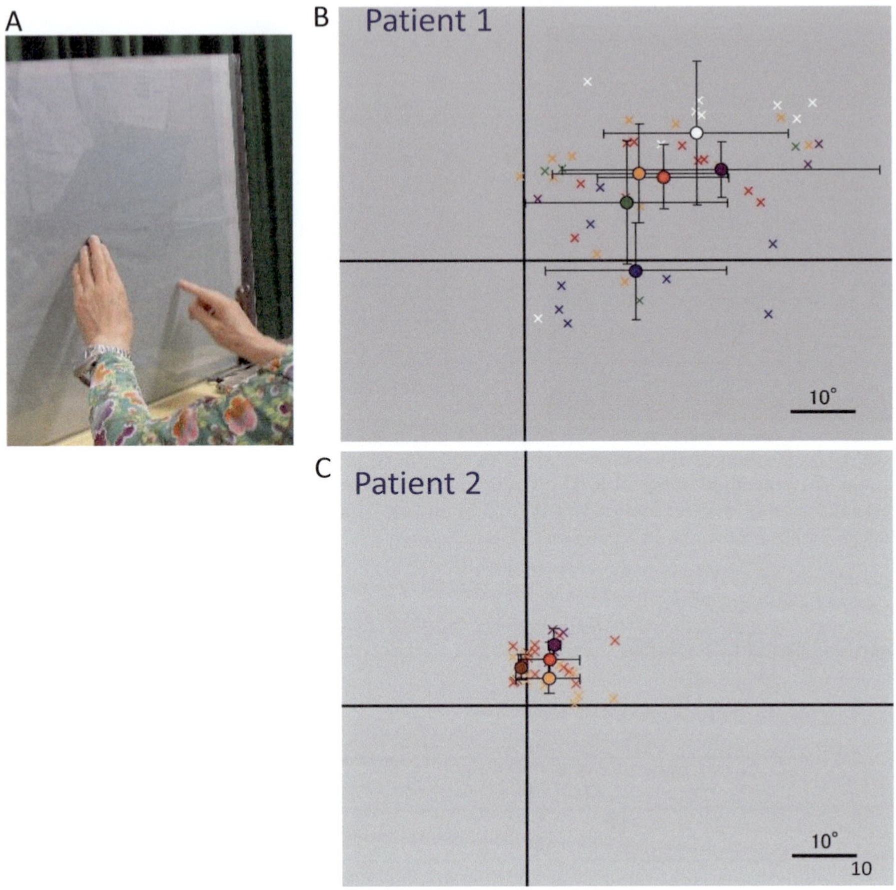


**Fig. 10.10** Map of perceived phosphenes in response to suprachoroidal transretinal stimulation of individual electrodes. (**a**) Method to record the position of a phosphene in relation to the center of the body. The left index finger is positioned at the center of the board and the right index finger is placed at the position of a perceived phosphene (Modified from (Fujikado et al. 2011), with kind permission from Association for Research in Vision and Ophthalmology). (**b**) The phosphene map of Patient 1. The results of the trials are superimposed. The *red*, *orange*, *green*, *dark blue*, *green*, *purple*, and *white Xs* indicate the position of the perceived phosphene in response to the separate stimulation of Ch 2, 3, 5, 6, 7, and 8, respectively. The *colored circles* indicate the gravitational center of the responses to the stimulation of the individual channels. The *bars* indicate the standard deviations. (**c**) The phosphene map of Patient 2. The *brown*, *red*, *orange*, and *purple Xs* indicate the position of the perceived phosphene in response to the stimulation of Ch 1, 2, 3, and 7

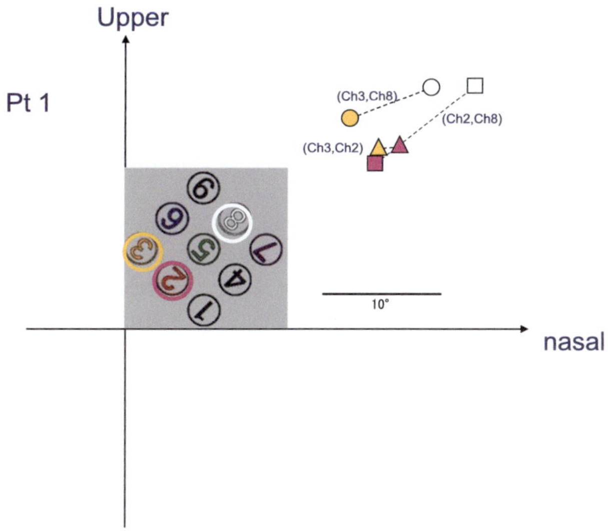


**Fig. 10.11** The positions of the perceived phosphenes in response to simultaneous activation of 2 electrodes in Patient 1 (single trial). The *circles* show the positions of the phosphene in response to simultaneous stimulation of Chs 3 and 8, *triangles* those in response to simultaneous stimulation of Chs 2 and 3, and *squares* those in response to simultaneous stimulation of Chs 2 and 8 (Modified from (Fujikado et al. 2011), with kind permission from Association for Research in Vision and Ophthalmology)

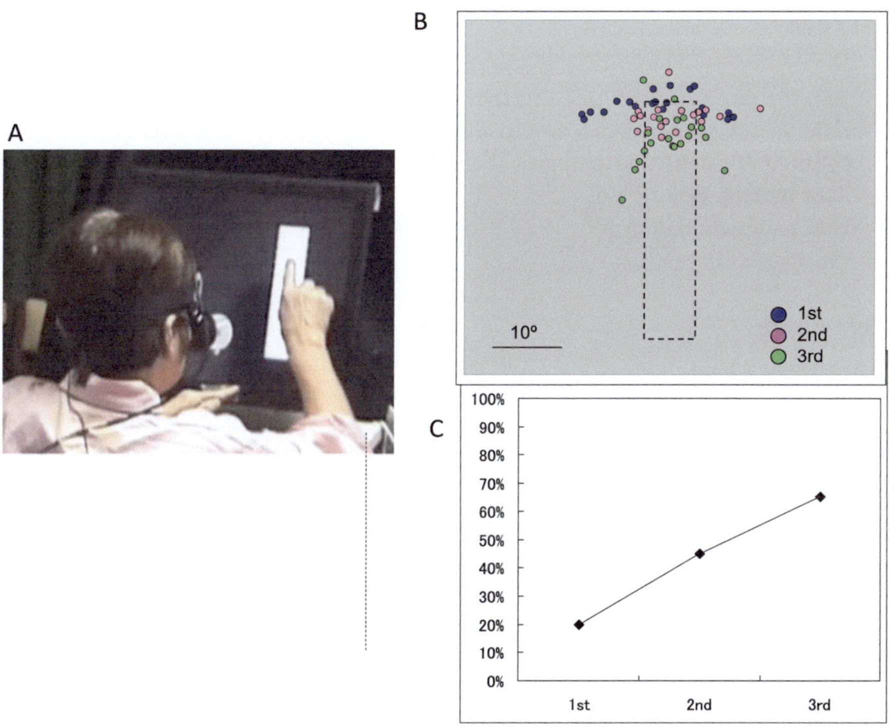


**Fig. 10.12** Bar-Identification task for Patient 2. (**a**) Patients are instructed to touch the bar presented on the panel. (**b**) Points touched by the patient when a *white bar* was presented on the panel. (**c**) The success rate after repeated examination. The *dashed rectangular* area in **a**–**c** represents the position of the *white bar*. The *blue dots* represent the first trial, the *pink dots* the second trial, and the *green dots* the third trial (Modified from (Fujikado et al. 2011), with kind permission from Association for Research in Vision and Ophthalmology)

## 10.4 Conclusion

The history and current status of sensory BMI have been reviewed in this chapter. The auditory prosthesis has reached a nearly completed stage for speech recovery in deaf patients. The visual prosthesis is at the clinically feasible stage but still has a lot of room for improvement.

To achieve an improvement of the quality of life for blind patients implanted with a retinal prosthesis, not only an improvement of surgical procedures and engineering advancement but also a development of effective rehabilitation are necessary. Even though visual acuity is still poor, patients with implants can read large letters.

The challenge is how to continue to improve the resolution of images to the level where completely blind and legally blind (20/200) patients can benefit from the use of retinal prostheses in their daily life. Relatedly, another challenge is how to enlarge the visual field so as to enable patients to walk without the aid of a cane.

## Exercise

1. From the point of view of cognitive neuroscience, in which patient is it appropriate to implant a cochlear prosthesis, a prelingually deafened patient or a postlingually deafened patient? If there is any condition for implantation for either patient, describe it.
2. What parameters does the visual prosthesis use to reconstruct the vision?

## References

Brelen, M.E., Vince, V., Gerard, B., et al.: Measurement of evoked potentials after electrical stimulation of the human optic nerve. Invest. Ophthalmol. Vis. Sci. **51**, 5351–5355 (2010)

Brindley, G., Rushton, D.: Implanted stimulators of the visual cortex as visual prosthetic devices. Trans. Am. Acad. Ophthalmol. Otolaryngol. **78**, 741–745 (1974)

da Cruz, L., Coley, B.F., Dorn, J., et al.: The Argus II epiretinal prosthesis system allows letter and word reading and long-term function in patients with profound vision loss. Br. J. Ophthalmol. **97**, 632–636 (2013)

Dobelle, W.H.: Artificial vision for the blind by connecting a television camera to the visual cortex. ASAIO J. **46**(1), 3–9 (2000)

Dobelle, W.H., Mladejovsky, M.G.: Phosphenes produced by electrical stimulation of human occipital cortex: and their application to the development of a prosthesis for the blind. J. Physiol. **243**(2), 553–576 (1974)

Fujikado, T., Morimoto, T., Kanda, H.: Evaluation of phosphenes elicited by extraocular stimulation in normals and by suprachoroidal-transretinal stimulation in patients with retinitis pigmentosa. Graefes Arch. Clin. Exp. Ophthalmol. **245**, 1411–1419 (2007)

Fujikado, T., Kamei, M., Sakaguchi, H., et al.: Testing of semi-chronically implanted retinal prosthesis by suprachoroidal-transretinal stimulation in patients with retinitis pigmentosa. Invest. Ophthalmol. Vis. Sci. **52**, 4726–4733 (2011)

Helms, J., Mueller, J., Schon, F., et al.: Evaluation of performance with the COMBI 40 cochlear implant in adults: a multicentric clinical study. ORL J. Oto-Rhino-Laryngol. Relat. Spec. **59**, 23–35 (1997)

Humayun, M.S., de Juan, E.J., Dagnelie, G., et al.: Visual perception elicited by electrical stimulation of retina in blind humans. Arch. Ophthalmol. **114**(1), 40–46 (1996)

Humayun, M.S., Dorn, J.D., da Cruz, L., et al.: Interim results from the international trial of Second Sight's visual prosthesis. Ophthalmology **119**(4), 779–788 (2012)

Maynard, E.M., Nordhausen, C.T., Normann, R.A.: The Utah intracortical electrode array: a recording structure for potential brain–computer interfaces. Electroencephalogr. Clin. Neurophysiol. **102**(3), 228–239 (1997)

Morimoto, T., Kanda, H., Kondo, M., et al.: Transcorneal electrical stimulation promotes survival of photoreceptors and improves retinal function in rhodopsin P347L transgenic rabbits. Invest. Ophthalmol. Vis. Sci. **28**(53), 4254–4261 (2012)

Niparko, J.K., Tobey, E.A., Thal, D.J., et al.: Spoken language development in children following cochlear implantation. J. Am. Med. Assoc. **303**, 1498–1506 (2010)

Rubinstein, J.T.: How cochlear implants encode speech. Curr. Opin. Otolaryngol. Head Neck Surg. **12**, 444–448 (2004)

Schatz, A., Röck, T., Naycheva, L., et al.: Transcorneal electrical stimulation for patients with retinitis pigmentosa: a prospective, randomized, sham-controlled exploratory study. Invest. Ophthalmol. Vis. Sci. **52**(7), 4485–4496 (2011)

Schmidt, E.M., Bak, M.J., Hambrecht, F.T., et al.: Feasibility of a visual prosthesis for the blind based on intracortical microstimulation of the visual cortex. Brain **119**(Pt. 2), 507–522 (1996)

Stingl, K., Bartz-Schmidt, K.U., Besch, D., et al.: Artificial vision with wirelessly powered subretinal electronic implant alpha-IMS. Proceed. R. Soc. Biol. Sci. **280**(1757), 20130077 (2013)

Veraart, C., Raftopoulos, C., Mortimer, J.T., et al.: Visual sensations produced by optic nerve stimulation using an implanted self-sizing spiral cuff electrode. Brain Res. **813**(1), 181–186 (1998)

Villalobos, J., Nayagam, D.A., Allen, P.J., et al.: A wide-field suprachoroidalretinal prosthesis is stable and well tolerated following chronic implantation. Invest. Ophthalmol. Vis. Sci. **54**, 3751–3762 (2013)

Wilson, B.S., Dorman, M.F., Woldorff, M.G., et al.: Cochlear implants matching the prosthesis to the brain and facilitating desired plastic changes in brain function. In: Schollellbor, J., Ganviczallli, M., Donielsell, N. (eds.) Progress in Brain Research, Elsevier, Amsterdam, vol. 194, pp. 117–129 (2011)

Zrenner, E.: Will retinal implants restore vision? Science **295**(5557), 1022–1025 (2002)

# Chapter 11
# Brain Machine-Interfaces for Motor and Communication Control

**Masayuki Hirata**

**Abstract** The brain-machine interface (BMI) is a technology which enables our brains to control external devices or to obtain sensory information by directly communicating with computers. BMIs are classified into two types: invasive type which uses intracranial electrodes, such as needle microelectrodes and brain surface electrodes, and noninvasive type which uses skin electrodes, near infrared spectroscopy, and etc. Noninvasive BMIs are promising for neurorehabilitation, while invasive BMIs are promising for neural prostheses for severely disabled people. BMIs using needle microelectrodes are characterized by high performance utilizing its detailed neural information, while BMIs using brain surface electrodes are noted for the high feasibility for clinical application based on its long term stability. We describe our development of a BMI using brain surface electrodes. It enables real time control of a robotic arm and a fully-implantable wireless system.

**Keywords** Brain machine interface • Functional restoration • Electrocorticogram • Robot • Implant

## 11.1 Introduction

Brain-machine interface (BMI) is a technology which enables our brains to control external devices or to obtain sensory information by directly communicating with computers. BMI is expected to be a functional assistive technology for improving functional disturbance in severely disabled people. In this chapter, we describe the basics of BMIs and the trends of BMI research, as well as the history of its development. We mainly discuss motor BMIs and our own BMI system.

BMI is also called 'brain-computer interface' (BCI). The term 'BMI' is mainly used for the device that uses neuronal spike activities, whereas the term 'BCI' is used for the device that uses electroencephalography (EEG). Nevertheless, there is no clear distinction between these two. In this chapter, we use only 'BMI.'

BMIs are classified into two types: motor BMIs and sensory BMIs (Fig. 11.1). Motor BMIs record brain signals, decode the behavioral information that the brain

M. Hirata (✉)
Graduate School of Medicine, Osaka University, Suita, Osaka, Japan
e-mail: mhirata@nsurg.med.osaka-u.ac.jp



M. Kasaki et al. (eds.), *Cognitive Neuroscience Robotics B*,
DOI 10.1007/978-4-431-54598-9_11

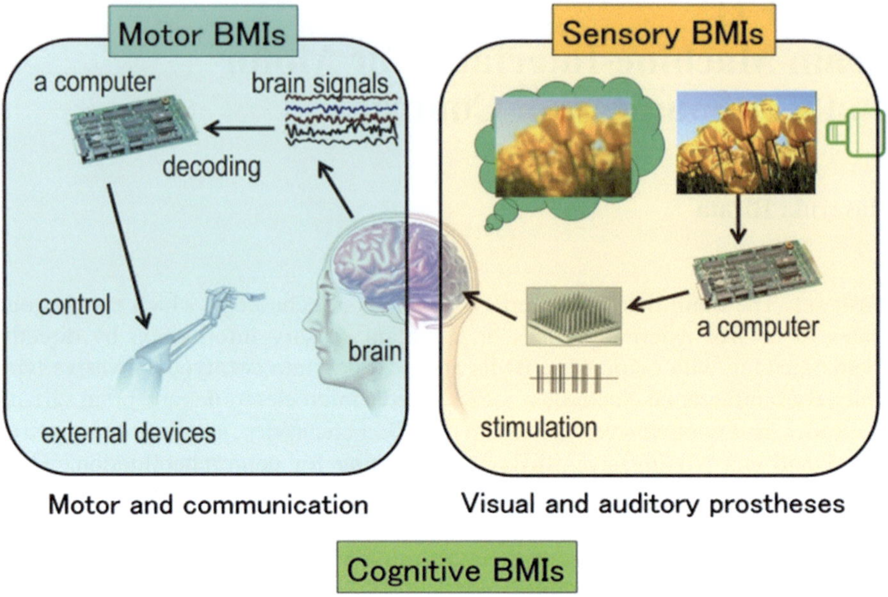


**Fig. 11.1** Classification of invasive BMIs: motor BMIs, sensory BMIs, and cognitive BMIs

signals encode or represent (neural decoding), and manipulate external devices according to decoded signals. Motor BMIs are expected to be an assistive technology to restore motor function and communication for patients who are disabled due to amyotrophic lateral sclerosis, spinal cord injury, or brain stroke. Sensory BMIs obtain environmental information from sensors, translate it into appropriate stimuli (neural encoding), and input them to the brain by stimulating the nervous system. Brain stem auditory implants and artificial visual prostheses are representative of sensory BMIs. The former are already clinically used, but not widely. In addition to motor and sensory BMIs, cognitive BMIs are developed recently. They directly decode cognitive activities in the brain, and therefore they are expected to restore communication function for severely disabled people.

BMIs are classified into two types in terms of invasiveness: invasive BMIs and noninvasive BMIs (Fig. 11.2). Invasive BMIs use intracranial electrodes, such as micro-needle electrodes and brain surface electrodes, and noninvasive BMIs use skin electrodes, near infrared spectroscopy, and etc. Invasive BMIs are further classified into two types: less invasive BMIs and highly-invasive BMIs. Less invasive BMIs record electrocorticograms (ECoGs) from brain surface electrodes, and highly-invasive BMIs record neuronal spiking activities from micro-needle electrodes.

There are various brain signals used for BMIs, as they are listed in Table 11.1. It is important to understand the characteristics of each brain signal. The

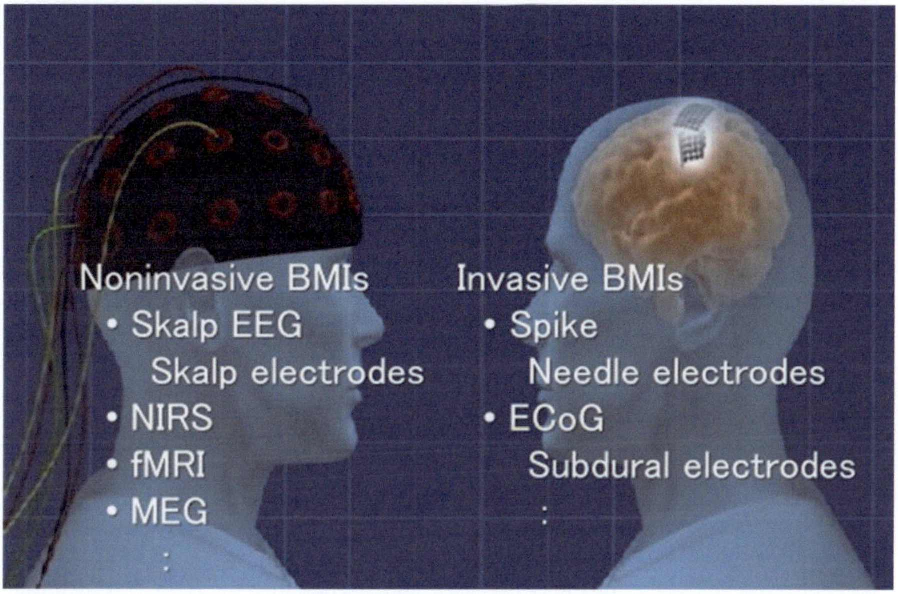


**Fig. 11.2** Classification of BMIs: noninvasive BMIs and invasive BMIs

**Table 11.1** Brain signals used for BMIs

| | Measured physiological phenomena | Spatial coverage | Spatial resolution | Temporal resolution | Time delay | Invasive ness | Long term recording stability | Portability |
|---|---|---|---|---|---|---|---|---|
| fMRI | CBF | whole brain | ○ 3–5 mm | × 4–5 s | × 4–5S | ◎ | ○ | × |
| NIRS | CBF | cortex | × 2 cm | × 4–5 s | × 4–5S | ◎ | ○ | ○ |
| EEG | Neural activities | whole brain | × 3–4 cm | ○ 1 ms | ◎ 0 | ◎ | ○ | ○ |
| MEG | Neural activities | cortex | △ 5–10 mm | ◎ 0.1 ms | ◎ 0 | ◎ | ○ | × |
| ECoG | Neural activities | 100 $cm^2$ | ○ 2–3 mm | ◎ 0.1 ms | ◎ 0 | △ | ◎ | ◎ |
| LFP | Neural activities | 1 $cm^2$ | ○ 1 mm | ◎ $<$ 0.1 ms | ◎ 0 | × | △ | ◎ |
| spike | Neuronal activities | 1 $cm^2$ | ◎ 0.2 mm | ◎ $<$ 0.1 ms | ◎ 0 | × | × | ◎ |

*fMRI* functional magnetic resonance imaging, *NIRS* near infrared spectroscopy, *EEG* electroencephalogram, *MEG* magnetoencephalography, and *LFP* local field potential characteristics include recording area, spatial and temporal resolution, time delay, invasiveness, long term recording stability, and portability. It is also important to utilize them as they are intended to be used. For example, electroencephalograms (EEGs) are non-invasive and have high temporal resolution, and therefore have been widely used in BMI research. But they are limited in performance due to their low spatial resolution. Both functional MRI and near infrared spectroscopy (NIRS) are noninvasive but not suitable for real time recording, because they measure the cerebral blood flow and the cerebral blood flow is delayed for 4–5 s from cerebral functional activities. Functional MRIs and MEGs are not portable but good in spatial resolution; so they are mainly used for research purposes. Micro-needle electrodes record spike activities of a single neuron and local field potentials (LFPs), i.e., electric potentials in the extracellular space around neurons. Although micro-needle electrodes have the highest spatiotemporal resolution, they are invasive to brain tissues and low in long term stability, because their recording efficacy gradually decreases in months due to chronic inflammation caused by tissue micro-injury. Electrocorticograms (ECoGs) are intracranial electroencephalograms recorded by electrodes directly placed on the brain surface. They are low-noise and capable of detecting high frequency activities, as compared with scalp electroencephalograms. ECoGs are less invasive and higher in long term stability than micro-needle electrodes (Chao et al. 2010). They are well-balanced methods, although they require a surgical intervention.

Basic technologies, such as neural decoding, are needed for functional restoration using BMIs. In particular, the systematic and sophisticated integration of numerous basic technologies listed in Table 11.2 enable invasive BMIs to be usable for this purpose. Integrative research and development based on the collaboration of medicine and engineering and academic-industrial alliance are indispensable for the clinical application of BMIs.

**Table 11.2** Basic technologies and issues for invasive BMIs

| |
|---|
| (1) Neural recording with high spatiotemporal resolution |
| (2) High speed transfer and processing of neural signals |
| (3) Optimal extraction of neurophysiological features |
| (4) Neural decoding |
| (5) Control of external devices such as robot arms |
| (6) Downsizing, integration, and implantation of electronic devices, and the use of wireless technology. |
| (7) Non-invasive evaluations for appropriate surgical indications. |
| (8) On-target survey and analysis of patient needs |
| (9) Addressing neuroethical issues |

## 11.2 History and Recent Trends of BMI Researches

### *11.2.1 Early BMI Researches*

#### 11.2.1.1 P300 and Mu Rhythm

Motor BMI research has become active in the past 10 years or so. The first BMI research was reported in 1988 by Farwell and Donchin (1988). They developed a communication device that displayed intended characters by using scalp EEG to detect P300 potentials when they were evoked by randomly-flashed characters (Farwell and Donchin 1988). P300 evoked potentials are the responses generated over the parietal area when irregular stimuli are recognized. This device is presently well known as the P300 speller. The device is able to display 2–3 alphabetic characters per minute. In this sense, the first BMI was a cognitive BMI.

BMI using mu rhythms were reported 3 years after the P300 speller was reported. The mu rhythm is a rhythmic brain wave in the 8–12 Hz band, and it is found in the center area related to voluntary movement. Wolpaw et al developed a BMI device in 1991, such that a cursor moves up and down in accordance with an increase and decrease in mu rhythm power (Wolpaw et al. 1991). This device usually requires a certain amount of training for users to voluntarily control the increase and decrease in mu rhythm power. Since the development of these devices, P300-based and mu-rhythm-based BMIs have become standard EEG-based BMIs, and many researchers have been using them.

#### 11.2.1.2 Electrocorticograms

With the initial success of EEG-based BMI, it is easily expected that electrocorticograms (ECoGs), i.e., intracranial electroencephalograms recorded from brain surface electrodes, can advance the performance of BMIs. Brain surface electrodes are called ‘subdural electrodes,’ and have been clinically used in the field of neurosurgery for more than 30 years to identify the epileptic foci of intractable epilepsy (Ojemann 1983). The first ECoG study was reported by Levine and Huggins in 1999; it provided a basis for ECoG-based BMIs (Huggins et al. 1999; Levine et al. 1999). The first quantitative assessment of an ECoG-based BMI reported that one-dimensional cursor control had been achieved at a rate of 74–100 % (Leuthardt et al. 2004).

#### 11.2.1.3 Neuronal Spiking Activities and Directional Tuning

A rat experiment reported by Chapin and Nicolelis was noted for the first real-time control of a robot arm using neuronal spiking activities in the motor cortex (Chapin et al. 1999). They trained rats to control and position a robot arm in a search of water by pressing a lever. Then, by recording the neuronal spiking activities from micro-needle electrodes inserted into the motor cortex, they modified the control of the robot arm in such a way that it was able to be controlled by the neuronal spiking occurred just before pressing the lever, as opposed to actually pressing it. The majority of rats controlled the robot arm by brain-derived signals and obtained water. Interestingly, when rats were given more training for this neuronal robotic control, they were less disposed to move the lever or stopped doing it. Although the control is very simple (turning on and off a robot arm movement), this study is significant in two respects: first, it controls a robot arm using spiking activities of multiple neurons in the motor cortex, and second, it suggests a possible means for movement restoration in paralysis patients.

In 2000, Wessburg and Nicholelis reported that accurate real-time predictions of three-dimensional arm movement trajectories were obtained by the cortical neuronal activity recorded from two monkeys (Wessberg et al. 2000). In 2002, Serruya and Donoghue reported that a cursor could be controlled two-dimensionally by a relatively small number of neuronal spiking activities (7–30 neurons) (Serruya et al. 2002). Tayler and Schwartz reported on three-dimensional control of a robot arm in monkey experiments (Taylor et al. 2002).

All of these studies use the "directional tuning" property of neurons in the motor cortex (Fig. 11.3). It is the property that motor neurons of the upper extremities respond preferentially to a specific direction of movement of the upper extremities (Georgopoulos et al. 1982). Significantly, because of this property, the movement direction of the upper extremities is accurately predicted. Spiking patterns of a few dozen of motor neurons are used for two-dimensional prediction, and those of a few

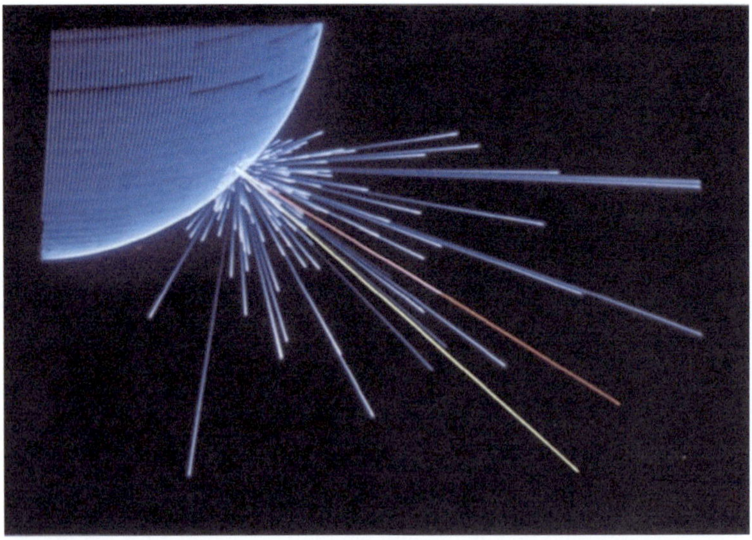

**Fig. 11.3** An example of directional tuning of movement direction (Referred from Georgopoulos et al. 1988) (Permission obtained to copy from SfN)

hundred of motor neurons are used for three-dimensional prediction (Wessberg et al. 2000; Serruya et al. 2002).

To summarize these early BMI studies, BMIs are initially applied to humans by using scalp EEGs in the 1990s, and become an active area of research in the 2000s after American neurophysiologists succeeded in real time control of computer cursors and robot arms in animal experiments.

## *11.2.2 Noninvasive BMIs*

### 11.2.2.1 EEG and MEG

Scalp EEG (hereafter simply described as EEG) is one of the most basic brain signals. EEGs are relatively easy to record, and have the advantages of noninvasiveness and high temporal resolution. For these reasons, they have been widely used in BMI research. P300 evoked potentials, cerebral oscillatory changes (power changes of mu rhythms), and visual evoked potentials are used as neurophysiological features for BMIs (Wolpaw et al. 2002). However, the spatial resolution of EEGs is low, because EEGs are damped and distorted by the tissues between the brain and electrodes, such as cerebrospinal fluids, skull bones, and skins. It is difficult to detect high frequency components of EEGs because of the filtering effects of these intervening tissues. For these reasons, current EEG-based BMIs are limited in performance, but they take advantage of noninvasiveness. Neurorehabilitation is one of the most promising applications of EEG-based BMI (Birbaumer et al. 2009; Ramos-Murguialday et al. 2013). A clinical study of stroke patients with chronic complete paralyses reported that activities of the paralyzed finger-extensor muscles were restored in a half of the patients by neurorehabilitation, which used a hand prosthesis. The hand prosthesis was controlled by cerebral oscillatory changes which were, in turn, induced by an imaginary movement of the paralyzed hand. This result suggests that EEG-based BMIs are a promising tool to enhance the efficacy of rehabilitation (Shindo et al. 2011). P300 spellers are one of the most classical and popular BMI methods. Various improvements in practical utility have been reported (Ikegami et al. 2012).

Magnetoencephalograms (MEGs) have been recently used in scientific research, because they are superior in spatial resolution to EEGs, albeit their low portability. Toda and Imamizu reported that they decoded the fingertip position during wrist movement at an average accuracy of 15 mm, by using MEG data combined with functional MRI data as prior information regarding activation sites (Toda et al. 2011). We decoded three types of unilateral hand and arm movements on a single trial basis at an accuracy of 60–70 % (Sugata et al. 2012a). We also demonstrated that there was a positive correlation between amplitudes of movement-related neuromagnetic fields and decoding accuracies (Sugata et al. 2012b). These results may contribute to the establishment of presurgical evaluation for invasive BMI treatments and to the application of MEG to neurorehabilitation.

#### 11.2.2.2 Near Infrared Spectroscopy and Functional MRIs

Many recent BMI studies use near infrared spectroscopies (NIRSs) and functional MRIs. Both methods measure cerebral hemodynamic changes noninvasively. However, they are less suitable for real time recording than EEGs and MEGs, because cerebral hemodynamic changes lag behind electrical neural activities by several seconds. NIRSs have low spatial resolution but can be used even during movement. They are promising for practical use in neurorehabilitation (Sitaram et al. 2009).

Functional MRIs are mainly used for scientific research because they have high spatial resolution and thereby show relatively high performance, although they lack portability. Kamitani et al pioneered neural decoding studies using fMRI (Kamitani and Tong 2005). They visually presented two types of lines in different directions to subjects, and successfully decoded the directions only from subtle differences in spatial distribution of activation in the visual cortex. They also decoded visually-presented words by improving the method of information extraction; they used a modular decoding method: they divided region-of-interests into multiple small image bases and applied decoders to each of them (Miyawaki et al. 2008). Most recently, they successfully decoded visual imagery during sleep (Horikawa et al. 2013).

#### 11.2.2.3 Decoded Neurofeedback

Shibata and Kawato used an online-feedback method with decoded fMRI signals to induce activity patterns only in early visual cortex corresponding to an orientation without stimulus presentation or participants' awareness of what was to be learned, resulting in improvement of visual perceptual performance (Shibata et al. 2011). They called the method 'decoded neurofeedback.' Decoded neurofeedback is expected to be a new therapeutic modality, although its efficacy and safety still need to be validated.

### *11.2.3 Invasive BMIs*

#### 11.2.3.1 Microneedle Electrodes

Microneedle electrodes record neuronal action potentials and local field potentials (LFPs); LEFs reflect the activities of neuronal population. As mentioned above, if the directional tuning property of the motor neurons in the motor cortex is used, a relatively small number of neuronal activities are sufficient for predicting the trajectory of the upper limb (Wessberg et al. 2000; Serruya et al. 2002). More recently, Velliste et al provided a method for real-time three-dimensional control of a robot arm, and showed that monkeys were able to use it (Velliste et al. 2008). Hochberg et al succeeded in developing a system for real time control of a computer cursor based on the neuronal spiking activities recorded from 100 channel microneedle electrodes that were inserted into the hand motor area. They tested a system in a patient with complete tetraplegia caused by injury to the spinal cord (Hochberg et al. 2006). They also reported that a patient with tetraplegia caused by a brain stem stroke used a robotic arm to drink coffee from a bottle (Hochberg et al. 2012). Collinger et al reported that a patient with tetraplegia caused by spinocerebellar degeneration was able to use a robot arm more skillfully and coordinate reaching and grasping in 13 weeks (Collinger et al. 2013). A problem is, however, that the insertion of microneedle electrodes may damage brain tissues and cause chronic inflammatory reactions. As a result, the efficiency of electrode recording gradually deteriorates over several months.

#### 11.2.3.2 Electrocorticograms

ECoGs are brain waves recorded from plate electrodes directly placed on the brain surface. They have a higher signal-to-noise ratio and detect higher frequency activities, as compared to scalp EEGs. Moreover, they are less invasive to brain tissues and better in long-term recording stability than scalp EEGs. Chao et al implanted ECoG electrodes in monkeys, and they were able to accurately decode the three-dimensional position of the upper arm of monkeys over a year (Chao et al. 2010); for 6 months, the arm position was able to be decoded without additional decoder learning. These results indicate that ECoGs allow for stable long-term recording. This is one of the most important factors for clinical application. There have been an increasing amount of published ECoG-based BMI studies since Chao et al reported on the experiment for one-dimensional decoding in 2004. Several groups including ours reported on experiments for two-dimensional decoding (Schalk et al. 2007; Pistohl et al. 2008), in one of which a cursor was controlled two-dimensionally (Schalk et al. 2008). Our group demonstrated that ECoGs within the central sulcus (the brain groove where the sensorimotor function are located) were useful for decoding upper limb movements (Yanagisawa et al. 2009). We succeeded in controlling a robot hand in real time using high frequency activities (Yanagisawa et al. 2011). Furthermore, we showed that, even in patients with severe motor disturbance, movement types were decoded by using high frequency activities during imaginary movement. We describe our findings in detail later.

## 11.3 Cerebral Oscillatory Changes

Cerebral cortical potentials change according to movements. They are called 'movement-related cortical potentials' (MRCP) (Fig. 11.4). MRCPs occur before the onset of movement. They consist of four main components. Readiness potential (RP) is a slow negative potential, and starts 1500–1800 ms before a movement. Negative slope potential (NS) is a prominent negative potential, and occurs predominantly in contralateral electrodes 400–500 ms before the onset of movement. NS is followed by premotor positivity (PMP) and motor potential (MP).

Basic rhythmic waves, such as α waves in 8–10 Hz over the parietal area during resting states, are called 'cerebral oscillations.' Brain activation induces local oscillatory changes in specific frequency bands. The increase in oscillatory change is called 'event-related synchronization' (ERS), and the decrease is called 'event-related desynchronization' (ERD) (Pfurtscheller and Neuper 2006). For example, hand grasping induces ERD mainly in the hand motor area bilaterally with contralateral predominancy in the α (8–13 Hz) and β (13–25 Hz) bands 500–1000 ms before the movement onset, and ERD peaks after the movement onset. In the high γ band (more than 50 Hz), ERS occurs a few hundred ms before the movement onset and peaks after it. These oscillatory changes are called 'movement-related cerebral oscillatory changes.' Cerebral oscillatory changes are observed not only during movements (Taniguchi et al. 2000) but also during language activities, somatosensory processing and mental concentration (Hirata et al. 2002, 2004; Ishii et al. 2009).

Both movement-related oscillatory changes and MRCPs reflect macroscopic electrical cortical activities related to movement, and they have a close relationship with each other. In order to decode movement intentions and movement types, appropriate features of movement-related oscillatory changes and potentials must be selected as input information for neural decoding on the basis of neurophysiological spatiotemporal properties related to movement.

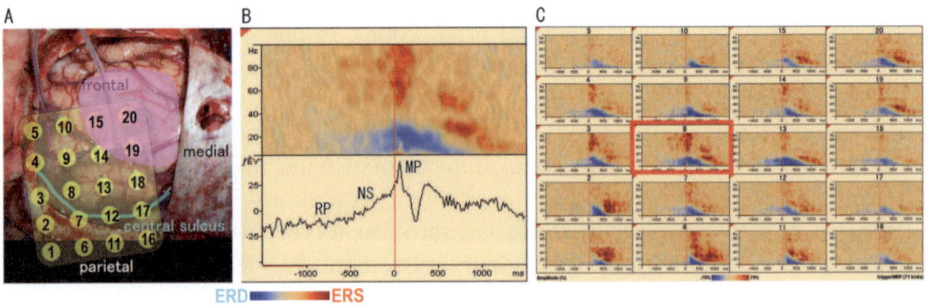


**Fig. 11.4** Movement-related cortical potentials and cerebral oscillatory changes

## 11.4 Neural Decoding Based on ECoGs

In neurosurgical treatments, grid electrodes are sometimes implanted directly on the brain surface for about 2 weeks. We performed this implantation for the purposes of identifying optimal stimulation sites in the motor cortex for intractable pain and identifying epileptic foci for intractable epilepsy. In addition, we placed strip electrodes within the central sulcus for the purpose of obtaining more efficient pain relief by motor cortex stimulation for patients with intractable pain (Hosomi et al. 2008). With the approval of the institutional ethical committee, we have been investigating BMIs using human ECoGs recorded from electrodes implanted in more than 30 patients.

Neural decoding is a key technology of BMIs, and there are many decoding methods. We mainly used a support vector machine (SVM). A SVM is a learning machine algorithm for classification. It has high classification accuracy by adjusting weight parameters and maximizing the margin between groups (Kamitani and Tong 2005) (Fig. 11.5).

We measured ECoGs during two or three types of simple hand or arm motor tasks, such as grasping, pinching and elbow flexion. We used a SVM to predict movement types on the basis of analysis of single trial ECoGs. We managed to predict movement types on a single trial basis with an accuracy rate of 70–90 %. More specifically, we had demonstrated for the first time that ECoGs from the anterior wall of the central sulcus (the groove in the brain where most of the primary motor cortex lies) are useful for the accurate and early decoding of movement types. Most of the primary motor cortex, which is responsible for the final output of motor commands, lies within the anterior wall of the central sulcus. The anterior wall of the central sulcus has many neurons directly projecting to the spinal anterior horn cells, especially in humans. These neurons are regarded to be related to fine movement control (Rathelot and Strick 2009). As far as our results are concerned, we suppose that appropriate neurophysiological feature extraction from the central sulcus has contributed to accurate movement decoding.

We investigated what frequency bands of oscillatory changes decoded movement types most accurately. We found that the normalized power in the high $\gamma$ band (80–150 Hz) gave the highest decoding accuracy of all the frequency bands from

**Fig. 11.5** Support vector machine

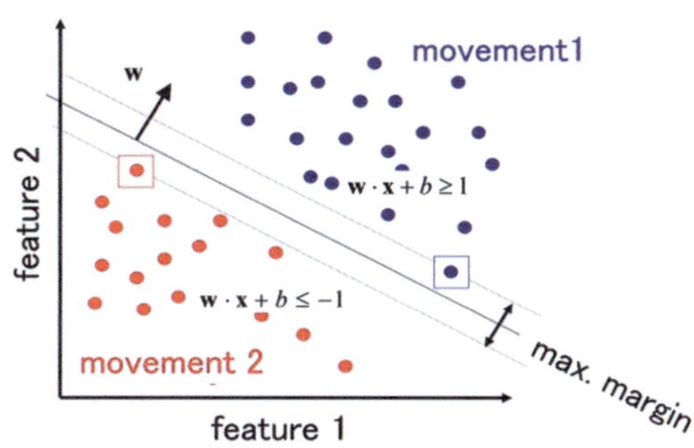

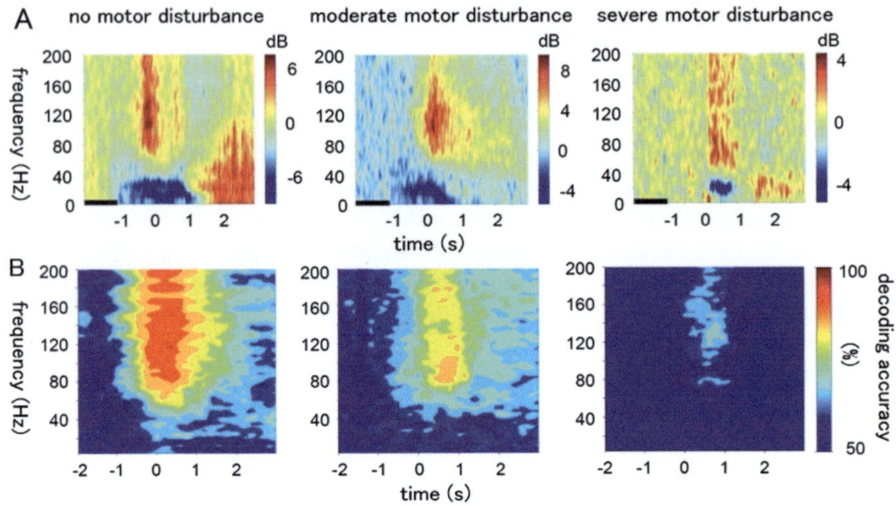


**Fig. 11.6** The relationship between motor disturbance and the decoding accuracy of unilateral upper limb movements. (**a**) The time-frequency spectrograms of the high γ band power during hand movements in patients without motor disturbance (*left*), with moderate motor disturbance (*middle*), and with severe motor disturbance (*right*). (**b**) The decoding accuracy by SVM of hand movements (chance level = 50 %)

3 to 150 Hz (Yanagisawa et al. 2011). We also found that, even in the case of severely-paralyzed patients, imageries of hand movements induced clear high γ band responses similar to those that real movements induced (Fig. 11.6).

There was a difference in spatial distribution of the high γ band activity between the patients who could clearly differentiate the imageries of upper limb movements and those who could not. In the patients who could clearly differentiate the imageries, a clear difference was observed in spatial distribution of the high γ band activities between hand grasping and elbow flexion, whereas in the patients who could not differentiate the imageries clearly, there was no significant difference (Yanagisawa et al. 2012b). This result suggests that the ability to differentiate motor imageries reflects differentiated cerebral neural representations of corresponding movements, and provides a new insight into functional reorganization.

Our most recent finding has revealed the mechanism of motor control. The power amplitude of the high γ band activities couples with the phase of α band activities before movement onset, and the coupling is attenuated just before movement onset (Yanagisawa et al. 2012a). This is called the 'cross-frequency phase-amplitude coupling.' It might play an important role in control of movement initiation and postures.

## 11.5 Real Time Robot Control and Communication

We applied our decoding method to an ECoG-based BMI system for real time control of a robot arm (Fig. 11.7). In particular, we measured ECoGs by using a 128-channel digital EEG system (EEG 2000; Nihon Koden Corporation, Tokyo, Japan) and digitized them at a sampling rate of 1000 Hz. We performed successive decoding every 200 ms by a hierarchical decoding and control method: first, Gaussian process regression was used to estimate the likelihood of the decoding feature for movements, and when the likelihood was above the threshold level, SVM was used to classify the types of hand and arm movements; second, "transitional states" were introduced into the robot posture, and the robot arm was gradually controlled from the initial posture to the target posture. The hierarchical decoding and control contributed to the smooth motion of the robot arm, even if decoding accuracy was not perfect (decoding accuracy on a single trial basis was approximately 70 %). As well as the hierarchical decoding and control method, we used a method of "modular decoding": different decoders were used for hand and elbow movements. This allowed for simultaneously and independently decoding and controlling the hand and elbow. The robot arm was an experimental anthropomorphic hand (Yokoi et al. 2009) and similar to the human arm in general movement mechanism and degrees of freedom. The robot hand was equipped with 8 DC motors, and they independently actuate 8 individual tendons in it. The 8 tendons work in a coordinated manner and produce finger flexion or extension. We succeeded in voluntary controlling the robot arm so as to grasp and release objects (Yanagisawa et al. 2012b). The robot arm was successfully controlled for 4 days after the initial decoder training without additional decoder training, although electrode implantation is limited for only a few weeks in clinical situation. This result is evidence that ECoG-based BMIs have long term stability.

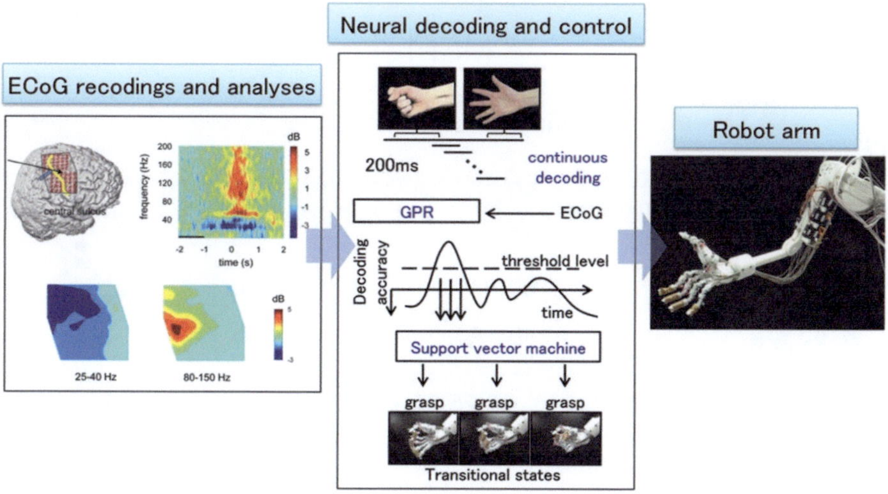


Fig. 11.7 The ECoG-based BMI system for real time robot control

We developed not only the method of classification-based decoding of hand posture but also a method of regression-based decoding of the reaching motion of arm. It uses a Bayesian algorithm called 'sparse linear regression' (SLR) (Nakanishi et al. 2013). If we combine both decoding methods, SVM for hand posture and SLR for the reaching motion of arm, we are able to decode hand and arm movements.

## 11.6 Noninvasive Neural Decoding Using Neuromagnetic Recording

A noninvasive evaluation of individual BMI performance is necessary for determining the surgical indication of invasive BMI treatment. A magnetoencephalography (MEG) is a potentially noninvasive method for evaluating individual BMI performance; it has high spatiotemporal resolution and neurophysiological compatibility with ECoGs. By using an MEG, we investigated the neural decoding performances of three types of unilateral hand and arm movements on a single trial basis (Sugata et al. 2012a, b). We used an SVM to decode movement types, with the peak amplitudes of the first movement-related cortical component that is induced after the movement onset (MEFI in Fig. 11.8a) as decoding features. The neural

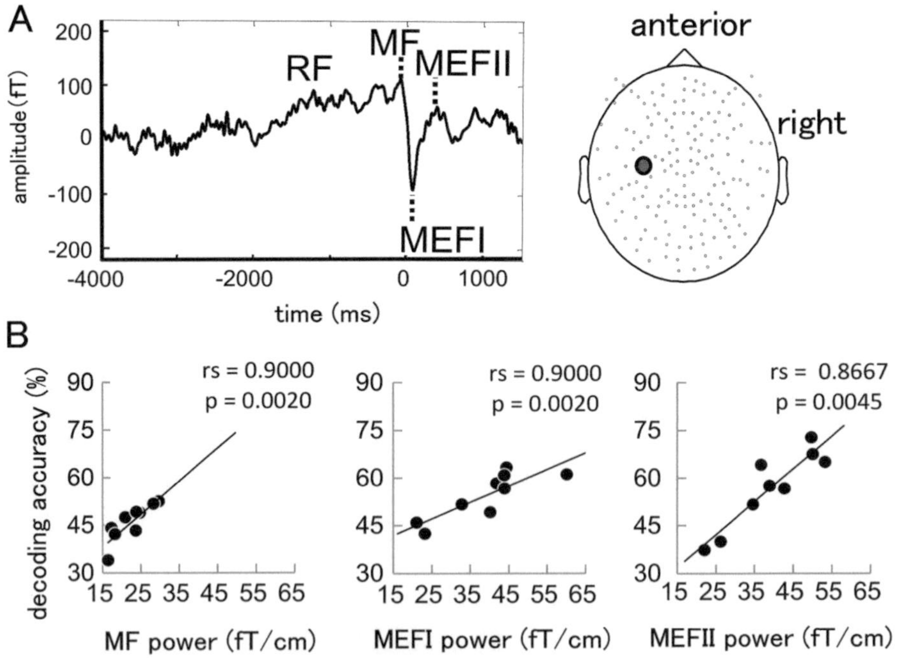


**Fig. 11.8** Neural decoding using magnetoencephalography. (**a**) A typical averaged waveform of a movement-related cortical field. (**b**) The relationships between neural decoding accuracies and the peak amplitudes of the first movement-related component that is induced after the movement onset. There were significant positive correlations between the amplitudes and the decoding accuracies for all of the three components

decoding accuracies largely exceeded the chance level in all of the nine healthy subjects. The MEFI and decoding accuracies were strongly correlated ($rs = 0.900$, $p = 0.002$) (Fig. 11.8). These results suggest that neurophysiological profiles might serve as predictors of individual BMI performances and assist in the improvement of BMI technology in general.

## 11.7 A Fully Implantable Wireless Device

### *11.7.1 Needs for Implantable Devices for BMIs*

The BMI system we employ for clinical research uses a wired lead, and it penetrates the skin and connect electrodes with an external EEG recording system. This wired lead poses a higher risk of infection as the implanted period is longer. Therefore, wired electrode implantation needs to be temporary, and electrodes need to be removed within a few weeks, in order to reduce the infection risk. It is necessary for making the infection risk low to fully implant both electrodes and the EEG recording system within the body. If these devices are implanted, indeed, there is another benefit: the BMI system will be more convenient for use because patients do not have to wear or remove the system anymore. However, there have been few reported implantable BMI devices in clinical use. There is only one report regarding clinical implantation in which only a 2-channel device was used (Guenther et al. 2009). For this reason, we have developed two prototypes of a fully implantable 128-channel ECoG recording system. This is a wireless system. By integrating it into a real-time BMI system, we ultimately aim to develop a Wireless Human ECoG-based Real-time BMI System (W-HERBS) (Hirata et al. 2011; Matsushita et al. 2013).

### *11.7.2 System Overview*

The second prototype we have developed is shown in Fig. 11.9. This system is fully implantable, and includes many new technologies: the head part consists of tissue conformable brain surface microelectrodes, a titanium head casing that functions as an artificial skull, and a 128-channel amplifier unit with two 64-channel chips; and the body part consists of a wireless data transfer unit, a microchip data controller, and a wireless rechargeable unit.

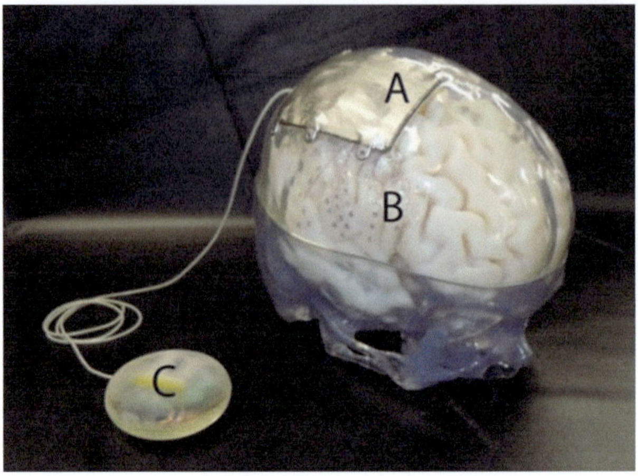


**Fig. 11.9** The second prototype of a fully implantable wireless system for the W-HERBS: Titanium head casing/ artificial skull bone (**a**). Brain surface microelectrodes conformable to the outer surface of the individual brain (**b**). Body part (**c**)

### *11.7.3 Tissue Conformable Brain Surface Microelectrodes*

To record the ECoGs with a higher spatiotemporal resolution, we developed 3-dimensional high-density grid electrodes, which were designed to fit individual's brain surface (Hirata et al. 2010). First, we extracted 3-dimensional (3D) surface data of the brain surface and brain groove from patient's individual magnetic resonance (MR) images. We used an automatic brain groove extraction software program (Brain VISA, http://brainvisa.info/). Next, we designed male and female molds for the grid electrodes using 3D CAD software (Mimics and 3-matic, Materialize Japan, Tokyo, Japan) (Fig. 11.10). Next, the molds were made by rapid manufacturing using a 3D printer (Polyjet ® 3D printer, Objet Geometries Ltd., Israel). Silicon sheets fitting the brain surface were subsequently produced from these molds. In addition, the location of each platinum electrode (1.0 mm in diameter) was decided with the 3D CAD software by taking account of individual's anatomical information. The inter-electrode spacing was up to 2.5 mm, and the brain groove grid electrodes were located on both sides of the electrode sheet. These 3D grid electrodes were fitted onto the brain surface with a minimal compression of the brain tissue, and they generated high ECoGs yields due to their close contact with the brain surface.

### *11.7.4 Integrated Analog Amplifier Unit*

ECoGs are characterized as signals in low frequency bands from 0.1 to 500 Hz, and they produce amplitudes from 1 μV to 1 mV. It is necessary for recording ECoG signals to reduce the input-referred noise of the amplifier (Yoshida et al. 2010). It is also important to have a variable bandwidth and wide dynamic range, because

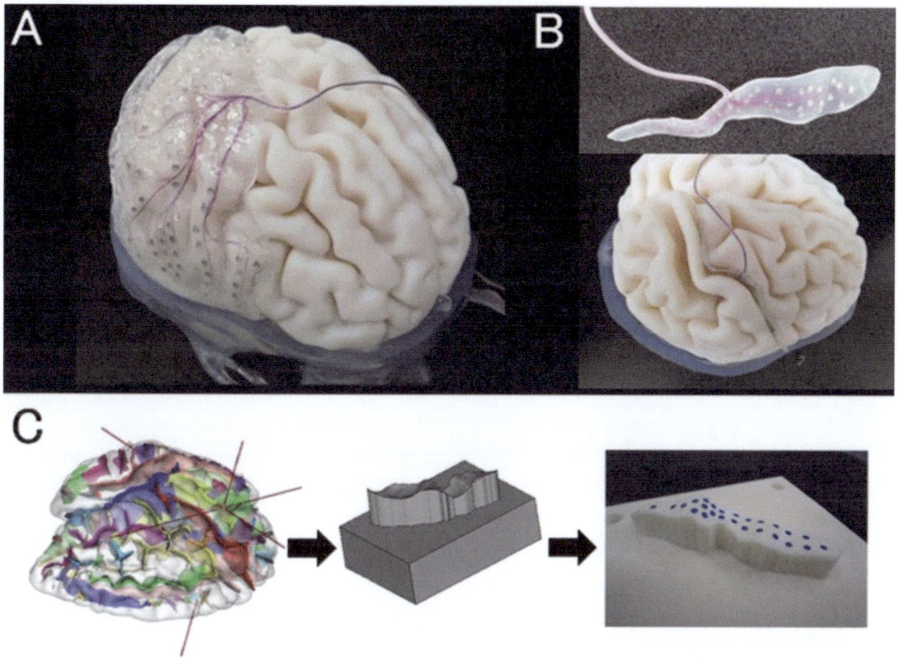


**Fig. 11.10** Tissue conformable brain surface microelectrodes fitted onto the individual brain surfaces. (**a**) Gyral (brain surface) electrodes. (**b**) Sulcal (brain groove) electrodes. (**c**) The mold of the electrode sheet is made by a 3D printer with rapid manufacturing technique

commercial AC noise can easily contaminate ECoG signals; they share the frequency bands. Thus, a high-linearity low noise amplifier with a variable bandwidth was developed to cover the frequency bands and voltage gains that are needed for recording ECoG signals (Yoshida et al. 2011). It is the low noise amplifier with a 0.1 Hz roll-off frequency implemented with core differential amplifiers using large-sized MOSFETs and a capacitor feedback scheme biased by ultrahigh resistors of cascade 12 MOSFETs. A VLSI chip was fabricated using CMOS 0.18 μm process technology in the chip fabrication program of the VLSI Design and Education Center (VDEC) at the University of Tokyo.

The main characteristics of the chip are listed in Table 11.3.

A 128-channel analog amplifier board consists of two chips, and they are mounted on two high-density printed sub-boards that are bridged by flexible printed wiring (Fig. 11.11). The size of the board is 20 mm × 30 mm × 2.5 mm, and it is small enough to be placed within a head casing, which will be described later in the text.

**Table 11.3** Main characteristics of the integrated analog amplifier chip

| | |
|---|---|
| Number of channels: | 64 channels |
| A/D converter: | 12 bits |
| Voltage gain: | 40–80 dB |
| Signal frequency bands: | 0.1–1000 Hz |
| Input referred noise: | 2.8 μV |
| Power consumption: | 4.9 mW |
| Chip size: | 5.0 mm × 5.0 mm |
| Master/slave function for a 128-channel system | |

**Fig. 11.11** A 128-channel integrated analog amplifier board

### *11.7.5 Wireless Data Transfer Unit*

We used a Bluetooth protocol communication (Class 2) for the first prototype because it was easy to use. However, it had several problems: first, its data transfer speed was not sufficient (data transmission rates of max. 400 kbps); second, it was too big in size (60 mm × 60 mm × 8 mm); third, its average power consumption was as high as 300 mW, and most of the system power was consumed by wireless data transfers. Therefore, we employed a Wi-Fi protocol communication for the second prototype. Our Wi-Fi chip achieves an effective data transmission rate of 1.6 Mbps, and transfers 128-channel × 12-bit ECoG data × 1 kHz in real time. The max power consumption of the second prototype is approximately 200 mW, and this means that most of the system power is still consumed by wireless data transfers. The size is reduced to 40 mm × 40 mm × 5 mm.

### 11.7.6 Wireless Rechargeable Unit

The wireless battery charging system of the second prototype consists of two parts. One is a transmitter to be positioned outside the human body, and the other is a receiver to be located inside the human body. The system exhibits a wireless charging power of 400 mW at a distance of 20 mm, and this is sufficient for running the entire system. The coil size of the abdominal portion is 40 mm in diameter and 8 mm in thickness. The size may be scaled down more if power consumption can be reduced.

### 11.7.7 Head Casing and Artificial Skull Bone

We have developed a titanium head casing which contained a 128-channel amplifier unit. This casing functions as both a head casing and an artificial skull bone. It was designed to fit a patient's skull by using the 3D CAD (3 matic, Materialize Japan, Tokyo) and 3D CAM (Gibbs CAM, Gibbs and Associates, USA) software programs (Fig. 11.12). The shape of the head casing has cosmetic advantages over and is safer than other convex shapes; they pose a higher risk of cutaneous fistula.

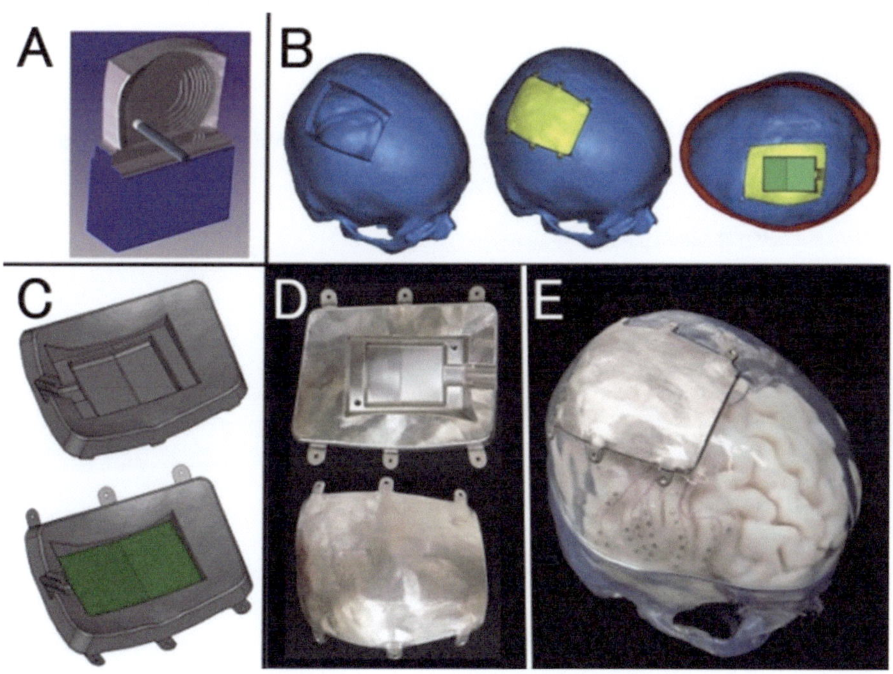


**Fig. 11.12** A titanium head casing/artificial skull bone. (**a**) A computer simulation of machining path using 3D CAM software. (**b**) A computer simulation of a head casing fitting the skull. *Left*: skull opening. *Middle*: Outer side view of a head casing fitting the skull. *Right*: Inner side view. (**c**) A head casing designed using 3D CAD software. *Upper*: A head casing without an electronic circuit board. *Lower*: A head casing with an electronic circuit board. (**d**) A prototype casing. *Upper*: inner side view. *Lower*: outer side view. (**e**) A prototype casing attached to a skull model

### *11.7.8 Animal Experiments*

To test the system, we conducted acute and chronic animal experiments with monkeys. We performed an acute experiment of implanting the first prototype for 10 days to evaluate its recording performance. Cortical somatosensory evoked potentials were clearly recorded (Fig. 11.13). To test the second prototype, we built its variant for animal experiments and chronically implanted it in two monkeys for 6 months. Since monkey's head and body space are small, the variant had 64-channel flat electrodes, a 64 channel amplifier chip, and a titanium head casing/an artificial skull bone. The abdominal part was implanted on the back, because monkeys tended to remove implanted devices. For one monkey, the device stopped working 3 months after the implantation, most probably due to incomplete hermeticity. For another monkey, the device worked throughout 6 months without major troubles. A head bulge was formed due to the titanium head casing/artificial skull bone, but it is not distinctly visible (Fig. 11.14).

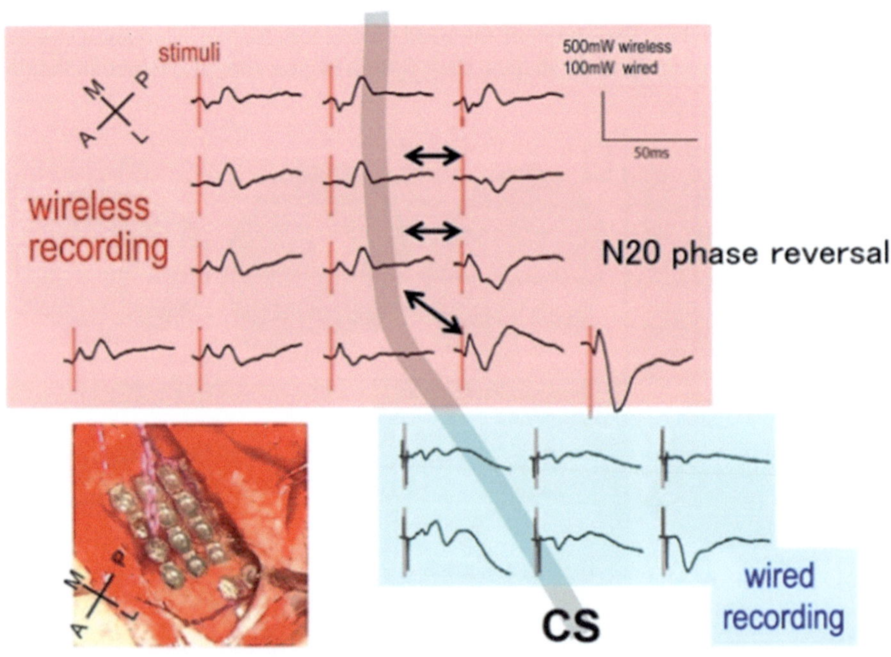


**Fig. 11.13** Cortical somatosensory evoked potentials (SEPs) recorded by a fully implantable device. Upper graphs show wirelessly transferred data recorded by an implanted device, while lower graphs show SEPs recorded by a wired EEG recorder

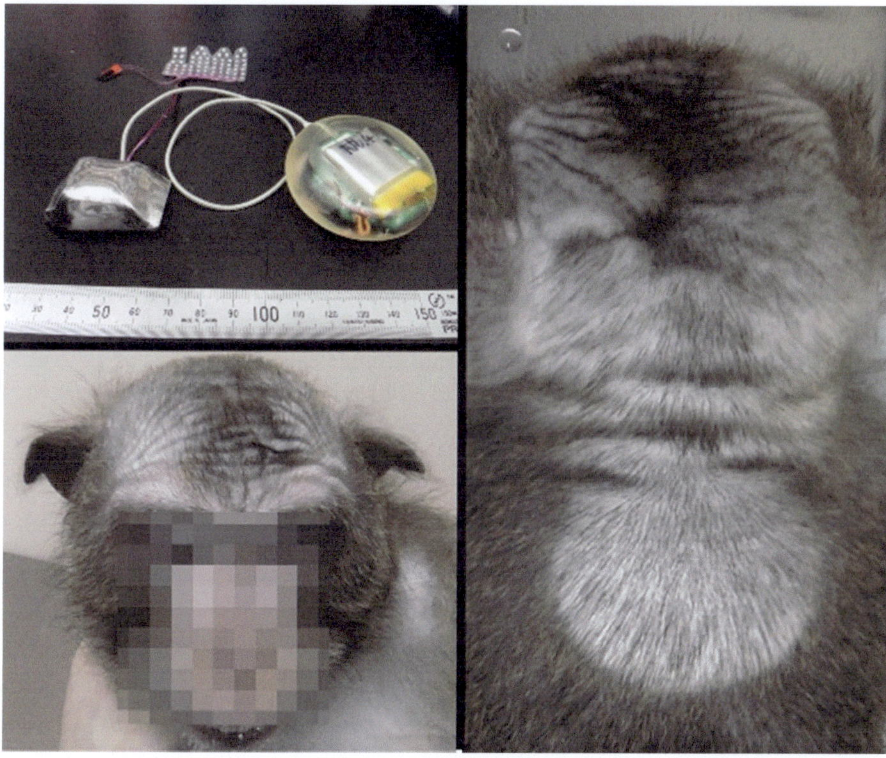

**Fig. 11.14** Monkey with a wireless device implanted in the left head and on the back. The head skin implant bulge is indistinct

## 11.8 Summary

In this chapter, we have described the history and trends of BMI research, and explained our research and development of ECoG-based BMI. Using a wired BMI system, patients have successfully controlled a robot arm in real time. We have developed the first prototype of a fully-implantable wireless system. A fully-implantable wireless system is indispensable for the clinical application of invasive BMI with a low risk of infection. BMIs are promising for restoring motor control in severely disabled patients.

**Acknowledgments** This work was partly supported by a grant for "Brain Machine Interface Development" from the Strategic Research Program for Brain Sciences of MEXT, and a Health Labour Sciences Research Grant (23100101) from the Ministry of Health, Labour, and Welfare of Japan. We would like to thank Toshiki Yoshimine, Takufumi Yanagisawa, Hisato Sugata, Morris Shayne, Takashi Moriwaki (Osaka Univ.), Yukiyasu Kamitani (ATR), and Hiroshi Yokoi (Univ. Electro. Com.). We would also like to thank Takafumi Suzuki, Hiroshi Ando (NICT), Takeshi Yoshida (Hiroshima Univ.), Fumihiro Sato (Tohoku Univ.), Yukio Nishimura, Tatsuya Umeda (NINS), Atsushi Iwata (A-R-Tec Corp), Shinichi Morikawa (Unique Medical), Naohiro Hayaishi

(Keisuugiken Corp), Shinichi Yoshimura, Shuhei Kosaka (Aska Electric Co Ltd), and Hirofumi Itoh (Junkosha Inc), for prototype manufacturing of our implantable system. Our appreciation also goes to VLSI Design and Education Center (VDEC), University of Tokyo, for offering the chip fabrication program.

## Exercises

1. Briefly review the classifications of BMIs in term of invasiveness and neural information.
2. Briefly review the characteristics of brain signals used for BMIs.
3. What neurophysiological features are useful for neural decoding? How are they useful?
4. Briefly describe the issues related to the clinical application of invasive BMIs. How do you think we are able to overcome those issues?

## References

Birbaumer, N., Ramos Murguialday, A., Weber, C., Montoya, P.: Neurofeedback and brain-computer interface clinical applications. Int. Rev. Neurobiol. **86**, 107–117 (2009)

Chao, Z.C., Nagasaka, Y., Fujii, N.: Long-term asynchronous decoding of arm motion using electrocorticographic signals in monkeys. Front. Neuroeng. **3**, 3 (2010)

Chapin, J.K., Moxon, K.A., Markowitz, R.S., Nicolelis, M.A.: Real-time control of a robot arm using simultaneously recorded neurons in the motor cortex. Nat. Neurosci. **2**, 664–670 (1999)

Collinger, J.L., Wodlinger, B., Downey, J.E., Wang, W., Tyler-Kabara, E.C., Weber, D.J., McMorland, A.J., Velliste, M., Boninger, M.L., Schwartz, A.B.: High-performance neuroprosthetic control by an individual with tetraplegia. Lancet **381**, 557–564 (2013)

Farwell, L.A., Donchin, E.: Talking off the top of your head: toward a mental prosthesis utilizing event-related brain potentials. Electroencephalogr. Clin. Neurophysiol. **70**, 510–523 (1988)

Georgopoulos, A.P., Kalaska, J.F., Caminiti, R., Massey, J.T.: On the relations between the direction of two-dimensional arm movements and cell discharge in primate motor cortex. J. Neurosci. **2**, 1527–1537 (1982)

Georgopoulos, A.P., Kettner, R.E., Schwartz, A.B.: Primate motor cortex and free arm movements to visual targets in three-dimensional space. II. Coding of the direction of movement by a neuronal population. J. Neurosci. **8**, 2928–2937 (1988)

Guenther, F.H., Brumberg, J.S., Wright, E.J., Nieto-Castanon, A., Tourville, J.A., Panko, M., Law, R., Siebert, S.A., Bartels, J.L., Andreasen, D.S., Ehirim, P., Mao, H., Kennedy, P.R.: A wireless brain-machine interface for real-time speech synthesis. PLoS One **4**, e8218 (2009)

Hirata, M., Kato, A., Taniguchi, M., Ninomiya, H., Cheyne, D., Robinson, S.E., Maruno, M., Kumura, E., Ishii, R., Hirabuki, N., Nakamura, H., Yoshimine, T.: Frequency-dependent spatial distribution of human somatosensory evoked neuromagnetic fields. Neurosci. Lett. **318**, 73–76 (2002)

Hirata, M., Kato, A., Taniguchi, M., Saitoh, Y., Ninomiya, H., Ihara, A., Kishima, H., Oshino, S., Baba, T., Yorifuji, S., Yoshimine, T.: Determination of language dominance with synthetic aperture magnetometry: comparison with the Wada test. Neuroimage **23**, 46–53 (2004)

Hirata, M., Yoshimine, T., Saitoh, Y., Yanagisawa, T., Goto, T., Watanabe, Y., Saito, T.: Intracranial electrode and method for producing same. US Patent 7,860,577, 2010

Hirata, M., Matsushita, K., Suzuki, T., Yoshida, T., Sato, F., Morris, S., Yanagisawa, M., Goto, T., Kawato, M., Yoshimine, T.: A fully-implantable wireless system for human brain-machine interfaces using brain surface electrodes: W-HERBS. IEICE Trans. Commun. **E94-B**, 2448–2453 (2011)
Hochberg, L.R., Serruya, M.D., Friehs, G.M., Mukand, J.A., Saleh, M., Caplan, A.H., Branner, A., Chen, D., Penn, R.D., Donoghue, J.P.: Neuronal ensemble control of prosthetic devices by a human with tetraplegia. Nature **442**, 164–171 (2006)
Hochberg, L.R., Bacher, D., Jarosiewicz, B., Masse, N.Y., Simeral, J.D., Vogel, J., Haddadin, S., Liu, J., Cash, S.S., van der Smagt, P., Donoghue, J.P.: Reach and grasp by people with tetraplegia using a neurally controlled robotic arm. Nature **485**, 372–375 (2012)
Horikawa, T., Tamaki, M., Miyawaki, Y., Kamitani, Y.: Neural decoding of visual imagery during sleep. Science **340**, 639–642 (2013)
Hosomi, K., Saitoh, Y., Kishima, H., Oshino, S., Hirata, M., Tani, N., Shimokawa, T., Yoshimine, T.: Electrical stimulation of primary motor cortex within the central sulcus for intractable neuropathic pain. Clin. Neurophysiol. **119**, 993–1001 (2008)
Huggins, J.E., Levine, S.P., BeMent, S.L., Kushwaha, R.K., Schuh, L.A., Passaro, E.A., Rohde, M. M., Ross, D.A., Elisevich, K.V., Smith, B.J.: Detection of event-related potentials for development of a direct brain interface. J. Clin. Neurophysiol. **16**, 448–455 (1999)
Ikegami, S., Takano, K., Wada, M., Saeki, N., Kansaku, K.: Effect of the green/blue flicker matrix for P300-based brain-computer interface: an EEG-fMRI study. Front. Neurol. **3**, 113 (2012)
Ishii, R., Canuet, L., Herdman, A., Gunji, A., Iwase, M., Takahashi, H., Nakahachi, T., Hirata, M., Robinson, S.E., Pantev, C., Takeda, M.: Cortical oscillatory power changes during auditory oddball task revealed by spatially filtered magnetoencephalography. Clin. Neurophysiol. **120**, 497–504 (2009)
Kamitani, Y., Tong, F.: Decoding the visual and subjective contents of the human brain. Nat. Neurosci. **8**, 679–685 (2005)
Leuthardt, E.C., Schalk, G., Wolpaw, J.R., Ojemann, J.G., Moran, D.W.: A brain-computer interface using electrocorticographic signals in humans. J. Neural Eng. **1**, 63–71 (2004)
Levine, S.P., Huggins, J.E., BeMent, S.L., Kushwaha, R.K., Schuh, L.A., Passaro, E.A., Rohde, M. M., Ross, D.A.: Identification of electrocorticogram patterns as the basis for a direct brain interface. J. Clin. Neurophysiol. **16**, 439–447 (1999)
Matsushita, K., Hirata, M., Suzuki, T., Ando, H., Ota, Y., Sato, F., Morris, S., Yoshida, T., Matsuki, H., Yoshimine, T.: Development of an implantable wireless ECoG 128ch recording device for clinical brain machine interface. Conf. Proc. IEEE Eng. Med. Biol. Soc. **2013**, 1867–1870 (2013)
Miyawaki, Y., Uchida, H., Yamashita, O., Sato, M.A., Morito, Y., Tanabe, H.C., Sadato, N., Kamitani, Y.: Visual image reconstruction from human brain activity using a combination of multiscale local image decoders. Neuron **60**, 915–929 (2008)
Nakanishi, Y., Yanagisawa, T., Shin, D., Fukuma, R., Chen, C., Kambara, H., Yoshimura, N., Hirata, M., Yoshimine, T., Koike, Y.: Prediction of three-dimensional arm trajectories based on ECoG signals recorded from human sensorimotor cortex. PLoS One **8**, e72085 (2013)
Ojemann, G.A.: Neurosurgical management of epilepsy: a personal perspective in 1983. Appl. Neurophysiol. **46**, 11–18 (1983)
Pfurtscheller, G., Neuper, C.: Future prospects of ERD/ERS in the context of brain-computer interface (BCI) developments. Prog. Brain Res. **159**, 433–437 (2006)
Pistohl, T., Ball, T., Schulze-Bonhage, A., Aertsen, A., Mehring, C.: Prediction of arm movement trajectories from ECoG-recordings in humans. J. Neurosci. Methods **167**, 105–114 (2008)
Ramos-Murguialday, A., Broetz, D., Rea, M., Laer, L., Yilmaz, O., Brasil, F.L., Liberati, G., Curado, M.R., Garcia-Cossio, E., Vyziotis, A., Cho, W., Agostini, M., Soares, E., Soekadar, S., Caria, A., Cohen, L.G., Birbaumer, N.: Brain-machine interface in chronic stroke rehabilitation: a controlled study. Ann. Neurol. **74**, 100–108 (2013)
Rathelot, J.A., Strick, P.L.: Subdivisions of primary motor cortex based on cortico-motoneuronal cells. Proc. Natl. Acad. Sci. U. S. A. **106**, 918–923 (2009)

Schalk, G., Kubanek, J., Miller, K.J., Anderson, N.R., Leuthardt, E.C., Ojemann, J.G., Limbrick, D., Moran, D., Gerhardt, L.A., Wolpaw, J.R.: Decoding two-dimensional movement trajectories using electrocorticographic signals in humans. J. Neural Eng. **4**, 264–275 (2007)
Schalk, G., Miller, K.J., Anderson, N.R., Wilson, J.A., Smyth, M.D., Ojemann, J.G., Moran, D.W., Wolpaw, J.R., Leuthardt, E.C.: Two-dimensional movement control using electrocorticographic signals in humans. J. Neural Eng. **5**, 75–84 (2008)
Serruya, M.D., Hatsopoulos, N.G., Paninski, L., Fellows, M.R., Donoghue, J.P.: Instant neural control of a movement signal. Nature **416**, 141–142 (2002)
Shibata, K., Watanabe, T., Sasaki, Y., Kawato, M.: Perceptual learning incepted by decoded fMRI neurofeedback without stimulus presentation. Science **334**, 1413–1415 (2011)
Shindo, K., Kawashima, K., Ushiba, J., Ota, N., Ito, M., Ota, T., Kimura, A., Liu, M.: Effects of neurofeedback training with an electroencephalogram-based brain-computer interface for hand paralysis in patients with chronic stroke: a preliminary case series study. J. Rehabil. Med. **43**, 951–957 (2011)
Sitaram, R., Caria, A., Birbaumer, N.: Hemodynamic brain-computer interfaces for communication and rehabilitation. Neural Netw. **22**, 1320–1328 (2009)
Sugata, H., Goto, T., Hirata, M., Yanagisawa, T., Shayne, M., Matsushita, K., Yoshimine, T., Yorifuji, S.: Neural decoding of unilateral upper limb movements using single trial MEG signals. Brain Res. **1468**, 29–37 (2012a)
Sugata, H., Goto, T., Hirata, M., Yanagisawa, T., Shayne, M., Matsushita, K., Yoshimine, T., Yorifuji, S.: Movement-related neuromagnetic fields and performances of single trial classifications. Neuroreport **23**, 16–20 (2012b)
Taniguchi, M., Kato, A., Fujita, N., Hirata, M., Tanaka, H., Kihara, T., Ninomiya, H., Hirabuki, N., Nakamura, H., Robinson, S.E., Cheyne, D., Yoshimine, T.: Movement-related desynchronization of the cerebral cortex studied with spatially filtered magnetoencephalography. Neuroimage **12**, 298–306 (2000)
Taylor, D.M., Tillery, S.I., Schwartz, A.B.: Direct cortical control of 3D neuroprosthetic devices. Science **296**, 1829–1832 (2002)
Toda, A., Imamizu, H., Kawato, M., Sato, M.A.: Reconstruction of two-dimensional movement trajectories from selected magnetoencephalography cortical currents by combined sparse Bayesian methods. Neuroimage **54**, 892–905 (2011)
Velliste, M., Perel, S., Spalding, M.C., Whitford, A.S., Schwartz, A.B.: Cortical control of a prosthetic arm for self-feeding. Nature **453**, 1098–1101 (2008)
Wessberg, J., Stambaugh, C.R., Kralik, J.D., Beck, P.D., Laubach, M., Chapin, J.K., Kim, J., Biggs, S.J., Srinivasan, M.A., Nicolelis, M.A.: Real-time prediction of hand trajectory by ensembles of cortical neurons in primates. Nature **408**, 361–365 (2000)
Wolpaw, J.R., McFarland, D.J., Neat, G.W., Forneris, C.A.: An EEG-based brain-computer interface for cursor control. Electroencephalogr. Clin. Neurophysiol. **78**, 252–259 (1991)
Wolpaw, J.R., Birbaumer, N., McFarland, D.J., Pfurtscheller, G., Vaughan, T.M.: Brain-computer interfaces for communication and control. Clin. Neurophysiol. **113**, 767–791 (2002)
Yanagisawa, T., Hirata, M., Saitoh, Y., Kato, A., Shibuya, D., Kamitani, Y., Yoshimine, T.: Neural decoding using gyral and intrasulcal electrocorticograms. Neuroimage **45**, 1099–1106 (2009)
Yanagisawa, T., Hirata, M., Saitoh, Y., Goto, T., Kishima, H., Fukuma, R., Yokoi, H., Kamitani, Y., Yoshimine, T.: Real-time control of a prosthetic hand using human electrocorticography signals. J. Neurosurg. **114**, 1715–1722 (2011)
Yanagisawa, T., Yamashita, O., Hirata, M., Kishima, H., Saitoh, Y., Goto, T., Yoshimine, T., Kamitani, Y.: Regulation of motor representation by phase-amplitude coupling in the sensorimotor cortex. J. Neurosci. **32**, 15467–15475 (2012a)
Yanagisawa, T., Hirata, M., Saitoh, Y., Kishima, H., Matsushita, K., Goto, T., Fukuma, R., Yokoi, H., Kamitani, Y., Yoshimine, T.: Electrocorticographic control of a prosthetic arm in paralyzed patients. Ann. Neurol. **71**, 353–361 (2012b)
Yokoi, H., Kita, K., Nakamura, T.: Mutually adaptable EMG devices for prosthetic hand. Int. J. Fac. Autom. Robot. Soft Comput. **1**, 74–83 (2009)

Yoshida, T., Masui, Y., Eki, R., Iwata, A., Yoshida, M., Uematsu, K.: A neural recording amplifier with low-frequency noise suppression. IEICE Trans. Electrons **E93-C**, 849–854 (2010)
Yoshida, T., Sueishi, K., Iwata, A., Matsushita, K., Hirata, M., Suzuki, T.: A high-linearity low-noise amplifier with variable bandwidth for neural recording systems. Jpn. J. Appl. Phys. **50**, 04DE07 (2011)

# Chapter 12
# Norms and Games as Integrating Components of Social Organizations

**Yasuo Nakayama**

**Abstract** Previous chapters in this book mainly discuss individual humans from the viewpoint of psychology or neuroscience. In this chapter, we discuss how mental states and attitudes of individuals are related to the society from a philosophical viewpoint. In this discussion, we are particularly concerned with the question of what kinds of roles norms play when individuals establish social connections. Then, we discuss how various kinds of human capacities are related to each other and how people build and maintain social organizations. In the end, we see that humans are beings that have cognitive capacities to form their own lives by behaving in accordance with the accepted norms and changing the norms for better lives.

**Keywords** Social organizations • Collective beliefs • Social norms • Obligation • Normative systems • Action space • Gamification • Language game • Information update • Knowledge representation

## 12.1 What Are Norms?

What are norms? Why are norms important for explaining human behaviors? One of the main aims of this chapter is to answer these questions. Let us start with the first question. Our preliminary answer is that norms are *what normative sentences express*. Then, what are normative sentences? We distinguish between *assertive* and *normative* sentences. They both belong to declarative sentences, but express different kinds of contents.[1] Assertive sentences simply describe states or events in the world. Examples of assertive sentences are the following:

(1a) There are three tables in this room.

[1] Different kinds of speech acts may be distinguished (Seale 1979). Our view is that the distinction between assertive and normative sentences is fundamental, and speech acts can be expressed in terms of this distinction. See Nakayama (2014).

Y. Nakayama (✉)
Graduate School of Human Sciences, Osaka University, Suita, Osaka, Japan
e-mail: nakayama@hus.osaka-u.ac.jp



M. Kasaki et al. (eds.), *Cognitive Neuroscience Robotics B*,
DOI 10.1007/978-4-431-54598-9_12

(1b) A presidential election took place in the USA in 2013.

The first sentence describes a physical fact, and the second sentence does a social fact. Unlike physical facts, social facts are those the existence of which is based on collective beliefs (Searle 1995, 2010; Nakayama 2011). Collective beliefs are closely related to social norms (See Sect. 12.4 and Appendix).

There are three kinds of normative sentences. They express *obligation*, *prohibition*, and *permission*, respectively:

(2a) 'You ought to be quiet in the lecture' expresses an obligation.
(2b) 'It is forbidden to sleep in the lecture' expresses a prohibition.
(2c) 'It is permitted to ask questions anytime when you do not completely understand' expresses a permission.

It is difficult to say of a normative sentence whether it is true or false. Then, it may be more appropriate to ask whether it is collectively accepted or not. This difficulty of assigning a truth value reveals a fundamental difference between normative and assertive sentences. For an assertive sentence is, in principle, true or false.

Sentences (2a) to (2c) are not purely normative but involve certain assertive components. For example, sentence (2a) can be paraphrased as a conditional obligation in the following way:

(2a*) If you are sitting with other students in the lecture, you ought to be quiet.

The second clause expresses an obligation, but the first clause expresses the condition in which this obligation is imposed. Suppose a female student is sitting in the lecture. Then, we can check by observation whether she fulfills the obligation. If she has been quiet during the whole lecture, we say that she has fulfilled it. The fulfillment of an obligation is often observable. This suggests that facts and norms are different but closely related.

Assertive and normative sentences are not the only kinds of sentences. There are other types of sentences, such as *imperative* and *interrogative* sentences. But there is a good reason to claim that assertive and normative sentences are semantically more fundamental than other types of sentences. In fact, Nakayama (2014) shows that all types of speech acts entail assertive or normative components.

There are logical relations among three kinds of normative modalities: *obligation*, *prohibition*, and *permission*.[2] If something is *permitted*, it is not *forbidden*. And if something is *forbidden*, its negation is an *obligation*.[3] In Sect. 12.4, we construct a logical framework that describes logical properties of normative modalities, but we cannot go into full detail and only illustrate some main principles of the

[2] These normative modalities are also called *deontic modalities*.

[3] These two conditional propositions are expressed as $\mathbf{P}p \rightarrow \neg\mathbf{F}p$ and $\mathbf{F}p \rightarrow \mathbf{O}\neg p$ in the *standard deontic logic* (SDL). SDL is a modal logic proposed by von Wright. In SDL, *prohibition* and *permission* are definable through *obligation*: $\mathbf{F}p \Leftrightarrow_{\mathrm{def}} \mathbf{O}\neg p$ and $\mathbf{P}p \Leftrightarrow_{\mathrm{def}} \neg\mathbf{O}\neg p$, where $\neg$ is the sign of negation. See Åqvist (2002).

logical framework for normative systems (See Sects. 12.4, 12.5, and Appendix of this chapter).

## 12.2 Why Norms Are so Important? – Plays, Games, and Norms

In this section, we give the first answer to the question why norms are important for human life. The answer is that a child becomes a member of a society *through accepting common beliefs and sharing social norms*. Sharing social norms means that people in a social organization agree on what they should do and what they are forbidden to do; or in other words, that they plan what to do (individually) in accordance with social norms.

A social organization might be complex, but a game can be used as a toy model for it. Games illustrate in a simple way how social interactions work. Thus, children can learn social behaviors through playing games. In fact, some developmental psychologists, such as Jean Piaget (1896–1980) and Lev Vygotsky (1896–1934), did extensive researches on plays of children. They held that the mastery of games is crucial for learning social behaviors. For example, Piaget proposed three broad stages of play activity (Piaget 1951; Birch 1997: p. 55f):

(3a) **Mastery play** corresponds to the sensori-motor stage of development (from birth to two years, approximately). Important things at this stage are practice, control of movements, and exploration of objects through sight and touch. Children's play activity contains repetitive movements, and children enjoy them for the simple pleasure of demonstrating their developing mastery of the relevant skills.

(3b) **Symbolic play** coincides with the pre-operational stage (from two to seven years, approximately). Children employ fantasy, make believe in play, and delight in using one object to symbolize another—a chair may become a motor car, and a sheet a fashionable dress.

(3c) **Play with rules** characterizes the operational stage (from seven years onwards, approximately). Children's thought processes are developed to be more logical, and their play involves the use of rules and procedures.

This description of each stage demonstrates how plays and games in a broad sense are crucial components of the human life from its beginning. Plays usually contain some normative components; there are actions you should do and actions you must not do. You are required to make your choice and what you can choose is restricted by the rules of the play. Such normative constraints define your *action space*. The learning of correct normative behaviors through plays prepares children to behave socially in the real life. When children go to school, they learn how to deal with social norms. For example, they ought to listen to the teacher during the lesson; they ought to stay in the school until school hours are over, and so on.

Recent successful applications of gamification also indicate how attractive games are for humans. *Gamification* means the use of game thinking and game mechanics in a non-game context to engage users and solve problems (Zichermann and Cunningham 2011). Boring activities can be made more attractive by employing gamification techniques, such as giving rewards, transforming a problem into a cooperative game, or visualizing each development stage of competition in real time.

## 12.3 Language Games

*Language game* is one of the crucial ideas that characterize the philosophy of Ludwig Wittgenstein (1889–1951). There are two interesting facts about language games. First, there are *primitive* language games, i.e., those that can be played without presupposing other language games. Children can learn primitive language games directly through playing them. Primitive language games are simpler and easier to learn than non-primitive language games. Second, language games have dynamic aspects. A new language game could emerge anytime and an old one could be forgotten. There is a kind of metabolism in the system of language games.

The second paragraph of Wittgenstein's *Philosophical Investigations* (*PI*) describes an example of primitive language game:

> That philosophical concept of meaning has its place in a primitive idea of the way language functions. But one can also say that it is the idea of a language more primitive than ours.
>
> Let us imagine a language for which the description given by Augustine is right. The language is meant to serve for communication between a builder A and an assistant B. A is building with buildingstones: there are blocks, pillars, slabs and beams. B has to pass the stones, and that in the order in which A needs them. For this purpose they use a language consisting of the words "block", "pillar", "slab", "beam". A calls them out;—B brings the stone which he has learnt to bring at such-and-such a call.—Conceive this as a complete primitive language. (Wittgenstein 1958: *PI* I.§2)

This primitive language game can be interpreted as a two player game. The game consists of two main rules:

[Rule 1] The builder is permitted to call out one of the four nouns "block", "pillar", "slab", "beam", when the assistant is waiting for a call.

[Rule 2] When the builder calls out a noun, the assistant is obligated to bring a stone denoted by the noun.

The builder makes a move of calling out one of the four nouns, and the assistant makes a move of delivering an appropriate stone to the builder. The builder and the assistant alternate moves. The game continues until the builder decides to end it.

Wittgenstein also suggests how the total system of language games develops.

> Do not be troubled by the fact that languages (2) and (8) consist only of orders. If you want to say that this shows them to be incomplete, ask yourself whether our language is complete;—whether it was so before the symbolism of chemistry and the notation of the

> infinitesimal calculus were incorporated in it; for these are, so to speak, suburbs of our language. (And how many houses or streets does it take before a town begins to be a town?) Our language can be seen as an ancient city: a maze of little streets and squares, of old and new houses, and of houses with additions from various periods; and this surrounded by a multitude of new boroughs with straight regular streets and uniform houses. (Wittgenstein 1958: *PI* I.§18)

Here, Wittgenstein talks about *our language*, i.e., the language of our linguistic community. However, this description is applicable to the development of the language of one particular person as well, because she has a personal history and has learned different sorts of language games in various situations.

If we take these two paragraphs from *PI* together, we obtain a picture of how a child learns a language. She starts playing various kinds of primitive language games. When she becomes adult, she has experienced many language games. Some of them are her favorite ones, and some of them are forgotten. She plays the former every day, and she no longer plays the latter. She plays language games through her life time.

## 12.4 Logic for Normative Systems

An aim of this chapter is to describe social norms rigorously and systematically. For this purpose, we use *Logic for Normative Systems* (LNS) that is proposed in Nakayama (2010, 2011). LNS is initially proposed to elucidate normative inferences and clarify the notion of *normativity*. As we discussed in Sect. 12.1, norms are closely related with facts. LNS is a framework that can describe context-dependent interactions between norms and facts. At first, we begin with clarifying what normative systems are.

Nakayama (2010) defines *Logic for Normative Systems* (LNS) in a certain way, and the following is a modification of the definition proposed there.[4]

(4a) A *normative system NS* consists of *belief base T* and *normative base N*. A normative base consists of *obligation base OB* and *permission base PER*. Thus, $NS = \langle T, \langle OB, PER \rangle \rangle$.[5] Each of *T*, *OB*, and *PER* is a set of sentences (more precisely, a set of sentences in *First-Order Logic*[6]).

(4b) A normative system *NS* is consistent ⇔ *union*(*NS*) is consistent, where *union* $(NS) = T \cup OB \cup PER$.

[4] We use *not*, &, ⇒, and ⇔ as meta-semantic abbreviation for *not*, *and*, *if . . . then*, and *if and only if*. For more precise definition and characterization of LNS, see Appendix of this chapter. The reader may consult Nakayama (2013, 2014) as well.

[5] Formally, we can avoid the use of *PER*, because what is permitted is uniquely determined when we determine *T* and *OB*. However, in this chapter, we include *PER* into the structure of normative systems, because we allow information updates of normative systems with respect to permission, in order to take ordinary language practices into consideration.

[6] First-Order Logic is the *standard logic* that is usually used for proving mathematical theorems.

(4c) A sentence $q$ belongs to the *belief set* of $NS \Leftrightarrow q$ follows from $T$.

(4d) A sentence $q$ belongs to the *obligation set* of $NS \Leftrightarrow$ ($NS$ is consistent & $q$ follows from $T \cup OB$ & $q$ does not belong to the *belief set* of $NS$).

(4e) A sentence $q$ belongs to the *prohibition set* of $NS \Leftrightarrow$ the negation of $q$ belongs to the *obligation set* of $NS$.

(4f) A sentence $q$ *belongs to the permission set of* $NS \Leftrightarrow$ ($union(NS) \cup \{q\}$ is consistent & $q$ does not belong to the *belief set* of $NS$).

(4g) We interpret that $NS$ represents a normative system accepted by a person or a group at a particular time. Thus, we allocate *what a person* (or *a group*) *believes to be true* to the *belief base*, *what he believes to ought to be done* to the *obligation base*, and *what he believes to be permitted* to the *permission base*.

According to this definition, a normative system presupposes certain non-normative theories and facts. The contents of beliefs are expressed in terms of assertive sentences, and hence non-normative theories and facts are treated as beliefs in $NS$ (see (4g)). Beliefs are separated from norms. In other words, they include no common element (see (4d) to (4f)). If you perform an action that is your obligation, then it becomes a fact and a part of your beliefs that you have performed this action. As a result, the obligation ceases to be your obligation. Beliefs and normative requirements are closely related this way. According to (4f), permission is understood in a wide sense: everything compatible with a given normative system is considered to be permissible. The *set of all actions that are permitted for you*, we call *your action space*. When you fully accept a given normative system, you would always plan your action within your current action space.

In this chapter, we use the following abbreviations (Table 12.1).

We now understand how a social organization $G$ is connected with $G$'s members. Suppose that $ns(G)$ is the normative system of $G$. Member's sharing of a normative system can be easily defined as follows, where a normative system comprises their normative bases, i.e., a belief base, an obligation base, and a permission base:

(5) All members of $G$ share the normative system $ns(G) \Leftrightarrow$ For every member $a$ of $G$, each of $a$'s normative bases contains a corresponding normative base of $G$ (i.e., $bel(G) \subseteq bel(a)$ & $ob(G) \subseteq ob(a)$ & $per(G) \subseteq per(a)$ & $ns(G) = \langle bel(G), \langle ob(G), per(G) \rangle \rangle$ & $ns(a) = \langle bel(a), \langle ob(a), per(a) \rangle \rangle$).

When all members of $G$ share a normative system, they share their beliefs and norms (obligations and permissions). Only those whose normative systems are

**Table 12.1** Abbreviations in LNS

| Abbreviation | Meaning | How to read |
|---|---|---|
| $\mathbf{B}_{NS}\, p$ | $p$ belongs to the *belief set* of *NS* | It is *believed* in *NS* that $p$ |
| $\mathbf{O}_{NS}\, p$ | $p$ belongs to the *obligation set* of *NS* | It is *obligated* in *NS* that $p$ |
| $\mathbf{F}_{NS}\, p$ | $p$ belongs to the *prohibition set* of *NS* | It is *forbidden* in *NS* that $p$ |
| $\mathbf{P}_{NS}\, p$ | $p$ belongs to the *permission set* of *NS* | It is *permitted* in *NS* that $p$ |

consistent and who believe that the normative requirements in question have not been fulfilled can share *G*'s norms (for an exact formulation, consult (A3a) to (A3d) in the Appendix). The shared normative system of *G* constitutes a part of the personal normative system of each of *G*'s members. It influences behaviors of *G*'s members, because it partially determines their action spaces.

## 12.5 Normative Systems in Reality

In this section, we consider two examples of normative systems and show how to represent them in LNS. LNS is designed to deal with normative systems in the ordinary life. We then choose two natural examples from our daily life.

The first example is from the Constitution of the United States (*CUS*). In principle, any law can be treated in a similar way. Now, look at Article 1, Section 1 of *CUS*:

[*CUS* Art. 1. Sect. 1] All legislative Powers herein granted shall be vested in a Congress of the United States, which shall consist of a Senate and House of Representatives.

This sentence can be divided into two components:

(6a) A Congress of the USA should be composed of Senate and House of Representatives.
(6b) Only the Congress of the USA has legislative Powers. That means that it is *permitted* for the Congress of the USA to make laws in the USA, while it is *forbidden* for any other organization to do so.

We can transform these sentences in order to express them in LNS. The transformation process involves two parts. First, we eliminate normative expressions from (6a) and (6b). Second, we separate permission components from prohibition components in (6b). After these operations, we obtain the following three sentences:

(7a) A Congress of the USA is composed of Senate and House of Representatives.
(7b) The Congress of the USA makes laws in the USA.
(7c) For any *x*, (if *x* is an organization that is not the Congress of the USA, then *x* does not make laws in the USA).

We put each sentence in an appropriate normative context, so that the meaning of (6a) and (6b) can be recaptured. We then obtain:

Sentences (7a) and (7c) belong to the *obligation base* of *CUS*, and (7b) belongs to the *permission base* of *CUS* (i.e., $\{(7a), (7c)\} \subseteq ob(CUS)$ & $\{(7b)\} \subseteq per(CUS)$).[7]

[7] Actually, we should translate sentences (7a), (7b), (7c) into sentences of First-Order Logic. For simplicity, we omit this translation step.

The *normative system of the Constitution of the United States* can be defined as follows: $ns(CUS) = \langle bel(CUS), \langle ob(CUS), per(CUS)\rangle\rangle$.[8] With this definition, we obtain three conclusions that roughly capture the content of (6a) and (6b)[9]:

(8a) It is an *obligation* in *ns*(*CUS*) that a Congress of the USA is composed of Senate and House of Representatives.[10]
(8b) It is *permitted* in *ns*(*CUS*) that the Congress of the USA makes laws in the USA.
(8c) It is *forbidden* in *ns*(*CUS*) that an organization other than the Congress of the USA makes laws in the USA.

The second example is taken from submission rules of a journal. We extract only a small fragment from the submission rules of *Annals of the Japan Association for Philosophy of Science*.

(9a) Manuscripts considered for publication in this journal should be original and related to philosophy of science.
(9b) Manuscripts should be written in English, German, or French, within 8000 words.

As we did for the first example, first, we omit normative expressions from the original sentences, and then we put them in the normative contexts that are indicated by the original sentences.

(10a) If author $x$ considers manuscript $y$ for publication in *AJAPS*, then $y$ is original and related to philosophy of science.
(10b) If author $x$ considers manuscript $y$ for publication in *AJAPS*, then $y$ is written in English, German, or French within 8000 words.
(10c) (10a) and (10b) belong to the obligation base of *ns*(*AJAPS*) (i.e., $\{(10a), (10b)\} \subseteq ob(AJAPS)$).

From this description, it follows in *ns*(*AJAPS*) that any manuscript that is considered for publication in *AJAPS* should be original, should be related to philosophy of science, and should be written in English, German, or French, within 8000 words (Formally, $\mathbf{B}_{ns(AJAPS)}$ ($m$ is considered for publication in *AJAPS*) $\Rightarrow \mathbf{O}_{ns(AJAPS)}$ ($m$ is original $\wedge$ $m$ is related to philosophy of science $\wedge$ $m$ is written in English, German, or French $\wedge$ $m$ is within 8000 words)). If someone sends to *JSPS* a manuscript that violates some of the normative rules in *ns*(*AJAPS*), then it will be automatically rejected by *JAPS*.

As you have seen in this section, it is relatively easy to express ordinary normative systems in LNS.

---

[8] *bel*(*CUS*) means the *belief base* of *CUS*.

[9] For this derivation, please consult theorems in Appendix of this chapter.

[10] In LNS, once a Congress of the USA is built, it is no longer obligatory to construct the Congress, and its existence becomes a (social) fact.

## 12.6 Dynamic Normative Logic and Information Update

*Dynamic Normative Logic* (DNL) is proposed as an extension of LNS in Nakayama (2013). DNL is developed to describe information update of normative systems.[11] In this chapter, we use the expression $ns^2(a, t)$ to refer to the *normative state*[12] accepted by agent $a$ in period of time $t$, where this agent may be an individual or a group. We use the expressions $bel^2(a, t)$, $ob^2(a, t)$, and $per^2(a, t)$ as the *belief base of agent a in time t*, the *obligation base of a in t*, and the *permission base of a in t*, respectively. Thus, $ns^2(a, t) = \langle bel^2(a, t), \langle ob^2(a, t), per^2(a, t)\rangle\rangle$.

To emphasize the *part of information that is updated*, we distinguish between the elementary theory $et(a)$ and the set $fact^2(a, t)$ (facts for $a$ in $t$). In other words, the belief base $bel^2(a, t)$ is composed of mutually excluding two sets, $et(a)$ and $fact^2(a, t)$ (Formally, $bel^2(a, t) = et(a) \cup fact^2(a, t)$ & $et(a) \cap fact^2(a, t) = \emptyset$).[13] In many cases, only $fact^2(a, t)$ is updated and $ob^2(a, t)$ and $per^2(a, t)$ remain unchanged during the time interval under consideration. In such a case, we sometimes simply write: $ns^2(a, t) = \langle bel^2(a, t), \langle ob(a), per(a)\rangle\rangle$.

In DNL, we can describe information update of collective normative systems. Because we consider the change of normative states through time, we slightly modify the previous definition (5) of the sharing of normative systems as follows.

(11) All members of $G$ share the normative state $ns^2(G, t)$ in period of time $t \Leftrightarrow$ For every member $a$ of $G$, each of $a$'s normative bases in $t$ contains a corresponding normative base of $G$ in $t$ respectively (i.e., $bel^2(G, t) \subseteq bel^2(a, t)$ & $ob^2(G, t) \subseteq ob^2(a, t)$ & $per^2(G, t) \subseteq per^2(a, t)$ & $ns^2(G, t) = \langle bel^2(G, t), \langle ob^2(G, t), per^2(G, t)\rangle\rangle$ & $ns^2(a, t) = \langle bel^2(a, t), \langle ob^2(a, t), per^2(a, t)\rangle\rangle$.

Thus, the normative state of $G$ in $t$ represents the common part of the normative states of all members of $G$ in $t$. In many cases, the elementary theory and norms remain unchanged, and only the set of facts is updated.[14] When all members of $G$ share a normative state in $t$, they share their non-normative beliefs and their norms in $t$, to the extent that they believe in $t$ that the normative requirement in

[11] DNL is a framework for logical analysis of social interactions. Dynamic epistemic logic is a well-known framework developed for the same purpose. See van Benthem (2011). However, his description is restricted on propositional logic and does not discuss about quantification. In Sects. 12.7 and 12.8, we will see that typical games can be defined in DNL. A game is a kind of normative system that includes information update (Nakayama 2013). A move in a play changes the previous information state.

[12] We say sometimes *normative state* instead of *normative system*, when information update plays a role. In other words, a normative state is a normative system that an agent has for a period of time. The index 2 in $ns^2(a, t)$ indicates that $ns^2$ is a function with two arguments. Generally, we use $ns^2(a, \tau)$, when term $\tau$ involves a temporal component.

[13] $\emptyset$ denotes the empty set, i.e., the set that contains no element.

[14] In such a case, for every time $t$ and for every $G$'s member $a$, $ob^2(a, t) = ob^2(G, t)$ & $per^2(a, t) = per^2(G, t)$.

question has not been fulfilled, provided that their normative systems are consistent (for an exact formulation, consult (A4a) to (A4d) in the Appendix).

## 12.7 Dynamic Normative Logic and Language Games

We can describe primitive language games in DNL. Let us take the example discussed in Sect. 12.3 again (We refer to this example as *PI§2*). The elementary theory and norms do not change, and only the set of facts is updated in this example. Thus, the elementary theory of *PI§2*, the obligation base for *PI§2*, and the permission base for *PI§2* can be expressed as *et*(*PI§2*), *ob*(*PI§2*), and *per*(*PI§2*), respectively, independently from the temporal component.

We interpret this language game as that in which a builder A creates obligations for an assistant B, and B tries to fulfill each of the obligations at a time.[15] An obligation for B consists in bringing a piece of $int_A(\alpha)$ to A, when A calls out 'α.'[16] To fulfill this obligation successfully, B has to interpret α correctly. B can successfully play this language game, only if B's interpretation of α coincides with A's interpretation of α (Formally, $int_B(\alpha) = int_A(\alpha)$). If B's interpretation is wrong, then A would reject receiving $int_B(\alpha)$. After several trials, B would learn the right interpretation of α, so that $int_B(\alpha) = int_A(\alpha)$.[17] In other words, we can say that B has learned this language game, if B always brings a piece of $int_A(\alpha)$ for any α in {"block", "pillar", "slab", "beam"}.

We use $ns^2(\{A, B\}, n)$ as the normative state of group {A, B} at game stage *n*, where *n* is an integer and $ns^2(\{A, B\}, n) = \langle et(PI\S2) \cup fact^2(\{A, B\}, n), \langle ob(PI\S2), per(PI\S2)\rangle\rangle$. We use a function $t_f$ to express the period of time of a game stage; $t_f(n)$ means the period of time of game stage *n*. For example, when this language game starts at 2014 May 23 12:00 and the first A's call takes place at 2014 May 23 12:05, $t_f(0)$ is uniquely determined as the interval [2014 May 23 12:00, 2014 May 23 12:05).

We assume that elementary theory *et*(*PI§2*) contains the sentence 'If *m* is smaller than *n*, then the time of *m* is earlier than the time of *n*' (Formally, $\{\forall m\ \forall n\ (m \leq n \rightarrow t_f(m) \leq_t t_f(n))\} \subseteq et(PI\S2)$).[18] In this example, the permission base contains the sentence (12a), and the obligation base *ob*(*PI§2*) contains the sentence (12b) (i.e., *per*(*PI§2*) = {(12a)} and *ob*(*PI§2*) = {(12b)}).

---

[15] A command is a speech act that creates an obligation for the addressee. See Searle (1979), Yamada (2008), and Nakayama (2014).

[16] $int_A(\alpha)$ means A's interpretation of word α.

[17] This problem of finding the right interpretation for an uttered expression seems closely related to the *symbol grounding problem*. The latter problem is explained as follows: "abstract, arbitrary symbols such as words need to be grounded in something other than relations to more abstract arbitrary symbols if any of those symbols are to be meaningful" (Glenberg and Robertson 2000, p. 381). See also Shapiro (2011, pp. 95–98).

[18] *et*(*PI§2*) contains axioms of linear order for $\leq$ and $\leq_t$ as well.

(12a) For every word α in {"block", "pillar", "slab", "beam"} and for every game stage $n$, A calls out 'α' in $t_f(n+1)$, when the assistant is waiting for a call in $t_f(n)$.
(12b) For every word α in {"block", "pillar", "slab", "beam"} and for every game stage $n$, if A calls out 'α' in $t_f(n)$, then B brings a piece of $int_A(\alpha)$ to A in $t_f(n+1)$.

We assume that A and B share the normative state of {A, B} during the game. This assumption is justified by the story in *PI§2*. Now, a possible development of the language game in *PI§2* can be described as follows.

A calls out "block", when B is waiting for a call in $t_f(0)$. Then, in $ns^2(\{A,B\},1)$, B is obligated to bring a piece of $int_A$(block) to A in $t_f(2)$. In accordance with this obligation, B brings a piece of $int_A$(block) to A in $t_f(2)$. A calls out "pillar", when B is waiting for a call in $t_f(2)$. Then, in $ns^2(\{A,B\},3)$, B is obligated to bring a piece of $int_A$(pillar) to A in $t_f(4)$. In accordance with this obligation, B brings a piece of $int_A$(pillar) to A in $t_f(4)$. (The game continues in this way.)

This development can be more formally described as follows:

$fact^2(\{A, B\}, 0) = \{$B is waiting for a call in $t_f(0)\}$.

$\mathbf{P}_{ns2(\{A,B\},0)}$ (A calls out "block" in $t_f(1)$).[19]

$fact^2(\{A, B\}, 1) = fact^2(\{A, B\}, 0) \cup \{$A calls out "block" in $t_f(1)\}$.

$\mathbf{O}_{ns(\{A,B\},1)}$ (B brings a piece of $int_A$(block) to A in $t_f(2)$).[20]

$fact^2(\{A, B\}, 2) = fact^2(\{A, B\}, 1) \cup \{$B brings a piece of $int_A$(block) to A in $t_f(2)$, B is waiting for a call in $t_f(2)\}$.

$\mathbf{P}_{ns2(\{A,B\},2)}$ (A calls out "pillar" in $t_f(3)$).

$fact^2(\{A, B\}, 3) = fact^2(\{A, B\}, 2) \cup \{$A calls out "pillar" in $t_f(3)\}$.

$\mathbf{O}_{ns2(\{A,B\},3)}$ (B brings a piece of $int_A$(pillar) to A in $t_f(4)$).

$fact^2(\{A, B\}, 4) = fact^2(\{A, B\}, 3) \cup \{$B brings a piece of $int_A$(pillar) to A in $t_f(4)$, B is waiting for a call in $t_f(4)\}$.

...

As this example indicates, DNL can formally represent not only primitive language games but also typical games, such as chess and card games.

[19] The consistency of $union(ns^2(\{A, B\},0)) \cup \{$A calls out "block" in $t_f(1)\}$ can be shown by constructing a finite model.

[20] For the derivation of this sentence, you use theorem (A2e1) in Appendix.

## 12.8 Dynamic Normative Logic, Games, and Roles

In this section, we show how to describe baseball games in DNL. For this purpose, we consult the rules of *Major League Baseball* (Official Baseball Rules 2013, 2.00). The rules correspond to a normative system *ns*(*MLB*), where ⟨*et*(*MLB*), ⟨*ob*(*MLB*), *per*(*MLB*)⟩⟩. The rules are many and very complex, and so we can only describe a small fragment of them.

There are batters, a pitcher, a catcher, fielders, and umpires in a baseball game. They are distinguished in terms of the roles they play during a game.

[OBR*4.03] When the ball is put in play at the start of, or during a game, all fielders other than the catcher shall be on fair territory.

(a) The catcher shall station himself directly back of the plate. He may leave his position at any time to catch a pitch or make a play except that when the batter is being given an intentional base on balls, the catcher must stand with both feet within the lines of the catcher's box until the ball leaves the pitcher's hand.
(b) The pitcher, while in the act of delivering the ball to the batter, shall take his legal position.
(c) Except the pitcher and the catcher, any fielder may station himself anywhere in fair territory.

Notice that several roles are different in normative requirement. The differences can be reflected in the obligation base and the permission base in *ns*(*MLB*).

(13a) The catcher stations himself directly back of the plate.
(13b) The catcher leaves his position at any time to catch a pitch or make a play.
(13c) If the batter is being given an intentional base on balls, then the catcher stands with both feet within the lines of the catcher's box until the ball leaves the pitcher's hand.
(13d) If the pitcher is in the act of delivering the ball to the batter, then he takes his legal position.
(13e) Every fielder who is neither the pitcher nor the catcher stations himself anywhere in fair territory.
(13f) {(13a), (13c), (13d)} $\subseteq$ *ob*(*MLB*) & {(13b), (13e)} $\subseteq$ *per*(*MLB*).

With the help of this list of normative requirements, it is evident that a player needs to know about his own role in the baseball game in order to play it. For example, when you are registered as a catcher, if you mistake yourself as a fielder, you will station yourself in a wrong position and violate some rules of *MLB*.

Let us take as an example of information update a scene in a baseball game between teams *A* and *B*. The initial normative state can be represented as $ns^3(MLB, \{A, B\}, 0\blacktriangledown 0/3) = \langle et(MLB) \cup fact^3(MLB, \{A, B\}, 0\blacktriangledown 0/3), \langle ob(MLB), per(MLB)\rangle\rangle$. Suppose that team *A* has scored 5 runs and team *B* scored only 2 runs before the seventh inning; more simply, that *score* (*A*: 5, *B*: 2, 7▼0/3). Suppose also that an umpire now calls out a player in team *A*. Then, the score is updated and the number of outs is raised: *score* (*A*: 5, *B*: 2, 7▼1/3) (formally, $fact^3(MLB, \{A, B\}, 7\blacktriangledown 1/3) =$

*fact*$^3$(*MLB*, {*A*, *B*}, 7▼0/3)∪{*score* (*A*: 5, *B*: 2, 7▼1/3)}). We can describe the development of a baseball game in this way.

The game ends when the termination condition is satisfied. The precise termination condition of *MLB*-games is found in OBR.

[OBR*4.11] The score of a regulation game is the total number of runs scored by each team at the moment the game ends.

(a) The game ends when the visiting team completes its half of the ninth inning if the home team is ahead.
(b) The game ends when the ninth inning is completed, if the visiting team is ahead.
(c) If the home team scores the winning run in its half of the ninth inning (or its half of an extra inning after a tie), the game ends immediately when the winning run is scored.

These rules can be expressed as components of the elementary theory of *MLB* (Formally, {[OBR*4.11] (a) + (b) + (c)} ⊆ *et*(*MLB*)). They determine the end of *MLB*-games.

## 12.9 Games and Dynamic Knowledge Representation

As we have just seen, a baseball game has a rich and complex structure. In this section, we suggest that some real situations can be interpreted as games. Schank and Abelson (1977) proposed what they called the '*script*' as a useful framework for knowledge representation. They described a restaurant script from a customer's viewpoint. The restaurant script consists of four scenes, *entering*, *ordering*, *eating*, and *exiting* (these scenes may have further internal structures). Schank's theory treats every memory as episodic, i.e., as that which is organized around personal experiences. He regards a *script* as a generalized episode. For this reason, a script provides a passive and static way of knowledge representation. In contrast, DNL-representation can provide a dynamic knowledge representation of the relevant activities in a social place.[21]

Eating at a restaurant can be interpreted as a process in a game played by customers, waiters, cooks, a cashier, and an owner. These people play certain roles in a *theater* of a restaurant. In this game, the same person may sometimes play different roles in different situations. For example, a waiter becomes a cashier when a customer wants to pay the bill.

As we saw in Sect. 12.8, roles can be determined through normative requirements. For example, a restaurant customer is a person who may order a meal and has to pay for it. A waiter is a person who should ask customers what meals they

[21] Nakayama (2013) describes a small part of a restaurant scene in full detail in DNL.

want and should serve them with the meals they order. A cook is a person who should cook ordered meals. Different players of the game have different goals. A customer's goal is to eat a meal, and an owner's goal is to satisfy customers with good meals and get money from them.

As an example, we describe normative requirements for customers and waiters.

{A customer orders an available meal, A customer eats a meal she ordered} $\subseteq$ *per* (*RT*).
{A non-customer orders no meal} $\subseteq$ *ob*(*RT*).
{A customer pays the bill of all meals she ordered} $\subseteq$ *ob*(*RT*).
{A waiter asks customers about their orders, a waiter serves customers with meals they ordered when the meals have been cooked} $\subseteq$ *ob*(*RT*).

All of these descriptions can be explicitly written in DNL. We can use DNL to describe ongoing processes in a concrete situation. In a game, you may develop strategies to achieve a goal. For example, the owner might propose new menus to attract restaurant customers. In representing a game in a social place in DNL, phenomena pertaining to that game can be analyzed from different viewpoints of players.

In sum, DNL-representation is a uniform dynamic knowledge representation that is more complex and flexible than script representation. Because of this uniformity, it is easy to combine DNL-representations of activities in different contexts without giving rise to confusion.

## 12.10 Roles and Social Organizations

A baseball game is a kind of competition between two teams. Each team consists of more than nine members and has an internal structure that is determined by the roles of the team members. A member plays as a pitcher, a catcher, or one of seven fielders. Real social organizations have more complex structures than baseball teams, but the basic principles behind them are not very different.

Many social organizations, such as countries, cities, companies, and universities, are hierarchically structured. For example, a university consists of faculties, and each faculty divides into several departments. There are teaching staffs, office staffs, and students in each faculty. They are distinguished in terms of normative requirements pertinent to university activities.

Nakayama (2011: p. 153f) roughly characterizes social organization as follows.

(14a) [Social organization as a four-dimensional object] Social organization *O* is a four-dimensional object.[22] In other words, a social organization is an entity that has a temporal extension and an internal structure.

[22] For four-dimensionalism and mereology, see Sider (2001) and Nakayama (2009).

(14b) [Humans and artifacts as components of a social organization] A social organization is, in general, a mereological sum $G+A$, where $G$ is a group of rational agents (usually, humans) and $A$ is a mereological sum of certain artifacts. Thus, $O=G+A$. However, there are social organizations that are solely composed of humans. In such case: $O=G$.
(14c) [Self-knowledge of members] If social organization $O$ exists at $t$ and $O=G+A$ or $O=G$, then every member of $G$ knows at $t$ that she belongs to $O$.
(14d) [Member's knowledge about $O$'s property] Suppose that $O=G+A$ and $O$ exists at $t$. Then, there is at least one member in $G$ who believes at $t$ that for every part $x$ of $A$, $x$ belongs to $O$.
(14e) [Common belief about the existence of a social organization] The existence of social organization $O$ is a common belief of group $G$, where $O=G+A$ or $O=G$.
(14f) [Structure of social organization and its maintenance] $O$ is structured in the way in which the possibility of $O$'s future existence increases. In other words, if necessary, $O$'s structure is adjusted so that the possibility of $O$'s future existence increases.

As we already mentioned in this section, the structure of a social organization is determined by the roles and functions of its sub-organizations and their members. The characterization (14f) suggests that the internal structure of a social organization is determined and changed so as to realize its survival.[23]

## 12.11 Meta-games and Meta-norms

We can analyze a collective action of a social organization as a move in a game; for example, as a move in a survival game in which several organizations aim to maintain their existence, and perform collective actions to achieve this aim (Nakayama 2011).

Nakayama (2011) also suggests that a game can be composed of several sub-games. A tournament game is an example of a meta-game, i.e., a game that involves iteration of sub-games. A win in a sub-game is interpreted as a successful move in the meta-game. A tournament game is also a survival game. To remain in a tournament game, a team has to keep winning in a series of sub-games.

A legal philosopher Herbert L. A. Hart (1907–1992) distinguished between primary rules (rules of conduct) and secondary rules (rules addressed to officials and which set out to affect the operation of primary rules) (Hart 1961). According to Hart, a legal system is the union of primary and secondary rules. There are three types of secondary rules:

(15a) [Rules of recognition] They clarify what belongs to the primary rules.

[23] There are discussions on norms, roles, and plays in sociology. See Mead (1934), Parsons (1937), Merton (1949), Goffman (1959), and Habermas (1981).

(15b) [Rules of change] They are rules for changing the primary rules.
(15c) [Rules of adjudication] They are rules that give certain people the power to judge whether the primary rules are violated.

Hart's distinction of rules roughly corresponds to the distinction between norms and meta-norms in our terms. Meta-norms are norms at the meta-level and define or change social normative systems at a lower level. They should be socially approved. Article 1, Section 1 of *CUS*, as we analyzed it in Sect. 12.5, expresses a meta-norm that defines the structure and function of the Congress of the United States.

All Japanese laws are considered to be approved by the Japanese people, when they are approved in the Japanese parliament. The *Japanese Constitution* (*JS*) is a normative system that defines when a law is considered to be approved in the Japanese parliament. Thus, *JS* is a meta-normative system. There are a lot of meta-games and meta-norms.

## 12.12 Concluding Remarks

We have analyzed normative systems in this chapter. Laws and regulations are typical examples of normative systems. In addition, we have seen that a game can be interpreted as a dynamic system of players whose actions are restricted by a normative system. As we suggested, many social activities can be interpreted as games. In playing a game, players choose and perform their actions within the normative constrains given by the rules of the game and the roles they play in the game. An actual person lives in various normative systems and various games. So she always has to make a decision which game and what kind of role she will play. Her choices and performances give a social organization reality. In other words, when nobody follows social norms approved in a social organization, the organization is ignored and its reality is lost.

## Appendix

In this appendix, we describe the definitions and theorems mentioned in the main text in more exact manners.

The definitions (8a) to (8f) of LNS can be expressed more rigorously as follows:

(A1a) When each of $T$, $OB$, and $PER$ is a set of sentences in *First-Order Logic*, $NS = \langle T, \langle OB, PER \rangle \rangle$ is a *normative system*.
(A1b) A normative system $NS$ is consistent $\Leftrightarrow$ *union*($NS$) is consistent, where *union* ($NS$) $= T \cup OB \cup PER$.
(A1c) $\mathbf{B}_{NS}\, q \Leftrightarrow T \vdash q$.
(A1d) $\mathbf{O}_{NS}\, q \Leftrightarrow$ (*consistent*(*union*($NS$)) & $T \cup OB \vdash q$ & *not* $\mathbf{B}_{NS}\, q$).

(A1e) $\mathbf{F}_{NS}\, q \Leftrightarrow \mathbf{O}_{NS}\, \neg q$.
(A1f) $\mathbf{P}_{NS}\, q \Leftrightarrow (consistent(union(NS) \cup \{q\})\ \&\ not\ \mathbf{B}_{NS}\, q)$.

Based on these definitions, we can easily prove the main theorems of LNS, where *NS* ($NS = \langle T, \langle OB, PER\rangle\rangle$) indicates a normative system.

(A2a) $(\mathbf{B}_{NS}\, (p \rightarrow q)\ \&\ \mathbf{B}_{NS}\, p) \Rightarrow \mathbf{B}_{NS}\, q$.
(A2b1) $(\mathbf{O}_{NS}\, (p \rightarrow q)\ \&\ \mathbf{O}_{NS}\, p) \Rightarrow \mathbf{O}_{NS}\, q$.
(A2b2) $\mathbf{F}_{NS}\, p \Leftrightarrow \mathbf{O}_{NS}\, \neg p$.
(A2b3) $\mathbf{O}_{NS}\, p \Rightarrow \mathbf{P}_{NS}\, p$.
(A2b4) $\mathbf{F}_{NS}\, p \Rightarrow not\ \mathbf{P}_{NS}\, p$.
(A2c1) $\mathbf{P}_{NS}\, p \Rightarrow not\ \mathbf{B}_{NS}\, p$.
(A2c2) $\mathbf{B}_{NS}\, p \Rightarrow (not\ \mathbf{O}_{NS}\, p\ \&\ not\ \mathbf{F}_{NS}\, p\ \&\ not\ \mathbf{P}_{NS}\, \mathrm{p})$.
(A2d1) $(\mathbf{O}_{NS}\, (p \rightarrow q)\ \&\ \mathbf{B}_{NS}\, p) \Rightarrow \mathbf{O}_{NS}\, q$.
(A2d2) $(\mathbf{O}_{NS}\, (p \wedge q)\ \&\ \mathrm{not}\ \mathbf{B}_{NS}\, p) \Rightarrow \mathbf{O}_{NS}\, p$.
(A2d3) $(\mathbf{O}_{NS}\, (p \wedge q)\ \&\ \mathbf{B}_{NS}\, p) \Rightarrow \mathbf{O}_{NS}\, q$.
(A2d4) $(\mathbf{O}_{NS}\, (p \vee q)\ \&\ \mathbf{B}_{NS}\, \neg p) \Rightarrow \mathbf{O}_{NS}\, q$.
(A2d5) $(\mathbf{O}_{NS}\, (p \vee q)\ \&\ \mathbf{F}_{NS}\, p) \Rightarrow \mathbf{O}_{NS}\, q$.
(A2d6) $(\mathbf{O}_{NS}\, p\ \&\ not\ \mathbf{B}_{NS}\, q) \Rightarrow \mathbf{O}_{NS}\, (p \vee q)$.
(A2d7) $(\mathbf{B}_{NS}\, (p \rightarrow q)\ \&\ \mathbf{O}_{NS}\, p\ \&\ \mathbf{P}_{NS}\, q) \Rightarrow \mathbf{O}_{NS}\, q$.
(A2e1) $(\mathbf{O}_{NS}\, \forall x_{1\ldots}\, \forall x_{\mathrm{n}}\, (P(x_1, \ldots, x_{\mathrm{n}}) \rightarrow Q(x_1, \ldots, x_{\mathrm{n}}))\ \&\ \mathbf{B}_{NS}\, P(a_1, \ldots, a_{\mathrm{n}})\ \&\ not$ $\mathbf{B}_{NS}\, Q(a_1, \ldots, a_{\mathrm{n}})) \Rightarrow \mathbf{O}_{NS}\, Q(a_1, \ldots, a_{\mathrm{n}})$.
(A2e2) $(\mathbf{F}_{NS}\, \exists x_{1\ldots}\, \exists x_{\mathrm{n}}\, (P(x_1, \ldots, x_{\mathrm{n}}) \wedge Q(x_1, \ldots, x_{\mathrm{n}}))\ \&\ \mathbf{B}_{NS}\, P(a_1, \ldots, a_{\mathrm{n}})\ \&\ not$ $\mathbf{B}_{NS}\, \neg Q(a_1, \ldots, a_{\mathrm{n}})) \Rightarrow \mathbf{F}_{NS}\, Q(a_1, \ldots, a_{\mathrm{n}})$.

Two theorems express the relationship between the normative states of a group and those of its members:

[T1] If all members of *G* share the normative system *ns*(*G*), then the following sentences hold:

(A3a) For all $a \in G$, $(\mathbf{B}_{ns(G)}\, p \Rightarrow \mathbf{B}_{ns(a)}\, p)$.
(A3b) For all $a \in G$, $((\mathbf{O}_{ns(G)}\, p\ \&\ consistent(ns(a))\ \&\ not\ \mathbf{B}_{ns(a)}\, p) \Rightarrow \mathbf{O}_{ns(a)}\, p)$.
(A3c) For all $a \in G$, $((\mathbf{F}_{ns(G)}\, p\ \&\ consistent(ns(a))\ \&\ not\ \mathbf{B}_{ns(a)}\, p) \Rightarrow \mathbf{F}_{ns(a)}\, p)$.
(A3d) For all $a \in G$, $((\mathbf{P}_{ns(G)}\, p\ \&\ consistent(ns(a))\ \&\ not\ \mathbf{B}_{ns(a)}\, p) \Rightarrow \mathbf{P}_{ns(a)}\, p)$.

[T2] If all members of *G* share the normative state $ns^2(G, t)$ at *t*, then the following sentences hold:

(A4a) For all time *t* and for all $a \in G$, $(\mathbf{B}_{ns2(G,t)}\, p \Rightarrow \mathbf{B}_{ns2(a,t)}\, p)$.
(A4b) For all time *t* and for all $a \in G$, $((\mathbf{O}_{ns2(G,t)}\, p\ \&\ consistent(ns^2(a, t))\ \&\ not$ $\mathbf{B}_{ns2(a,\, t)}\, p) \Rightarrow \mathbf{O}_{ns2(a,\, t)}\, p)$.
(A4b) For all time *t* and for all $a \in G$, $((\mathbf{F}_{ns2(G,t)}\, p\ \&\ consistent(ns^2(a, t))\ \&\ not$ $\mathbf{B}_{ns2(a,\, t)}\, p) \Rightarrow \mathbf{F}_{ns2(a,\, t)}\, p)$.
(A4b) For all time *t* and for all $a \in G$, $((\mathbf{P}_{ns2(G,t)}\, p\ \&\ consistent(ns^2(a, t))\ \&\ not$ $\mathbf{B}_{ns2(a,\, t)}\, p) \Rightarrow \mathbf{P}_{ns2(a,\, t)}\, p)$.

T1 and T2 indicate the problem of consistency for a personal normative system. This problem is quite interesting in view of our real life experience. Suppose that you belong to groups $G_1$ and $G_2$ whose normative systems are mutually inconsistent, and that all members of $G_1$ and all members of $G_2$ share their normative systems $ns(G_1)$ and $ns(G_2)$. Then, your normative system becomes inconsistent, and, according to (A1d), (A1e), and (A1f), all of your normative requirements disappear. This is a case of social dilemma that many of us face in the real life.[24]

## Exercises

Discuss the following problems:

1. What is a *normative system*?
2. What is *Logic for Normative Systems*?
3. What is *Dynamic Normative Logic*?
4. How can normative powers be explained?
5. How can roles be explained?
6. How are individuals and social organizations related?

## References

Åqvist, L.: Deontic logic. In: Gabbay, D., Guenthner, F. (eds.) Handbook of Philosophical Logic, vol. 8, pp. 147–264. Kluwer Academic Pub, Dordrecht (2002)
Birch, A.: Developmental Psychology: From Infancy to Adulthood, 2nd edn. Macmillan Press, London (1997)
Glenberg, A., Robertson, D.: Symbol grounding and meaning: a comparison of high-dimensional and embodied theories of meaning. J. Mem. Lang. **43**, 379–401 (2000)
Goffman, E.: The Presentation of Self in Everyday Life. University of Edinburgh Social Sciences Research Centre, Edinburgh (1959)
Habermas, J.: Theorie des kommunikativen Handelns, Bde, 1-2. Suhrkamp, Frankfurt am Main (1981)
Hart, H.L.A.: The Concept of Law. Oxford University Press, Oxford (1961)
Mead, G.H.: Mind, Self, and Society. Edited by C. W. Morris. University of Chicago Press, Chicago (1934)
Merton, R.K.: Social Theory and Social Structure: Toward the Codification of Theory and Research. Free Press, New York (1949)
Nakayama, Y.: The Construction of Contemporary Nominalism: Its Application to Philosophy of Nominalism. Syunjyu-sha, Tokyo (2009), in Japanese
Nakayama, Y.: Logical framework for normative systems. In: SOCREAL 2010: Proceedings of the 2nd International Workshop on Philosophy and Ethics of Social Reality, 27–28 March 2010, Hokkaido University, Sapporo, Japan, pp. 19–24 (2010)
Nakayama, Y.: Norms and Games: An Introduction to the Philosophy of Society. Keiso shobo, Tokyo (2011), in Japanese

[24] This kind of social dilemmas is discussed in Nakayama (2011) chap. 7 sect. 3.

Nakayama, Y.: Dynamic normative logic and information update. In: Yamada, T. (ed.) SOCREAL 2013: 3rd International Workshop on Philosophy and Ethics of Social Reality, Abstracts, Hokkaido University, Sapporo, Japan, pp. 23–27 (2013)

Nakayama, Y.: Speech Acts, Normative Systems, and Local Information Update. In: Y. I. Nakano, et al. (eds.) New Frontiers in Artificial Intelligence, LNAI 8417, Springer, pp. 98–114 (2014)

Parsons, T.: The Structure of Social Actions. Free Press, New York (1937/1967)

Piaget, J.: Play, Dreams and Imitation in Childhood. Routledge and Kegan Paul, London (1951)

Schank, R.C., Abelson, R.P.: Scripts, Plans, Goals and Understanding: An Inquiry into Human Knowledge Structures. Lawrence Erlbaum Associates, Hillsdale (1977)

Searle, J.R.: Expression and Meaning – Studies in the Theory of Speech Acts. Cambridge University Press, Cambridge (1979)

Searle, J.R.: The Construction of Social Reality. Free Press, New York (1995)

Searle, J.R.: Making the Social World: The Structure of Human Civilization. Oxford University Press, Oxford (2010)

Shapiro, L.: Embodied Cognition. Routledge, London (2011)

Sider, T.: Four-Dimensionalism: An Ontology of Persistence and Time. Oxford University Press, Oxford (2001)

van Benthem, J.: Logical Dynamics of Information and Interaction. Cambridge University Press, Cambridge (2011)

Wittgenstein, L.: Philosophical Investigations. Translated by G. E. M. Anscomb, 2nd edn. Basil Blackwell, Oxford (1958)

Yamada, T.: Logical dynamics of some speech acts that affect obligations and preferences. Synthese **165**(2), 295–315 (2008)

Zichermann, G., Cunningham, C.: Gamification by Design. O'Reilly Media, Cambridge, MA (2011)

# Index




M. Kasaki et al. (eds.), *Cognitive Neuroscience Robotics B*,
DOI 10.1007/978-4-431-54598-9

MIX
Papier aus verantwortungsvollen Quellen
Paper from responsible sources
FSC® C105338

If you have any concerns about our products,
you can contact us on
**ProductSafety@springernature.com**

In case Publisher is established outside the EU,
the EU authorized representative is:
**Springer Nature Customer Service Center GmbH**
**Europaplatz 3, 69115 Heidelberg, Germany**

Printed by Libri Plureos GmbH
in Hamburg, Germany